Physics of Ice

Physics of Ice

VICTOR F. PETRENKO
Thayer School of Engineering, Dartmouth College, Hanover, New Hampshire

and

ROBERT W. WHITWORTH
School of Physics and Astronomy, The University of Birmingham

OXFORD
UNIVERSITY PRESS

Great Clarendon Street, Oxford OX2 6DP

Oxford University Press is a department of the University of Oxford.
It furthers the University's objective of excellence in research, scholarship,
and education by publishing worldwide in

Oxford New York
Athens Auckland Bangkok Bogotá Buenos Aires Calcutta
Cape Town Chennai Dar es Salaam Delhi Florence Hong Kong Istanbul
Karachi Kuala Lumpur Madrid Melbourne Mexico City Mumbai
Nairobi Paris São Paulo Singapore Taipei Tokyo Toronto Warsaw

with associated companies in Berlin Ibadan

Oxford is a registered trade mark of Oxford University Press
in the UK and in certain other countries

Published in the United States
by Oxford University Press Inc., New York

First published 1999

A catalogue record for this book is available from the British Library

Library of Congress Cataloging in Publication Data
Petrenko, Victor F.
Physics of ice/Victor F. Petrenko and Robert W. Whitworth.
Includes bibliographical references and index.
1. Ice. I. Whitworth, Robert W. II. Title.
GB2403.2.P44 1999 551.31–dc21 99–19984

ISBN 0 19 851895 1

Typeset by Newgen Imaging Systems (P) Ltd., Chennai, India
Printed in Great Britain
on acid-free paper by
Bookcraft (Bath) Ltd
Midsomer Norton, Avon

Preface

This book is addressed to the community of scientists who share a common interest in that one most abundant, most studied, and most fascinating crystalline solid material—ice. These scientists have a diversity of interests. There are geophysicists and glaciologists concerned with the large ice masses on the surface of the Earth. There are engineers concerned with structures built on, with, or in the vicinity of ice. There are meteorologists or cloud physicists concerned with the role of ice in our weather. There are chemists who see ice as the simplest example of a hydrogen-bonded solid. There are crystallographers concerned with the structural disorder that is specially characteristic of ice, and who are also concerned with the many different phases into which H_2O molecules can crystallize. There are physicists concerned with electrical properties, optical properties, lattice defects, lattice vibrations, and the unusual properties of the surface. For theoreticians ice is simple enough to think about, yet it presents unique problems, some of which have been solved and others that await solution.

Many readers will have had the experience of finding people, both scientists and others, express the view that 'surely there isn't anything new to be learnt about ice'. Because it is used to cool drinks or trodden underfoot in winter, ice is a commonly derided material. Anyone who has decided to read this preface will almost certainly realize that such attitudes are completely wrong. Ice is one of the most important and most studied materials on Earth, with a wealth of unique and intriguing physical properties. Moreover, the study of ice draws on and is applied in a wide range of physics. As new experimental and theoretical techniques have become available they have been applied to the study of ice, providing a continuous flow of new knowledge and understanding. This has been happening at an increasing rate throughout the last two centuries, and we anticipate that it will continue well into the next millennium.

Our book is about the *physics* of ice, by which we mean the properties of the material itself and the ways in which these properties are interpreted in terms of the crystalline structure and the properties of the water molecule. It therefore includes crystallography and some aspects that might be considered to be chemistry. The physics of ice is just one facet of the whole study of ice. We attempt to show how it relates to ice in the environment, but subjects like glaciology, ice mechanics, and cloud physics justify whole books and journals of their own.

In writing we have kept several aims in mind. First, we intend that the book should serve as an introduction to the subject for a graduate with some knowledge of physics and crystallography. In each chapter the subject is introduced from first principles and the reader is directed towards selected important papers on the topic. In some chapters we describe important practical techniques which are specific to working with ice. Second, we see each chapter as a critical review giving our assessment of current knowledge and understanding, and we hope that this will be of value to all workers in that area. Some issues remain controversial, and not everyone will agree with our point of view. We make no apology for devoting rather more attention to topics on which we have particular expertise and where we believe we have a larger contribution to make. Third, the book is intended to serve as a work of reference to which anyone can turn for factual information and references. However, the subject has become so large and the number of publications so enormous that our list of some 900 references is in no way comprehensive. Finally, although most chapters can stand alone, we wish to encourage cross-fertilization between areas of study. We hope that we have written a book in which readers will find their own speciality placed in a wider context, and that this will be found both fascinating and a stimulus to the development of fresh approaches to the study of ice.

There are two books about ice which have been of enormous value over the past 25 years. These are *The chemical physics of ice* by Neville H. Fletcher published in 1970, and *Ice physics* by Peter V. Hobbs published in 1974. The former provides an excellent introductory account of selected fundamental topics. The latter is a more comprehensive work of reference and includes an excellent bibliography of work done up to that time. A great deal of experimental work has been carried out and our theoretical understanding has developed considerably since these books were written. Although there have been specialized reviews on a few specific topics, no one has attempted a comprehensive account of the physics of ice since 1974. This has made life difficult for students entering the field, and tends to lead to a narrowness of view. We believe that the time is overdue for a book such as the one we have written, and we sincerely hope that it will be found useful to a wide range of people working with ice. We hope that we live to read a sequel written by someone else within less than 25 years time!

The idea for this book was initiated when in 1990 Victor Petrenko was commissioned by the US Army Cold Regions Research and Engineering Laboratory to write a series of six Special Reports on aspects of the physics of ice. The first two of these were published in 1993, and Robert Whitworth became co-author of two which were published in 1994. These reports formed the embryo of the book, though in the final product the reports are barely recognizable. The reports, and indeed the books by Fletcher and Hobbs, are rarely cited in the book itself. This is because the book largely supersedes them, and they are only referred to when they provide significant additional material or were the original source of some concept or understanding. For a listing of these reports and other general sources of reference the reader should refer to the Bibliography which appears immediately before the list of references.

The preparation of this book is the largest of a series of projects on which we have collaborated over the past twenty years. In it we have been greatly helped and encouraged by discussions with many members of the ice physics community, but certain people deserve particular acknowledgement. Dr John W. Glen has carried out extensive surveys of the literature on ice and made them available to us; he has also at different stages in their preparation read and commented on all of the chapters. Others who have reviewed particular sections for us are Professor Werner F. Kuhs (Chapters 2, 3, and 11), Dr Eric W. Wolff (Chapters 4, 5, and §12.5), Professor Erland M. Schulson (Chapter 8), Professor J. Paul Devlin (Chapter 10), Professor John F. Nagle (§§4.3 and 4.4), Dr Stephen F.J. Cox (§6.7), Professor John C. Dore (§11.9), Dr Clive P.R. Saunders (§12.3), and Dr David A. Rothery (§12.7). Mrs Anne Whitworth has been of great help in checking the manuscript. We also thank authors and publishers who have given permission to reproduce copyright material, which is identified in the captions to the relevant figures and tables. The following workers have kindly provided original photographs and other material for use in the figures or tables: Dr I-Ming Chou, Dr S.C. Colbeck, Dr Y. Furukawa, Dr L.W. Gold, Dr A.J. Gow, Professor W.F. Kuhs, Dr V.M. Nield, Dr M.I.J. Probert, Dr D.A. Rothery, Dr M. Sakata, Professor E.M. Schulson, Dr C.J.L. Wilson, and Dr E.W. Wolff.

Finally, we hope that in writing these chapters we are helping to establish a foundation for the fruitful study of ice in the coming years. Such study should include not only imaginative new ideas and experiments, but also a careful repetition of some of the classical measurements, taking full advantage of the improved instrumentation, specimens and understanding that are now available. We wish our readers, from whatever area they may come, every success in their study of this most fascinating material.

Hanover, New Hampshire, USA V.F.P.
Birmingham, UK R.W.W.
October 1998

Contents

1
Introduction

1.1 The importance of ice

The whole character of the planet Earth depends on the abundance of water and on the temperature being such that all phases—ice, liquid, and vapour—are present in significant quantities. At the present time the oceans cover 70% of the globe, and 10% of the land mass is covered with ice to depths of up to several kilometres. Depending on the time of year some 5% of the surface area of the oceans is covered with ice around one or other of the Poles. The comfortable situation for mankind at the present time depends on two delicately balanced equilibria. The first is that between the radiation received from the Sun and that reflected or re-radiated from the Earth. The resulting global average temperature is highly sensitive to the amount of snow, ice, and cloud cover. The second equilibrium concerns the evaporation of water from the oceans, leading to snowfall over the polar regions and subsequent flow of the ice sheets back into the oceans. Our environment is thus critically dependent on the properties of ice both in the meteorology of clouds and in the rheology of ice sheets. At earlier times in geological history ice has covered up to 30% of the land area and its motion has carved many features of the northern landscape.

In winter, snow and ice present hostile conditions for modern society, disrupting transport, causing damage by freezing pipes and ground, and producing icing of aircraft, ships and power lines. But in a different mood we marvel at the beauty of snow, frost formations, icebergs, and glaciers, and we take pleasure in winter sports. In summer people may forget about ice except for cooling drinks, but it remains important in the preservation of food, and its presence in the atmosphere is essential to the production of rain; it is the collision of ice particles in clouds that generates thunderstorm electricity.

Approaching the subject from another extreme, the water molecule is one of the simplest in chemistry, yet it forms a liquid with many features that are essential to life and the environment. These features can be attributed to the formation of hydrogen bonds, but that does not mean that they are fully understood, and much effort has been devoted to experimental and theoretical research on water. Ice should be an easier material to understand, because the molecules are arranged on a regular lattice, but the ice with which we are familiar (ice Ih) is just one of at least 13 crystalline phases which have been observed under different conditions of pressure and temperature. The crystal structures of ice Ih and of many of the other phases are unusual because, although the molecules lie on a regular crystal lattice, there is disorder in their orientations. This feature introduces a whole series of distinctive properties, of which the most significant are the electrical polarizability and conductivity. Ice can be described as a 'protonic semiconductor', and the theory of its electrical properties is now well developed. This understanding is highly relevant to more complicated systems in which proton transfer takes place along hydrogen-bonded chains in biologically important structures.

On the small scale ice is unusual in being a brittle material right up to its melting point, but on a larger scale we know that glaciers and ice sheets flow under their own weight. The plastic properties of crystalline materials have to be understood by the application of concepts developed in metallurgy, and this has been done very successfully for the mechanical properties of ice. Modelling the flow of ice sheets is an important part of glaciology, but engineers concerned with issues like building an aircraft runway on an ice sheet or the possible consequences of the impact of an iceberg on an oil drilling platform are more concerned with the strength of ice under sudden loading. Their field is called 'ice mechanics'.

These brief paragraphs show that ice is an important material both in our environment and as one of the simplest crystalline materials. It has many distinctive properties, and yet its structure is simple enough to be accessible to serious theoretical treatment. Great progress has been made in understanding the properties of ice, but the task is not and probably never will be complete.

1.2 The physics of ice and the structure of the book

The 'physics of ice' is the study of the properties of the material itself and the interpretation of these properties at the molecular level. Here we must recognise two stages. The crystallographic structure provides the link between individual molecules and the properties of single crystals, but on a large scale ice is polycrystalline and the mechanical properties in particular are dependent on this polycrystallinity.

In this section we provide an overview of the subject designed as a guide to the structure of the book. As is usual in condensed matter physics, the emphasis is on the interpretation of properties at the atomic level, but the chapters are designed so that up-to-date information on macroscopic physical properties is readily accessible to those who need it. A short list of the most basic parameters of ice will be found inside the back cover of the book, and inside the front cover there is a table of important conversion factors and a few relevant general physical constants.

This introductory chapter concludes with sections on the water molecule and hydrogen bonding, because these are the starting point for an account of the structure and properties of ice. Chapter 2 then focuses on the basic properties of ordinary ice Ih, including the preparation of monocrystalline specimens. The essential basis for work on the physics of ice is the crystal structure and in particular the 'proton disorder', or the disorder in the orientations of the H_2O

molecules. The crystallographic basics are developed with some care and are followed by the more specialized question of how the real structure differs from the average structure.

The theme of Chapter 3 is those properties which depend on small displacements of atoms or molecules from their equilibrium sites. The first such topic is elastic deformation, and this is followed by thermal properties arising from the lattice vibrations. Then at a more detailed level we describe the spectrum of the lattice vibrations as determined by optical measurements and neutron diffraction, and we discuss the modelling of these vibrations in terms of inter- and intramolecular forces.

Chapters 2 and 3 treat topics that can be explained in terms of the perfect crystal structure, with every atom remaining close to its equilibrium site. Many properties of ice involve the motion of atoms from one site to another. This requires the presence of crystal lattice defects, and these are fundamental to the next five chapters. In ice and related hydrogen-bonded crystals there is a unique class of protonic point defects which are responsible in particular for the electrical properties. A thorough treatment of these properties is perhaps the most important portion of the book. It occupies Chapter 4, in which the theory is developed, and Chapter 5 which deals with the experimental data, first on pure ice and then on ice doped with trace impurities. This chapter concludes with more specialised topics including the behaviour of metal electrodes on ice.

The more ordinary point defects and non-electrical properties of the protonic point defects are described in Chapter 6, which ends with a review of point defect characteristics. Chapter 7 treats dislocations, stacking faults, and grain boundaries. Ice is especially well suited to the study of dislocations by X-ray topography, so that this technique plays a dominant role in this chapter, yielding unusually detailed information about dislocations in this material. These dislocations are the defects which enable plastic flow to take place, so Chapter 7 leads naturally to an account of the mechanical properties of ice in Chapter 8, but that chapter has also to deal with the brittleness of ice, and for the first time with the dominant consequences of polycrystallinity.

Chapter 9 is a short chapter on the optical properties of ice, identifying the processes involved in the absorption of electromagnetic radiation across the whole range of the spectrum and the windows of transparency in the visible and radio-frequency ranges. Birefringence is treated from a practical point of view because of its usefulness in orienting specimens. A subject which is attracting growing interest, but which is probably the least well understood, is that of the surface of ice and the nature of a possible 'quasi-liquid layer' on the surface. This subject is reviewed in Chapter 10. Chapter 11 returns to crystallography to describe the variety of crystalline phases of ice which are formed under conditions of high pressure and low temperature, together with the amorphous forms of ice and the closely related clathrate hydrates.

In Chapter 12 we review the occurrence of various forms of ice on Earth and in the Solar System, and the ways in which the physics described earlier is of relevance to its properties. This chapter has a different character from the others. Each section will be trivial to those working in that field, but the chapter is included to show ice physicists the breadth of application of their studies and to introduce specialists in one topic to other areas of study. Some of these topics have been the subject of books in their own right, and the chapter points readers to these and other sources of further information. Finally, Chapter 13 deals with ice adhesion and friction.

Many aspects of ice in the environment depend only on simple macroscopic concepts like density, latent heat and heat capacity, coupled with an understanding of the processes of nucleation from the liquid or vapour and the converse processes of melting and sublimation. These are all applications of physics to ice, but the emphasis is not so much on the physics of the material itself as on a detailed treatment of the processes involved. We do not deal with these issues here. They form the core of a book entitled *The growth and decay of ice* by Lock (1990), and particular areas are dealt with more appropriately in the specific books and reviews referred to in Chapter 12.

In one sense ice is 'just another crystalline solid', and, although some people have devoted almost the whole of their working lives to the study of ice, others merely apply to ice techniques which they have previously used on other materials. Yet we believe that ice has sufficiently distinctive properties and areas of application for it to be useful to bring them together in a single volume. Indeed, it is important to do so for the benefit of those having just a single encounter with ice, because there is great danger of error from not appreciating the special peculiarities that arise from the proton disorder and the need for fractionally charged protonic point defects. Chapters 2 and 4 are especially important for such readers.

Since 1962 the essential coherence of the physics (or physics and chemistry) of ice and the need for the exchange of ideas between workers on different aspects of ice have been recognized by the organisation of a series of International Symposia on the Physics and Chemistry of Ice. These have been reviewed by Glen (1997) and are listed together with references to their proceedings on pp. 322–3.

1.3 The water molecule

On the atomic scale ice is made up from water molecules—H_2O, and we devote the last part of this introductory chapter to the properties of these molecules and of the hydrogen bonds which bind them together in a crystal of ice. This enormously important molecule in physics, chemistry, and biology has been, and still is, the subject of intensive theoretical and experimental study. Here we will summarize only those features that are relevant to the study of ice.

The positions of the oxygen and two hydrogen nuclei in the molecule are shown in Fig. 1.1. For a free molecule the equilibrium O—H distance is 0.9572 ± 0.0003 Å and the H—O—H angle is $104.52 \pm 0.05°$. These values, which determine the moments of inertia of the molecule about its three principal axes,

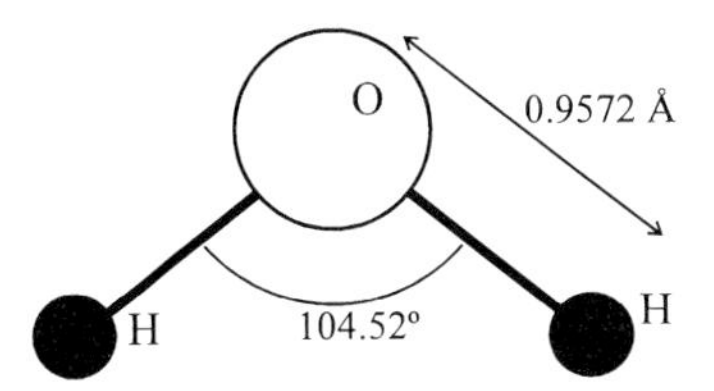

Fig. 1.1 The geometry of a free H_2O molecule.

have been deduced from the rotational components of the rotation–vibration spectrum (Benedict *et al.* 1956). It is of crucial importance to the properties of water and of ice that the molecule is bent rather than having the three atoms in a straight line. This bent form gives it a dipole moment and determines how the molecules can fit together in a crystal. The shape is a consequence of the nature of the ground state wave function of the electrons, and by way of contrast it is worth noting that, as we will see, the first electronically excited state of the molecule is linear.

The molecule contains ten electrons, eight from the oxygen atom and one from each of the hydrogen atoms. Of these ten electrons two occupy the 1s states tightly bound to the oxygen nucleus, and the remaining eight fill the eight states which correspond to the 2s and 2p states of the oxygen as perturbed by the presence of the hydrogen nuclei. Figure 1.2 shows the electron density distribution in the ground state of the molecule calculated from quantum mechanical first principles; such calculations are commonly called *ab initio* calculations. Because of the presence of the hydrogen nuclei the electronic charge is not distributed symmetrically around the oxygen nucleus. It is drawn towards the hydrogen nuclei but not sufficiently to neutralize their charge, and the molecule has an electric dipole moment oriented along the bisector of the H—O—H angle with its positive end on the hydrogen side of the molecule. For free molecules this dipole moment has been determined by Clough *et al.* (1973) to be $(6.186 \pm 0.001) \times 10^{-30}$ C m (often quoted as 1.8546 debye). This is in close agreement with the *ab initio* theoretical calculations of Xantheas and Dunning (1993). The 'centre of mass' of the electronic density distribution in Fig. 1.2 is slightly to the right of the oxygen nucleus, and this fact is incorporated in the electrostatic component of some of the empirical pair potentials to be introduced in §1.4.

The electronic excited states of the molecule have been described by Claydon *et al.* (1971). The first electronic excitation occurs under ultraviolet illumination at 7.5 eV (165 nm) to a state which, when relaxed, has a linear H—O—H singlet configuration with energy 5.79 eV above the ground state. A lower energy triplet state, which is also linear, exists at 4.76 eV but is not accessible by photo-absorption. Ultraviolet absorption at 7.5 eV can lead to photo-dissociation of the molecule. For a fuller treatment of the molecule itself see Kern and Karplus (1972). Its natural modes of vibration will be considered in §3.4.5. The important

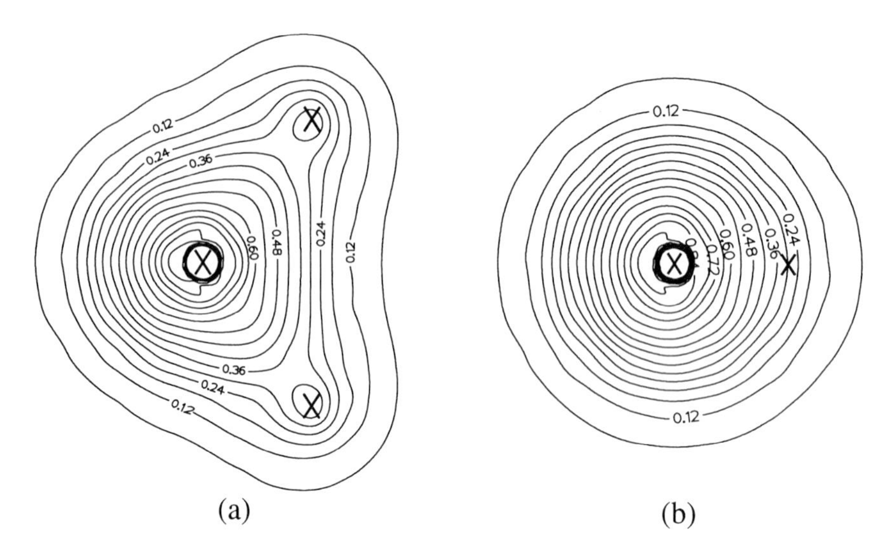

Fig. 1.2 Contour diagrams of the electron density in an isolated water molecule (a) in the plane containing the hydrogen nuclei, and (b) in the plane of symmetry perpendicular to this. The positions of the nuclei, or in (b) their projection on the plane, are marked by crosses. The plot is truncated so that the very high densities close to the oxygen nucleus are not shown. The densities were calculated by M.I.J. Probert, Cavendish Laboratory, Cambridge in 1997, with an *ab initio* density functional theory code, expanding the wave functions in a series of plane waves, using ionic pseudopotentials and the local density approximation for exchange and correlation. The units on the contour lines are electrons per (Bohr radius)^3, where the Bohr radius = 0.529 Å.

conclusion from these observations is that H_2O molecules are relatively stable entities. When they are bound together to form ice, they retain their main features, even though they are somewhat distorted and polarized by the interactions with their neighbours.

1.4 The hydrogen bond

Chemists have long recognised a distinct type of chemical bonding which arises when a hydrogen atom lies between two of the highly electronegative atoms F, O, or N (Pauling 1960, Chapter 12). This is the 'hydrogen bond'. It was proposed as being responsible for the binding of the molecules in water by Latimer and Rodebush (1920), and is now known to account well for the tetrahedral bonding of the molecules in ice. Its effectiveness for just the electronegative atoms in the first row of the periodic table is apparent from the fact that the boiling points of HF, H_2O, and NH_3 (19.5, 100, and −33.3 °C respectively) are much higher than those of the analogous compounds HCl, H_2S, and PH_3 in the next row of the table (−84.9, −60.7, and −87.4 °C) even though the latter have higher molecular weight.

In a hydrogen bond the hydrogen nucleus remains covalently bonded to one of the oxygen atoms, so that the bond is often represented diagrammatically as O—H···O, and the individual molecules remain intact. The H···O distance is much larger than the O—H covalent bond length. The molecule to which the proton is covalently bonded is sometimes called the 'proton donor' and the other molecule is the 'acceptor'. Each H_2O molecule can act as a donor for two hydrogen bonds and an acceptor for two others, with the acceptor sites located in directions tetrahedrally opposite the covalent O—H bonds. This produces tetrahedral bonding around the oxygen, geometrically similar to that in diamond. The hydrogen-bonded water molecule is often described as having 'lone pair' orbitals in the directions of the two bonds not occupied by the hydrogen nuclei, but these lone pairs are not apparent in the electron density distribution of the free molecule in Fig. 1.2. The strength of the hydrogen bond is intermediate between that of a covalent bond and the residual van der Waals interaction, giving ice a melting point mid-way between those of a covalent crystal like diamond and a rare gas like neon, both of similar molecular weight.

The simplest example of an O—H···O hydrogen bond is that in the water dimer $(H_2O)_2$, which occurs in the vapour phase and has been studied experimentally by Dyke *et al.* (1977) and Odutola and Dyke (1980). The lowest energy configuration of this dimer is shown in Fig. 1.3. It has a mirror plane of symmetry with the 'trans' configuration, in which the dipole moments of the two molecules point as far as possible in opposite directions. The proton on the hydrogen bond is thought to lie close to the line joining the oxygen atoms, though it is difficult to determine this. The H—O—H angles in the molecules are close to those of the free molecule, and the angle marked 57° shows how the hydrogen bond links to the lone pair position on the acceptor molecule even when only one molecule is attached to it. The experimental O—O distance R_{OO} is 2.976 Å, and the equilibrium point of minimum energy after correcting for anharmonic zero-point motion corresponds to $R_{OO} = 2.946$ Å.

There have been innumerable attempts at modelling the hydrogen bond interaction by quantum-mechanical calculation of the electron orbitals (Allen 1975; Finney *et al.* 1985), with considerable success in the most recent *ab initio*

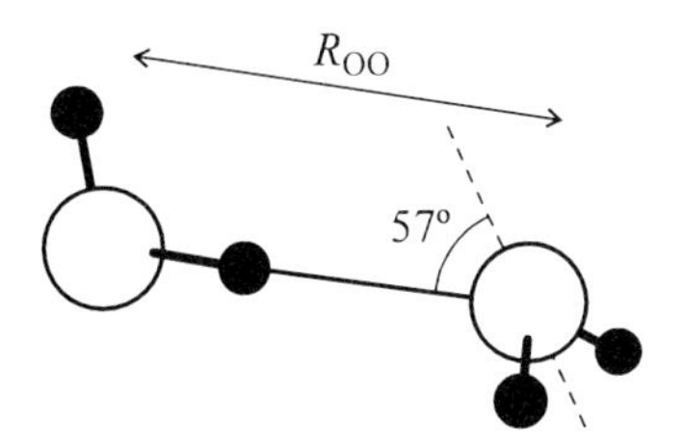

Fig. 1.3 Equilibrium configuration of the water dimer $(H_2O)_2$.

calculations of Xantheas and Dunning (1993). These calculations confirm the features of the dimer in Fig. 1.3. Because the potential minimum is rather flat, it is particularly difficult to calculate R_{OO}, but differing levels of approximation all agree with the experimental value to within 2%. The proton on the hydrogen bond moves slightly away from the oxygen nucleus of the donor molecule, but only by about 0.008 Å. The calculation and interpretation of the binding energy of the dimer presents some fundamental difficulties (Xantheas 1996*b*), but his value of 0.212 eV is consistent with the value of 0.24 ± 0.03 eV deduced from the experimental enthalpy of binding by Curtiss *et al.* (1979).

Experiments on larger hydrogen-bonded clusters (Liu *et al.* 1996) indicate that trimers, tetramers, and pentamers form closed rings, but the lowest energy form of hexamer is probably a cage. Ring clusters up to the hexamer have been modelled by Xantheas and Dunning (1993), who find that R_{OO} decreases with increasing size of the cluster towards a value of 2.8 Å which is typical of ice. However, while these calculations now show a good understanding of the nature of a hydrogen bond they apply to isolated clusters with dangling bonds on the outside, and they cannot be taken too far as models of the three-dimensional extended structure of ice. Xantheas (1996*a*) has shown that a cluster cannot be modelled accurately as a sum of pairwise interactions corresponding to single bonds. Each bond is affected by the presence of other molecules in the cluster.

Table 1.1 Examples of empirical pair potentials for H_2O molecules. All the examples below include a spherically symmetric Lennard-Jones potential of the form $A/r^{12} - C/r^6$. There are other potentials in which the spherical part is made up from exponentials or more complex shapes are used. The point 'δ' lies at a distance δ from O on the bisector of the H—O—H angle. e is the charge on a proton

Bernal and Fowler (1933)	$q = 0.49e$ on H atoms $-2q$ at point $\delta = 0.15$ Å
ST2 (Stillinger and Rahman 1974)	$q = 0.2357e$ on H atoms with H—O—H angle = 109.47° $-q$ on two sites tetrahedrally related to the H at 0.8 Å from O includes radially-dependent shielding function
TIP3P (Jorgensen *et al.* 1983)	$q = 0.417e$ on H $-2q$ on O
TIP4P (Jorgensen *et al.* 1983)	$q = 0.52e$ on H $-2q$ at $\delta = 0.15$ Å
Kozack and Jordan (1992)	$q = 0.6228e$ on H $+2q$ on O $-4q$ at $\delta = 0.138$ Å point dipole at δ of moment $\alpha\varepsilon_0\boldsymbol{E}$, where $\boldsymbol{E}$ is electric field and $\alpha = 1.47$ Å^3

This is referred to as 'hydrogen bond co-operativity' and can be described in theoretical terms by introducing 'many-body terms' in the interaction.

It is very difficult to extend *ab initio* calculations to liquid water, crystalline ice, or lattice defects in ice. For many years it has been necessary to introduce 'empirical pair potentials' (or 'effective pair potentials') to describe the hydrogen bonding in such circumstances, and such models are reviewed by Finney *et al.* (1985) (see also Morse and Rice 1982; Buch *et al.* 1998). By way of example only, a few of the commoner potentials that have been used in modelling ice are given in Table 1.1. Following the original idea of Bernal and Fowler (1933) these combine a van der Waals attraction and hard core repulsion with the electrostatic interaction arising from three, four, or five point charges suitably placed on each molecule. The parameters are chosen to fit experimental properties, usually of liquid water, or, as in the case of the more sophisticated MCY potential (Matsuoka *et al.* 1976), to correspond to theoretical predictions of *ab initio* modelling. It is beyond the scope of this book to discuss the properties of these potentials, but theoretical modelling has become an important aspect of the physics of ice, and we will at times refer to the results of such calculations. The limitations of using empirical pair potentials outside the context in which they were developed will be apparent. The direct use of *ab initio* methods is constrained by the size of the system being modelled, and the complications arising from many-body terms in the interaction present serious obstacles to the accurate modelling of the properties of ice.

2
Ice Ih

2.1 Introduction

Ice Ih is the normal form of ice obtained by freezing water at atmospheric pressure or by direct condensation from water vapour above about $-100\,^{\circ}C$. The number 'I' was assigned by Tammann (1900) following his discovery of the first of the high-pressure phases of ice, and the 'h' is commonly added to distinguish this normal hexagonal phase from a metastable cubic variant called ice Ic (see §11.8). After a brief account of how ice is formed and how specimens are produced in the laboratory, this chapter deals primarily with the crystal structure of ice Ih. The feature which is responsible for many of the special properties of ice is the disorder in the orientations of the water molecules, and special attention is given to the associated entropy and structural consequences of the disorder.

2.1.1 *The formation of ice*

The phase diagram for the equilibrium between ice Ih and the liquid and vapour phases is illustrated schematically in Fig. 2.1. Numerical data for the ice–vapour and ice–liquid boundary lines in this figure are given in Tables 2.1 and 2.2.

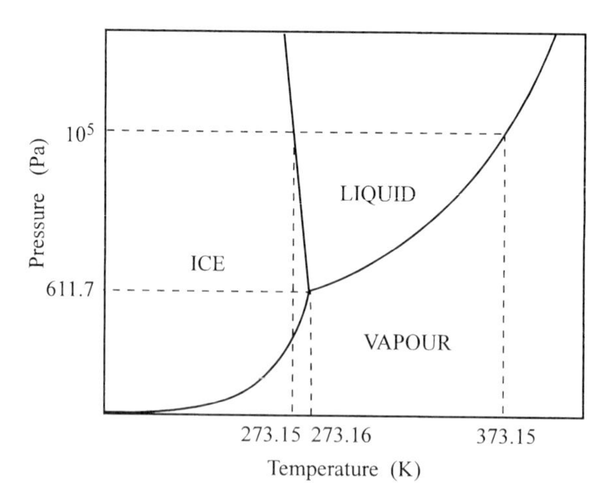

Fig. 2.1 Schematic phase diagram of water at low pressures (not to scale). Numerical data are given in Tables 2.1 and 2.2.

Table 2.1 The vapour pressure of ice Ih

Temperature (°C)	Vapour pressure (Pa)
0	611.1
−5	401.8
−10	259.9
−15	165.3
−20	103.3
−25	63.3
−30	38.0
−35	22.3
−40	12.8
−50	3.94
−60	1.08
−70	0.26
−80	0.055

Calculated from Wagner *et al.* (1994).

Table 2.2 The melting point of ice Ih as a function of pressure

Pressure (MPa)	Melting point (°C)
0.1	0.00
5	−0.36
10	−0.74
20	−1.52
50	−4.02
100	−8.80
150	−14.4
200	−20.7

Calculated from Wagner *et al.* (1994).

The triple point, where the three phases are in equilibrium, is, by definition of the Kelvin scale of temperature, at 273.16 K, and the corresponding pressure is 611.7 Pa. Water is unusual in that the melting curve has a negative slope with the melting point at atmospheric pressure being at 273.15 K, and this value is taken as the zero point of the Celsius scale of temperature. The negative slope of the melting curve is, by Le Chatelier's principle, a consequence of the fact that water expands on freezing, breaking vessels, bursting pipes and causing icebergs to float. This expansion is not unique to ice; it occurs also in silicon and germanium which have similar low-density structures in the solid state.

Table 2.3 Densities of ice and water at atmospheric pressure

	Temperature (°C)	Density ($Mg\,m^{-3}$)
Ice	0.0	0.91668
Water	0.0	0.999840
	2.0	0.999940
	4.0	0.999972
	6.0	0.999940
	8.0	0.999849
	10.0	0.999700

Data for ice from Ginnings and Corruccini (1947) and for water from Kell (1967).

Another unusual property of water is that the density of the liquid is a maximum slightly above the freezing point as shown in Table 2.3. A body of water is therefore stable against convection when freezing at the surface with water at about 4 °C at the bottom. This feature is very rare in liquids, but it is not unique; another example is In_2Te_3.

When ice condenses from the vapour it usually forms single crystals. These have a variety of shapes, including beautiful snow flakes, platelets, and less commonly needles depending on the conditions (Nakaya 1954; Kobayashi 1961). These crystals reveal the hexagonal symmetry of the crystal lattice of ice, and, in accordance with standard crystallographic convention, the hexagonal axis is denoted as the *c*-axis or [0001] in the Miller–Bravais notation appropriate for hexagonal structures. Under most conditions crystal growth is most rapid in directions perpendicular to the *c*-axis.

When liquid water freezes single crystals are nucleated initially, and these may be attached to the walls of the container or float on the surface. A few isolated nuclei form platelets lying on the surface with the *c*-axes vertical, and as these grow across the surface and then downward they form columnar grains with their *c*-axes parallel; this is called S1 ice (§12.1). If the ice is initially formed more rapidly, the surface is covered with a solid mass of randomly oriented grains, and as these spread into the liquid the grains which grow perpendicular to the *c*-axis grow most rapidly and predominate over the others. As a result a sheet of ice growing downwards from a free surface, as in a lake or at sea, will usually come to consist of long grains running perpendicular to the surface with their *c*-axes approximately horizontal (S2 ice). Polycrystalline ice with randomly oriented grains (T1 ice) can be produced in the laboratory by flooding compacted snow with water at 0 °C and then freezing (Cole 1979). Glacier ice is formed from compacted snow, but has a subsequent history of flow and recrystallization that leads to grains of complex shape with some preferred orientation. The formation of ice in the environment will be considered more fully in Chapter 12.

The solid solubility of most impurities in ice is very low, so that ice frozen from a solution usually contains liquid inclusions with a relatively high concentration of solute. This is an important feature of sea ice. On cooling the inclusions solidify at their eutectic compositions.

2.1.2 *Single crystals*

For many experiments single crystals are required. Crystals of good quality can be cut from large-grained columnar ice (Liu *et al.* 1992). At one time it was common to use large crystals which could be picked up from the Mendenhall Glacier in Alaska (Nakaya 1958). Single crystals can be formed in the laboratory simply by cooling an open vessel containing water under controlled conditions (Turner *et al.* 1987; Knight 1996*a*), or much more quickly and stably if surface cooling is achieved by rapid evaporation from the surface under reduced pressure (Khusnatdinov and Petrenko 1996). Bilgram *et al.* (1973) describe how a polycrystalline ingot can be converted to a large high-quality single crystal by zone refining. Single crystals have also been grown by variants of the Czochralski technique in which a cooled seed crystal is gradually lifted from the liquid held at 0 °C, but the highest quality crystals are produced by growing the ice through a capillary into a wider glass tube. The most stable conditions for growth are obtained when the ice forms above the liquid as in the Sapporo technique (Oguro 1988) shown in Fig. 2.2(a). This method requires a seed crystal, and, because of the expansion on freezing, it cannot be used with the water in a sealed container. These problems do not arise in the Birmingham technique where the ice grows from the bottom as in Fig. 2.2(b) (Ohtomo *et al.* 1987). As the growth tube is lowered into the column of cold antifreeze, polycrystalline ice is first formed in the nucleation bulb, and the ice then grows as a single crystal through the twisted capillary into the main part of the tube. However, in this method the liquid is unstable against convection and the temperature distribution has to be carefully controlled to eliminate a gradient in the upper part of the tube.

2.1.3 *Specimen preparation*

Ice can flow slowly in deformation experiments or in glaciers, but when specimens are being handled it behaves as a brittle material. There is no tendency for cleavage to follow any particular crystal planes or grain boundaries. In spite of its brittleness ice can be cut at around −10 °C with a circular saw, band saw, or lathe. It can be shaved with a razor blade or microtome, and polished on abrasive or ordinary paper sometimes moistened with alcohol. Any surface damage produced by such treatments anneals out within days at such temperatures.

Ice is a weakly birefringent uniaxial crystal, and this birefringence can be used to establish the orientation of the *c*-axis of a single crystal or of a grain within a

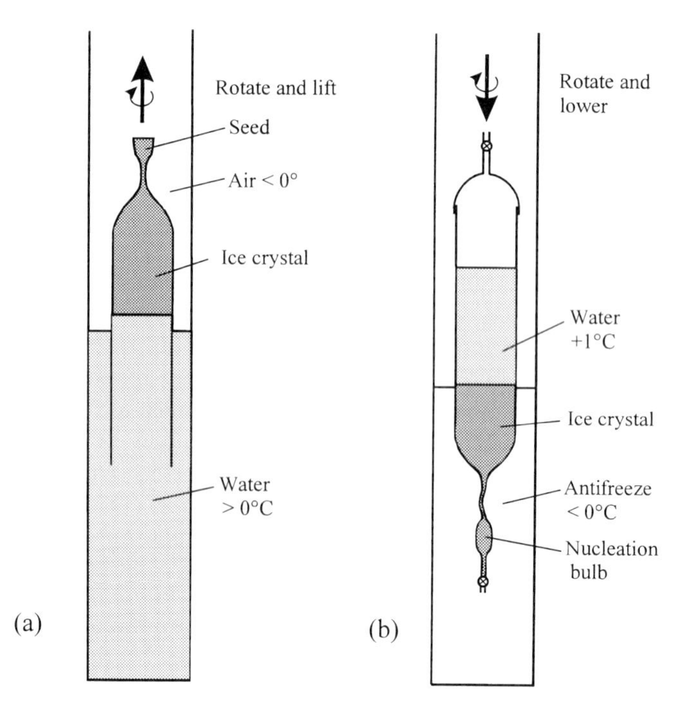

Fig. 2.2 Systems for the growth of high-quality single crystals of ice. (a) Sapporo technique in which the crystal is raised from the liquid. (b) Birmingham technique in which the growth tube is lowered into cold antifreeze.

section of polycrystalline ice (see §9.4). In approximately parallel light a crystal will appear dark between crossed polaroids if the projection of the *c*-axis lies parallel or perpendicular to the plane of polarization. When a thick crystal is viewed along the *c*-axis between crossed polaroids in convergent light (conoscopic illumination) a dark cross is seen superimposed on coloured circular fringes (Fig. 9.4). Together these features permit unique identification of the *c*-axis. The directions of the axes in the (0001) plane perpendicular to the *c*-axis are more difficult to establish. On a polished (0001) face hexagonal etch pits can be formed with their edges along the *a*-directions (Higuchi 1958). Alternatively one can observe the $\{1\bar{1}00\}$ facets on frost that grows on the surface, or the orientation of the 'Tyndall flowers' or 'melt figures' produced by internal melting under infrared radiation (S. Steinemann 1954). Ultimately of course absolute orientations are established by X-ray diffraction.

As shown in Table 2.1 ice has an appreciable vapour pressure, and specimens exposed to dry air in a typical cold room or refrigerator will deteriorate visibly by evaporation over a period of several hours. Specimens in sealed containers grow frost-like dendrites, though on a much slower time scale. Specimens are often preserved in kerosene or similar liquids.

Table 2.4 Effects of isotopic substitution

	H_2O	D_2O	T_2O	$H_2{}^{18}O$
Triple point (°C)	0.01	3.82[a]	4.49[b]	0.31[c]
Temperature of maximum density (°C)	3.9834[d]	11.2[a]	13.4[e]	4.30[f]

[a]Kirshenbaum (1951); [b]W.M. Jones (1952); [c]Nagano *et al.* (1993); [d]Watanabe (1991); [e]Goldblatt (1964); [f]Steckel and Szapiro (1963).

2.1.4 *Isotopic substitution*

The replacement of the hydrogen in ice by its heavy isotope deuterium (^2H, denoted here as D) has important effects on some properties and is essential for certain neutron diffraction experiments. Table 2.4 shows the effects of this and other isotopic substitutions on the melting point and the related temperature of maximum density of the liquid. $H_2{}^{18}O$ has the same molecular mass as D_2O but the change of H to D produces a more significant effect than the change of the oxygen isotope, reflecting the role of the hydrogen in the bonding. T_2O (i.e. ^3H_2O) continues the trend, but it is radioactive and prone to self heating, so that it is not used for ice physics experiments. Depending on its origin, normal water contains about 0.015% of deuterium and 0.2% of ^{18}O, but for our purposes it can be treated as pure $H_2{}^{16}O$. The geophysical significance of changes in the proportion of ^{18}O is described in §12.5.

2.2 Crystal structure

2.2.1 *The Pauling model*

The basic structure of ice Ih is well established to be that proposed by Pauling (1935) and illustrated in Fig. 2.3. The oxygen atoms, shown by the open circles, are arranged on a hexagonal lattice with a structure named after the mineral 'wurtzite' (the hexagonal form of ZnS). Each oxygen atom has four nearest neighbours at the corners of a regular tetrahedron. The hydrogen atoms, shown as the dark spots, are covalently bonded to the nearest oxygens to form H_2O molecules, and these molecules are linked to one another by hydrogen bonds, each molecule offering its hydrogens to two other molecules and accepting hydrogen bonds from another two as explained in §1.4. The essential feature of Pauling's model is that there is no long-range order in the orientations of the H_2O molecules or of the hydrogen bonds.

The disorder in the three-dimensional structure shown in Fig. 2.3 is difficult to visualize, and can be appreciated more easily in the larger two-dimensional layer shown in Fig. 2.4. There are two possible hydrogen sites on each bond and four of

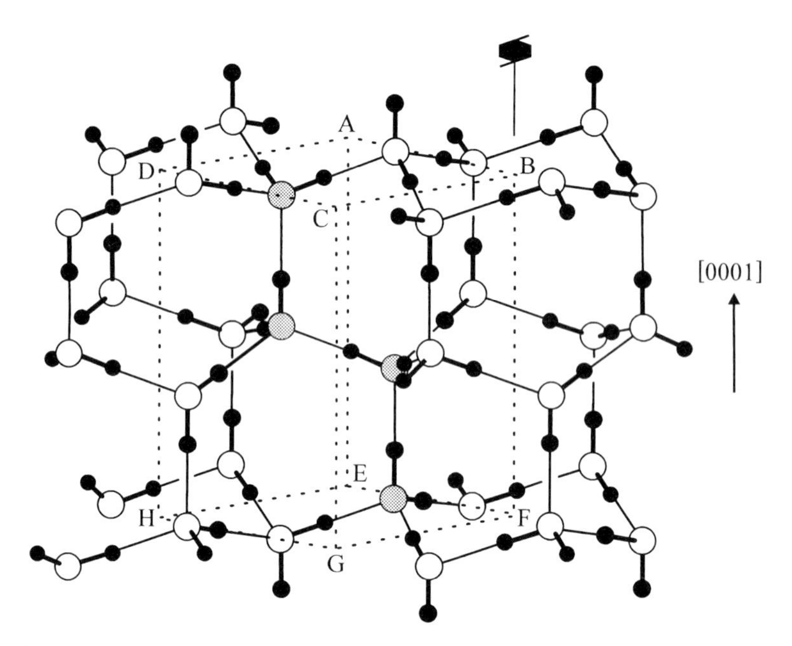

Fig. 2.3 The crystal structure of ice Ih, showing a particular disordered arrangement of the molecules, represented as in Fig. 1.1. The unit cell of the average structure is marked ABCDEFGH, and the four oxygen atoms contained within this cell are shown shaded. One 6_3 axis of symmetry of this structure lies along BF as indicated.

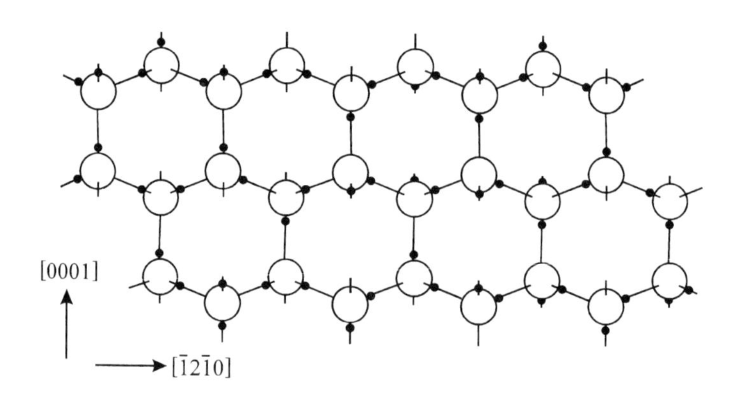

Fig. 2.4 A layer of the ice structure projected on the ($10\bar{1}0$) plane.

these sites adjacent to each oxygen. The disorder of the hydrogens over these sites satisfies the two *ice rules*, which are:

1. There are two hydrogens adjacent to each oxygen.
2. There is only one hydrogen per bond.

Pauling's hypothesis was that 'under ordinary conditions the interaction of non-adjacent molecules is not such as to stabilize any one of the many configurations satisfying [the above ice rules] with reference to the others'. Many of the properties of ice depend fundamentally on this limited degree of disorder within the crystallographically ordered arrangement of the oxygen atoms.

The historical background to Pauling's model and its subsequent confirmation are important to an understanding of the nature of ice. Ice was naturally an early material for study by X-ray crystallography, but X-rays are scattered by electrons and the electrons surround the oxygen nucleus with rather little perturbation to reveal the location of the hydrogen nuclei (see Fig. 1.2). Dennison (1921) was the first person to correctly determine the unit cell parameters, and Bragg (1922) used these parameters to propose a structure with oxygen ions at the sites shown in Fig. 2.3 and hydrogen ions midway between them. The hexagonal lattice and the locations of the oxygens were confirmed in detailed single-crystal diffraction experiments by W.H. Barnes (1929). Bernal and Fowler (1933) argued that the water molecules, which were known in the vapour phase to have the V-shape shown in Fig. 1.1, would remain intact in both water and ice, and this was supported by the similarity of the Raman spectra in all three phases (see §3.4.5). However, there is no way in which such molecules can be arranged on the four sites within the observed unit cell so as to preserve hexagonal symmetry. The simplest regular arrangement that retains hexagonal symmetry has a 12-molecule cell, but would still give the observed X-ray diffraction from the oxygens. This arrangement of the hydrogens gives a polar structure, but at the time that seemed plausible. Bernal and Fowler suggested that 'at temperatures just below the melting point the molecular arrangement is still partially or even largely irregular.' Pauling took the bolder step of supposing that the energies of all configurations of hydrogens satisfying the ice rules are so nearly equal that no particular ordering of the hydrogens will be stabilized at any ordinarily encountered temperature. The ice rules as we have stated them have often been referred to as the 'Bernal–Fowler rules', though their explicit formulation might better be attributed to Pauling.

The hydrogen disorder makes a contribution to the entropy of ice which will be calculated in §2.3. The fact that ice had an excess entropy had been established by Giauque and Ashley (1933), and Pauling devised his model to account for this; the idea is not unique to ice. According to the third law of thermodynamics the disordered state cannot remain the equilibrium state down to the lowest temperatures, and on cooling there should be a phase transition to a more ordered phase. If the energy gained in such an ordering transformation is ΔE and the entropy of disorder is ΔS, the transformation would occur at the temperature T at which the decrease in free energy $(\Delta E - T\Delta S) > 0$. Pauling's assumption is just that $\Delta E < T\Delta S$ for all ordinarily encountered temperatures. We now know that the temperature below which the ordered phase, called ice XI, is stable is 72 K; this phase will be described in §11.2. However, this ordered state is not easily attained, because, if the ice rules are totally satisfied, the orientation of any molecule is determined by the configuration of its neighbours and there is no mechanism by which ordering can occur. The transformation process requires defects that violate the ice rules, and in pure ice such defects have negligible concentration in thermal equilibrium at 72 K.

Another feature of Pauling's model is that it allows the dielectric permittivity of ice to be similar to that of liquid water. The water molecule has a dipole moment and this can be oriented in an electric field giving a permittivity varying with temperature as $1/T$. However, in ice the polarization process is many orders of magnitude slower than in water because the molecules are not free to reorient independently of one another. This opens up the whole question of dynamical processes in ice that depend on defects which violate the ice rules, and this is the subject of Chapter 4.

The location of the hydrogen atoms could not be determined crystallographically until the availability of neutron beams, which are scattered by the nuclei rather than the electrons in a crystal. The first such experiment on ice was that of Wollan *et al.* (1949). They used D_2O because neutron diffraction from ordinary H gives a large incoherent background which is almost absent for D. In Bragg diffraction the neutron wave is diffracted by the instantaneous configuration of a large number of cells, and what is deduced from such measurements is the average over these cells. Thus if in Pauling's model the hydrogen is at the top end of a bond in some cells and at the bottom end in an equal number of other cells, the diffraction pattern will be interpreted as coming from a single cell with the scattering power of half a hydrogen at each of the two sites. Wollan *et al.* observed six separate lines in the powder diffraction pattern and showed that their intensities corresponded more closely to this 'half-hydrogen model' than to other possibilities such as the ionic structure proposed by Bragg.

Definite confirmation of the model was provided by a detailed single-crystal neutron diffraction experiment on D_2O by Peterson and Levy (1957). From their data they constructed the Fourier map of the neutron scattering density shown in Fig. 2.5. This is a projection along the a or $[11\bar{2}0]$ direction equivalent to AC in Fig. 2.3. For the two O—O bonds in the plane of the figure the scattering clearly appears to come from two half-deuterons at sites part way along each bond, though in this projection the two sites on bonds coming out of the plane are not resolved. Goto *et al.* (1990) have performed an equivalent analysis using the most precise X-ray diffraction data. It is possible to detect the excess electron density associated with the half-hydrogen atoms, but the electronic distribution is not centred on the hydrogen sites determined by neutrons.

In addition to the Bragg scattering observed by Peterson and Levy, the disordered deuterons should give rise to elastic diffuse scattering into all angles with a distribution determined by the ice rules, in just the same way that neutron scattering reveals features about the structure of a liquid. This diffuse scattering from ice has been observed and theoretical calculations show that it has the general features expected from the Pauling model (Axe and Hamilton, quoted by Kamb 1973; Schneider and Zeyen 1980; Li *et al.* 1994). It also contains more information and we will return to this topic in §2.5.2.

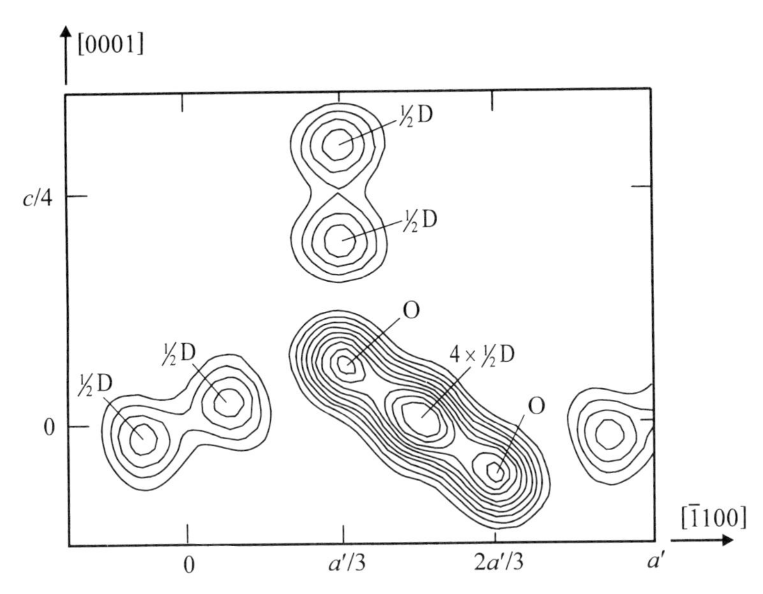

Fig. 2.5 Contour plot of neutron scattering density from D_2O ice, projected on the $(11\bar{2}0)$ plane DBFH of Fig. 2.3 with $a' = \frac{1}{2}$FH. (From Peterson and Levy 1957.)

A final question is whether the half-hydrogen model represents not just an average over many unit cells but a fundamental quantum-mechanical delocalization of each proton. An analogy would be the delocalization of the electrons in benzene, where one cannot distinguish between the single and double bonds in the Kekulé model (Pauling 1960, pp. 187–8). In ice the true eigenstates might be linear combinations of the supposedly degenerate Pauling configurations, leading to energy splitting arising from tunnelling between these states. Such speculation is believed to be inappropriate for ice Ih. As will be seen later (Table 2.7) the amplitude of the zero point motion of a proton on a hydrogen bond is only about 0.1 Å, so that the proton is well localized and individual molecules retain their identity. The small energy differences between the Pauling configurations are probably larger than any possible quantum-mechanical splitting. We can safely assume that the microstates described by the ice rules form an appropriate description of the system, and the whole of our treatment of the properties of ice Ih will be developed on this basis. In Chapter 11 we will see how this picture of ice being made up of distinct molecules breaks down for phases observed at extremely high pressures.

2.2.2 *The average structure*

As explained in §1.3 a free H_2O molecule has an H—O—H angle of 104.52°. This is different from the tetrahedral angle of 109.47° expected from Fig. 2.3 and there is nothing to require the molecule in ice to adopt the tetrahedral bond angle.

The molecule may also be displaced off its site in a direction that depends on the configuration of the hydrogens on the neighbouring molecules. However, crystallographic studies using Bragg diffraction reveal only the average atomic positions together with root-mean-square displacements from these sites which arise from the combined effects of disorder and thermal (including zero-point) motion. In this section we present the very precise information that is now available about this average structure, and we will consider what can be learnt about the actual structure in §2.5.

In the structure shown in Fig. 2.3 the oxygen atoms lie in layers perpendicular to the c-axis which are built up of puckered hexagonal rings. These layers are stacked in the sequence ABABAB... typical of hexagonal close-packed metals. This results in long open channels running parallel to the c-axis. The puckering of the rings forming these layers consists of displacements out of the plane following the sequence $+-+-+-$, and these rings are said to be 'chair-like'. The structure also contains hexagonal rings in planes perpendicular to these layers. They lie vertically in Fig. 2.3, and have the different puckering sequence $+--+--$; these rings are said to be 'boat-like'. In the metastable structure of ice Ic the stacking is ABCABC... and all hexagonal rings are chair-like. The question of stacking will be further discussed in §7.6.1.

The space group for the average structure with a half-hydrogen on each possible site is $P6_3/mmc$, with the result that the macroscopic properties of ice have the highest possible hexagonal point group symmetry P6/mmm. A single unit cell, located according to the standard conventions of the International Tables for Crystallography, is marked ABCDEFGH in Fig. 2.3, and it is drawn in isolation in Fig. 2.6 in which the hydrogen sites are shown as half-filled symbols. The 6_3 axis passes through the centre of the horizontal hexagonal rings and transposes one horizontal layer of molecules into the next by a rotation through $2\pi/6$ and translation by $3c/6$. There is a centre of symmetry at the centre of each horizontal ring and at the mid-point of each O—O$''$ bond (i.e. the bonds that form the horizontal puckered layers).

In a hexagonal crystal planes and directions are designated by the four-index Miller–Bravais system which reveals the true equivalence of different directions. This will be used throughout this book, and some typical directions are indicated this way in Fig. 2.6. The a-axis is one of type $\langle 2\bar{1}\bar{1}0\rangle$. For certain purposes it is more convenient to work with mutually perpendicular axes and in this case the C-centred eight-molecule orthorhombic cell shown in Fig. 2.7 is appropriate. The hexagonal symmetry requires that in this cell $b = a\sqrt{3}$.

The most precise measurements of the lattice parameters were made by Röttger *et al.* (1994) using synchrotron radiation, and were subsequently confirmed by Line and Whitworth (1996) using neutrons. These values for both H_2O and D_2O are tabulated as functions of temperature in Table 2.5 and plotted in Fig. 2.8. The thermal expansivity is isotropic, and has the property of being negative below 73 K (see §3.3.2). The c/a ratio is almost independent of temperature at 1.62806 ± 0.00009 for H_2O and 1.62828 ± 0.00012 for D_2O. This is close to the

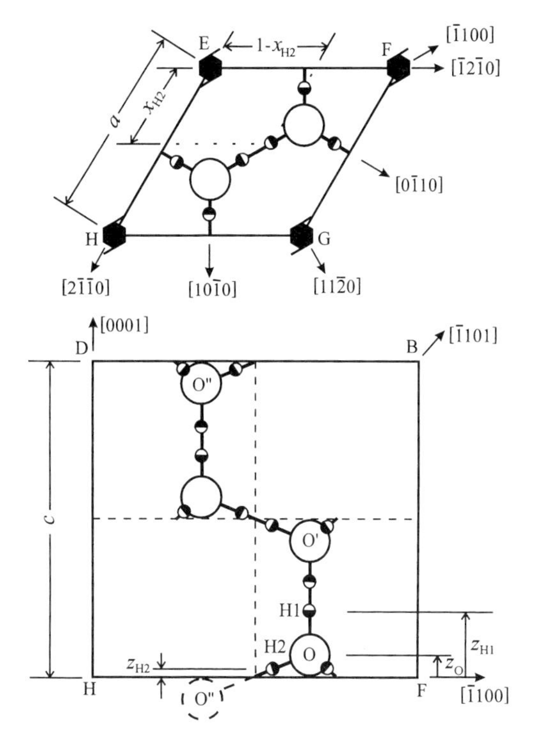

Fig. 2.6 Unit cell of the structure of ice Ih projected on (0001) and (11$\bar{2}$0) planes, and showing notation for atomic sites and co-ordinates. Half-filled circles represent possible sites for H in the half-hydrogen model.

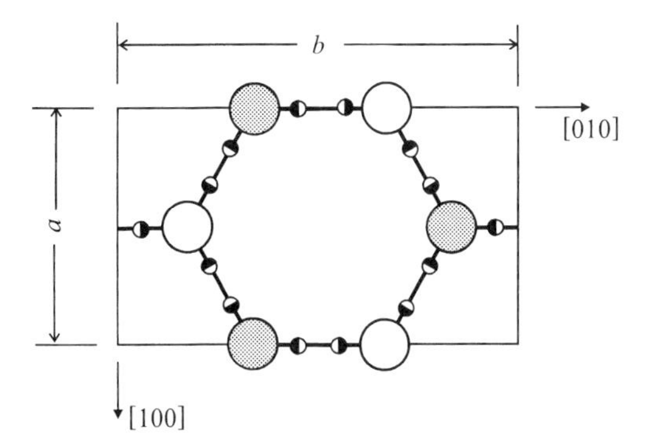

Fig. 2.7 C-centred eight-molecule orthorhombic cell for ice Ih, projected on (001). To retain hexagonal symmetry $b = a\sqrt{3}$. In a given layer the oxygen atoms shown by open and shaded circles are at levels $\pm z_O$.

Table 2.5 Lattice parameters of ice Ih

Temperature (K)	H_2O		D_2O	
	a (Å)	c (Å)	a (Å)	c (Å)
10	4.4969	7.3211	4.4982	7.3235
25	4.4967	7.3205	4.4980	7.3229
40	4.4967	7.3205	4.4977	7.3226
55	4.4964	7.3200	4.4973	7.3216
70	4.4959	7.3198	4.4969	7.3226
85	4.4961	7.3198	4.4972	7.3220
100	4.4966	7.3204	4.4977	7.3228
115	4.4975	7.3219	4.4986	7.3250
130	4.4988	7.3240	4.5001	7.3267
145	4.5002	7.3268	4.5014	7.3303
160	4.5021	7.3296	4.5035	7.3334
175	4.5042	7.3332	4.5057	7.3369
190	4.5063	7.3372	4.5083	7.3412
205	4.5088	7.3411	4.5114	7.3460
220	4.5117	7.3447	4.5146	7.3508
235	4.5148	7.3503	4.5180	7.3558
250	4.5181	7.3560	4.5216	7.3627
265	4.5214	7.3616	4.5266	7.3688

Data from Röttger *et al.* (1994). Precision typically 0.0002 Å in a and 0.0006 Å in c.

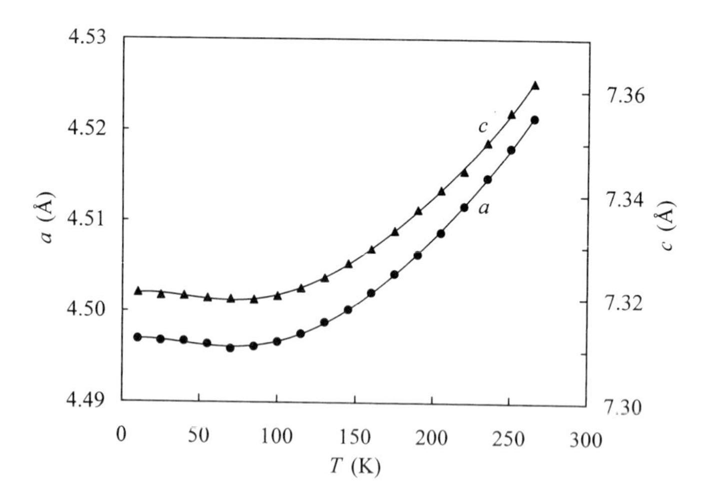

Fig. 2.8 Lattice parameters of H_2O ice Ih as a function of temperature (from Table 2.5).

ideal value of $2\sqrt{(2/3)} = 1.63299$ predicted for perfect tetrahedral symmetry around each oxygen, but as the symmetry permits a small difference it is not surprising that there is one.

For H_2O at a typical temperature of 253 K (−20 °C) the lattice parameters are $a = 4.519$ Å and $c = 7.357$ Å. With four molecules per unit cell these give a density of 3.074×10^{28} molecules per m^3, which is equivalent to $0.9197\ \mathrm{Mg\,m^{-3}}$. Allowing for thermal expansion (§3.3.2) this is in agreement with the experimental value at 0 °C given in Table 2.3. For perfect tetrahedral co-ordination all O—O bonds have the same length r_{OO}, and the number of molecules per unit volume N is related to r_{OO} by the equation

$$N = \frac{3\sqrt{3}}{8r_{OO}^3}. \tag{2.1}$$

The mean value of r_{OO} at 253 K is 2.764 Å.

Careful determinations of the atomic co-ordinates in the unit cell have been reported by Kuhs and Lehmann (1986) and values based on this work are summarized in Table 2.6. The site co-ordinates are expressed as fractions x and z of the lattice vectors a and c measured parallel to the corresponding edge of the unit cell as marked in Fig. 2.6. Within the unit cell there are four oxygen atoms at sites

$$\tfrac{1}{3}, \tfrac{2}{3}, z_O; \quad \tfrac{1}{3}, \tfrac{2}{3}, \tfrac{1}{2} - z_O; \quad \tfrac{2}{3}, \tfrac{1}{3}, \tfrac{1}{2} + z_O; \quad \tfrac{2}{3}, \tfrac{1}{3}, 1 - z_O.$$

The parameter z_O represents the puckering of the hexagonal rings lying in the plane perpendicular to the c-axis. For perfect tetrahedral symmetry around each oxygen site $z_O = \frac{1}{16} = 0.0625$, but in ice it is 0.0622 ± 0.0001, representing a slight flattening of the ring and an opening of the angle O″—O—O″ between

Table 2.6 Crystallographic parameters for ice Ih

	H_2O		D_2O	
	15 K	223 K	15 K	223 K
z_O	0.0622	0.0623(2)	0.0621	0.0621(2)
z_{H1}	0.1999(2)	0.1989(4)	0.1990	0.1983(2)
x_{H2}	0.4552	0.4540(2)	0.4545	0.4539
z_{H2}	0.0169	0.0167(3)	0.0169	0.0171
O—O′ (Å)	2.750	2.759(2)	2.752	2.763
O—O″ (Å)	2.751	2.761	2.752	2.762
O—H1 (Å)	1.008	1.004(3)	1.002	1.002(2)
O—H2 (Å)	1.006	1.000(2)	1.000	1.000
O′—O—O″ (deg)	109.33	109.36(4)	109.32	109.36(2)
O″—O—O″ (deg)	109.61	109.58(4)	109.63(2)	109.58(2)

Selected data based on Kuhs and Lehman (1986) with revisions by W.F. Kuhs in 1996. Precision (estimated std. dev.) is within ±1 in last digit except where larger figure is given in parentheses.

bonds lying in the ring from $2\sin^{-1}\sqrt{(2/3)} = 109.47°$ to $109.61°$. The angle O′—O—O″ between a vertical and an oblique bond is correspondingly reduced to 109.33°. Changes to the oxygen positions with temperature or isotope are barely significant.

The half-occupied hydrogen sites are of two kinds. There are four of type H1 on the O—O′ bonds parallel to the c-axis with co-ordinates like $\frac{1}{3}, \frac{2}{3}, z_{H1}$. Then there are 12 sites of type H2 on the oblique bonds O—O″. Those just below the oxygen site O have co-ordinates

$$x_{H2}, 1 - x_{H2}, z_{H2}; \quad x_{H2}, 2x_{H2}, z_{H2}; \quad 1 - 2x_{H2}, 1 - x_{H2}, z_{H2},$$

and there are equivalent sites with $z = \frac{1}{2} - z_{H2}$, $\frac{1}{2} + z_{H2}$, and $1 - z_{H2}$. The values of x_{H2} and z_{H2} are related so that these sites lie on the O—O″ bonds.

Table 2.6 gives values of the inter-site distances calculated from the fractional co-ordinates and the lattice parameters. There is no significant difference between the O—O′ and O—O″ or the O—H1 and O—H2 distances, which implies that all the hydrogen bonds are effectively the same, but we must remember that these distances are not actual bond lengths because of displacements arising from the disorder. The spread of the atomic positions about their average sites is described by the root mean square (r.m.s.) displacements $\langle u_O^2 \rangle^{1/2}$, $\langle u_{H\|}^2 \rangle^{1/2}$, $\langle u_{H\perp}^2 \rangle^{1/2}$, where the symbols $\|$ and $\perp$ refer to displacements along and perpendicular to the O—O bonds. Experimental values for these displacements at several temperatures are given in Table 2.7. To first order the oxygen displacements are isotropic, but for hydrogen the displacement perpendicular to the O—O bond is larger than that along it. At low temperatures these displacements are dominated by zero-point motion, but as the temperature rises thermal motion becomes apparent. As expected the displacements of the hydrogens are somewhat larger for H than for D. The contribution due to structural disorder will be considered in §2.5.2.

Table 2.7 Root-mean-square displacements $\langle u^2 \rangle^{1/2}$ (Å) for atoms in ice Ih

Temperature		15 K	123 K	233 K	275 K
$\langle u_O^2 \rangle^{1/2}$	H_2O	0.095	0.158	0.212	
	D_2O	0.093	0.153	0.208	0.232
$\langle u_{H\|}^2 \rangle^{1/2}$	H_2O	0.114	0.167	0.218	
	D_2O	0.104	0.159	0.210	0.226
$\langle u_{H\perp}^2 \rangle^{1/2}$	H_2O	0.167	0.206	0.243	
	D_2O	0.143	0.186	0.226	0.252

Data from refinement of neutron diffraction data in the harmonic approximation. The r.m.s. displacements are related to the crystallographic atomic displacement parameter B by the relation $B = 8\pi^2\langle u^2 \rangle$. (W.F. Kuhs, private communication 1991. See also Kuhs and Lehmann 1986; Kuhs and Knupfer 1992.)

2.3 Zero-point entropy

2.3.1 *The Pauling approximation*

Pauling's model was devised in part to account for the known zero-point entropy of ice. Boltzmann's relation, $S = k_B \ln W$, relates the zero-point entropy S_0 to the total number W of configurations of the hydrogen positions in the lattice that are consistent with the ice rules; k_B is Boltzmann's constant. W includes all possible arrangements, even those which exhibit macroscopic order or inhomogeneity, but, as is usual in the statistical mechanics of large numbers of particles, the result is completely dominated by arrangements which appear totally disordered. The entropy of ice can be calculated with remarkable precision by very simple arguments and we will give three different versions, each of which contributes to our understanding of the disorder in ice. The first two are based on the original methods of Pauling (1935).

Consider a crystal of N molecules large enough that the surfaces can be ignored. There will be $2N$ bonds between these molecules with one proton (or H atom) on each. Each of the molecules can be placed in the lattice in any one of the six orientations shown in Fig. 2.9, and if we ignore the second ice rule that there is only one proton on each bond there are 6^N possible arrangements of these molecules. Each bond within these possible arrangements will then have one of four equally probable states: HH, H–, –H, or – –, where H indicates the presence and – the absence of a proton. There is thus a probability of $\frac{1}{2}$ that each bond is correctly formed. The fraction of the 6^N arrangements in which all $2N$ bonds are correctly formed is therefore $(\frac{1}{2})^{2N}$, and the total number of acceptable configurations of the crystal is

$$W = 6^N \left(\tfrac{1}{2}\right)^{2N} = \left(\tfrac{3}{2}\right)^N, \tag{2.2}$$

giving

$$S_0 = Nk_B \ln\left(\tfrac{3}{2}\right). \tag{2.3}$$

Alternatively we may treat the crystal as an arrangement of $2N$ bonds each of which has two possible orientations. Ignoring the first ice rule that there are two protons adjacent to each oxygen, there are 2^{2N} possible arrangements of the bonds. At each oxygen site there are 16 possible arrangements with zero to four

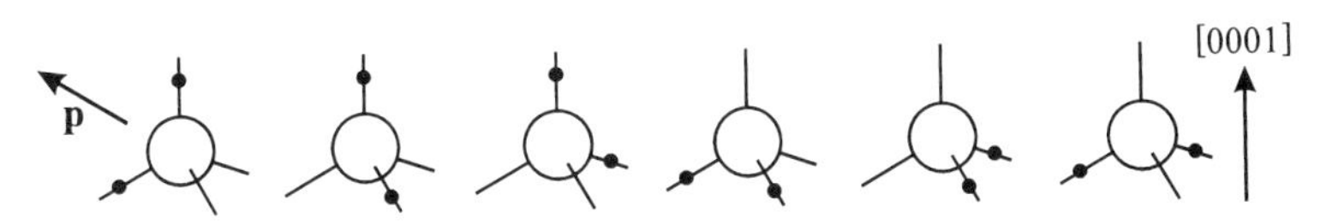

Fig. 2.9 The six possible orientations of an H_2O molecule at a given site in the lattice. The electric dipole moment of the first orientation is represented by the vector **p**.

protons on the four bonds. Of these arrangements only the six shown in Fig. 2.9 represent a correctly formed molecule, so the probability that all N molecules are correctly formed is $(\frac{6}{16})^N$, giving

$$W = 2^{2N}\left(\frac{6}{16}\right)^N = \left(\frac{3}{2}\right)^N, \tag{2.4}$$

which is the same as before. This second approach illustrates the fact that although energetically a crystal of ice should be thought of as an assembly of loosely bound molecules, it is topologically possible to consider it as an assembly of bonds, and this equivalence is important in understanding issues concerning dielectric polarization.

We give one further approach, based on that of Bjerrum (1951), because it shows how an ice crystal is built up in just one of its many possible configurations; it can only be changed into another configuration by processes which involve local violations of the ice rules. We imagine that a crystal of ice has been built up to the point illustrated in Fig. 2.10, where a layer growing from bottom left has formed on top of a lower (0001) layer up to the line X—Y. The next molecule M to be added to the second layer must make bonds to P in the top layer and to Q in the third layer. The following molecule N bonds to M and R, and so on. Ignoring edge effects the whole crystal can be built up in such a way that each added molecule forms just two bonds to an existing assembly. For the addition of molecule M there are four possibilities for the bonds to it from P and Q:

1. there are already protons adjacent to both P and Q;
2. there are no protons adjacent to P or Q;
3. there is a proton at P but not at Q;
4. there is a proton at Q but not at P.

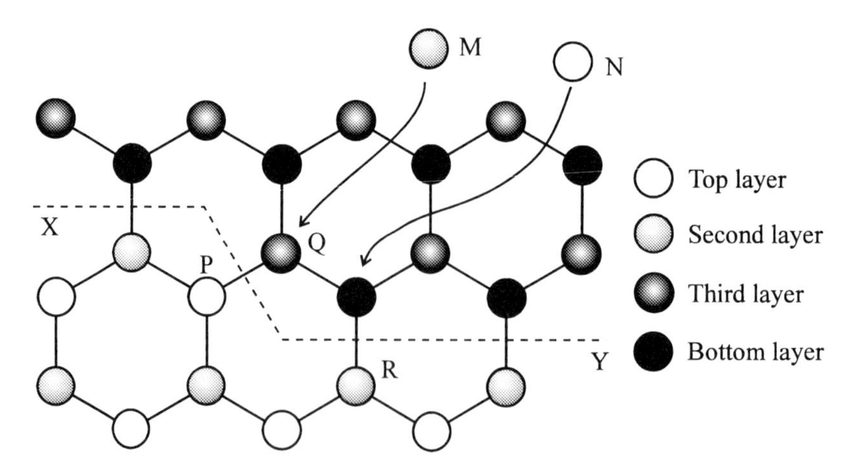

Fig. 2.10 Molecular positions for four layers projected on the (0001) plane, and used in calculating the entropy of disorder according to the method of Bjerrum (1951).

Let the total number of configurations of the rest of the crystal that result either in state 1 or in state 2 be A, and the number that result either in state 3 or in state 4 be B. There is only one way in which molecule M can be added to state 1 or to state 2, so this does not change A. For either of cases 3 or 4 there are two possible orientations of M, so that on adding M the total number of possible configurations of the crystal increases from $A + B$ to $A + 2B$. If the bonds presented to M by P and Q are uncorrelated, the probabilities of states 1 to 4 are all equal so that $A = B$. The addition of one molecule thus increases the number of configurations by the factor $(A + 2B)/(A + B) = \frac{3}{2}$, giving, as in Pauling's argument, $W = (\frac{3}{2})^N$.

In the simple Pauling approximation the zero-point entropy of one mole of ice is

$$S_0 = R \ln\left(\tfrac{3}{2}\right) = 3.371 \text{ J K}^{-1} \text{ mol}^{-1}. \tag{2.5}$$

2.3.2 *Theoretical refinements*

The first two calculations of the zero-point entropy given above would be exact if the structure were dendritic (i.e. all bonds spread out from a single point like the branches of a tree), but this is not the case, and the presence of closed loops in the structure introduces some error. Consider a given closed ring in our first calculation. When we have satisfied the conditions that five bonds around the ring are correctly formed with probability $\frac{1}{2}$, the probability that the sixth bond is also correct is slightly greater than $\frac{1}{2}$, and the calculation therefore underestimates W. Hollins (1964) showed that the correction factor for a single ring is $(1 + \frac{1}{729})$, and as the crystal contains $2N$ such rings the corrected result is

$$W = \left(\tfrac{3}{2}\right)^N \left(1 + \tfrac{1}{729}\right)^{2N} = 1.5041^N. \tag{2.6}$$

In Bjerrum's argument the approximation arises from the assumption that the bonds to P and Q are uncorrelated. The real problem is, however, more complicated because the rings have bonds in common. No exact solution has been found, but the best approximation is that of Nagle (1966), who used a series summation to give

$$W = (1.50685 \pm 0.00015)^N. \tag{2.7}$$

This gives

$$S_0 = 3.4091 \pm 0.0008 \text{ J K}^{-1} \text{ mol}^{-1},$$

which differs from the Pauling approximation (eqn (2.5)) by only about 1%.

A hypothetical structure which obeys the ice rules is 'square ice', in which the oxygen atoms lie on a two-dimensional square lattice, each atom linked by hydrogen bonds to its four neighbours. In the Pauling approximation square ice has the same entropy as ice Ih, but the correction for the four-membered rings is greater than that for the six-membered rings in ice Ih. In Nagle's theory $W = (1.540 \pm 0.001)^N$, but for this case Lieb (1967) has obtained an exact solution $W = (\frac{4}{3})^{3N/2} = (1.5396\ldots)^N$.

2.3.3 *Experimental*

Experimental evidence for the zero-point entropy of ice Ih comes from a comparison of the entropy of water vapour calculated from the properties of the vapour with that obtained from calorimetric measurements of the ice–water–vapour system. The absolute entropy of an ideal gas is given by the Sackur–Tetrode equation, and can be combined with spectroscopic data about molecular rotations and vibrations to give the entropy $S^{\ominus}$ of water vapour in its thermodynamic standard state (25 °C and 0.1 MPa). The calorimetric determination involves integrating $C_p\,dT/T$ from near 0 K and adding contributions for the latent heats of melting and vaporization. However, there is a problem because ice exhibits a small degree of slow partial ordering between 90 and 115 K (Haida *et al.* 1974). This effect is typical of a glass transition and occurs above the temperature (72 K) at which the transformation to ice XI occurs in suitably doped ice (§11.2). In determining the residual entropy for comparison with theory the contribution from this effect (<1% of the residual entropy) has to be excluded. The usually quoted treatment of this problem by Giauque and Stout (1936) has been updated by Haida *et al.* (1974) and their calculation is summarized in Table 2.8. The final value of $S_0 = 3.41 \pm 0.19\ \mathrm{J\,K^{-1}\,mol^{-1}}$ is in good agreement with the Pauling model, but it seems unlikely that experiments will ever be able to confirm that ice satisfies the ice rules to the level of accuracy of the theoretical calculations.

Table 2.8 Experimental determination of zero-point entropy of ice Ih

	ΔS $(\mathrm{J\,K^{-1}\,mol^{-1}})$
Heating ice from 0 to 15 K (Flubacher *et al.* 1960)	0.308
Heating ice from 15 to 273.16 K (fully disordered) (Haida *et al.* 1974)	37.530 ± 0.08
Melting—6006.8 $\mathrm{J\,mol^{-1}}$ at 273.16 K (Haida *et al.* 1974)	21.990 ± 0.04
Heating liquid from 273.16 to 298.15 K	6.615 ± 0.01
Vaporization to standard state (298.15 K and 0.1 MPa)	118.895 ± 0.05
Total	185.427 ± 0.18
Vapour in standard state	188.835 ± 0.010
Difference = zero-point entropy S_0	3.408 ± 0.19

Reprinted from *Journal of Chemical Thermodynamics*, **6**(9), O. Haida *et al.*, Calorimetric study of the glassy state X. Enthalpy relaxation at the glass-transition temperature of hexagonal ice, pp. 815–25, Copyright 1974, by permission of the publisher Academic Press, with revision of pressure at standard state from 1 atm to 0.1 MPa.

2.4 Lattice energy and hydrogen bonding

The same kind of calorimetric measurements as are used in Table 2.8 can be used to calculate the enthalpy required to dissociate a crystal of ice at 0 K into free H_2O molecules. This enthalpy of vaporization at 0 K was calculated by Whalley (1957) using the best data then available, giving for H_2O

$$\Delta H_{\mathrm{vap\ 0\,K}} = 47.34 \pm 0.02\,\mathrm{kJ\,mol^{-1}},$$

and for D_2O

$$\Delta H_{\mathrm{vap\ 0\,K}} = 49.89 \pm 0.04\,\mathrm{kJ\,mol^{-1}}.$$

At 0 K and zero pressure these experimental enthalpies are equivalent to energies. They include significant contributions from zero-point motion, which are different for H_2O and D_2O, but the zero-point energy is never included in modelling the bonding. Whalley (1957, 1976) has estimated the magnitude of this effect, using the known frequencies of the vibrational modes of the crystal and the free molecule, and hence deduced the lattice energy $\Delta E_{\mathrm{lattice}}$ required to convert a non-vibrating crystal at 0 K to non-vibrating molecules at the same temperature. For H_2O

$$\Delta E_{\mathrm{lattice}} = 58.95\,\mathrm{kJ\,mol^{-1}} = 0.6110\,\mathrm{eV\ per\ molecule},$$

and for D_2O

$$\Delta E_{\mathrm{lattice}} = 59.33\,\mathrm{kJ\,mol^{-1}} = 0.6149\,\mathrm{eV\ per\ molecule}.$$

These lattice energies are almost equal, though Whalley states that the residual difference is probably genuine.

The binding energy $\Delta E_{\mathrm{lattice}}$ arises from the hydrogen bonding and includes contributions from the van der Waals interaction and the closed shell repulsion. As there are two bonds per molecule it is equivalent to 0.306 eV per bond. The energy of the single hydrogen bond in the $(H_2O)_2$ dimer discussed in §1.4 is 0.24 eV and the observed O—O distance is 2.976 Å, compared with an inter-site distance of 2.750 Å in ice Ih at 0 K. The binding in ice is therefore stronger than that in the dimer, and in modelling ice it is clearly essential to consider the whole assembly of molecules and not just to sum the individual bonds.

The first attempt at modelling the structure and binding energy of ice in terms of the individual molecules was by Bernal and Fowler (1933). Bjerrum (1951) argued that the dipole–dipole interaction between adjacent molecules would favour the *trans* configuration shown for the dimer in Fig. 1.3, but Pitzer and Polissar (1956) showed that this effect would be much weaker when additional molecules were taken into account. All lattice energy calculations present a number of problems, and the first of these arises from the disorder in ice Ih. As we have seen in §1.4, the hydrogen bond interaction has a large electrostatic component which is represented in the empirical pair potentials by placing certain fixed charges on the H_2O molecule. These charges have long range interactions so that energy summations must be carried out over a sufficiently large number of molecules, but this is difficult to do in a disordered system. However, we know that the energies of the various Pauling configurations differ only slightly (the ordering energy is about 0.0025 eV compared with the lattice energy of 0.611 eV), and therefore for an initial calculation of the lattice energy the ordered or disordered structure chosen is unimportant. The difference between the ordered and disordered states cannot yet be accounted for theoretically (see §11.2). Morse and Rice (1982) calculated the lattice energy and density of an ordered ice structure using various empirical pair potentials, and more recently Kozack and Jordan (1992) have used different potentials for a 96 molecule unit cell within which the molecules are disordered with zero total dipole moment. The results are roughly consistent with the experimental values, but there are differences in energy or density of 10% or more between the various cases; such differences are comparable with the latent heat or the density change on melting! It must be remembered that these potentials have already been parametrized to fit experimental data on liquid water, and such calculations are just showing the limitations of such potentials for accurate modelling of ice.

Ab initio calculations, such as those of Pisani *et al.* (1996) and Casassa *et al.* (1997) determine the energy relative to separated nuclei and electrons. This is about 2500 eV per molecule, which is 4000 times larger than the lattice energy calculated relative to separate H_2O molecules, and the calculations therefore have to be very accurate for good values of the lattice energy to be determined. The present accuracy of such calculations is estimated to be of the order of 0.04 eV. One purpose of such calculations is to understand the stability of various structures under particular conditions of temperature and pressure. Lee *et al.* (1993) have shown from *ab initio* modelling that the Pauling structures consisting of identifiable molecules are stable relative to ones with the proton on the centre of the bond up to pressures far above those where ice Ih transforms to phases of higher density.

Handa *et al.* (1988) have shown from experiment that the energy of ice Ih is lower than that of the cubic variant ice Ic by not more than 0.0005 eV per molecule. This is much too small to be explained by absolute calculations, but by focusing on the dipole–dipole interaction energies in the two structures Huckaby *et al.* (1993) and Tanaka and Okabe (1996) have accounted for a difference of this order of magnitude.

We have already noted in §1.4 that it is not possible to treat the energy of a cluster or of ice as the sum of the energies of the separate hydrogen bonds. Some form of 'many-body' interaction has to be included, which is a theoretical way of saying that the interaction of one molecule with another is affected by the other molecules around it. One aspect of this problem can be considered as the polarization of a molecule in the electric field of its neighbours. Coulson and Eisenberg (1966) showed that in disordered ice Ih the average field at a particular molecule due to all the other molecules is oriented parallel to its dipole moment and is of sufficient size to increase its moment by about 50%; the actual effect fluctuates due to the disorder. To take some account of this problem the Kozack–Jordan potential (Table 1.1) includes a polarization term.

2.5 The actual structure

If we place an H_2O molecule at one of the oxygen sites in the half-hydrogen model the tetrahedral symmetry of that site will be destroyed. The molecule is a relatively rigid entity, and there is no requirement for the 104.52° bond angle of the free molecule to expand to the tetrahedral value of 109.47°. Moreover, the molecule can be displaced off its site along its own axis of symmetry, which is along the bisector of the two hydrogen bonds. As all molecules are subject to such displacements the tetrahedral symmetry of the neighbours of a given site will be lost, and every molecule will be subject to components of displacement off its own axis of symmetry. This disorder of the oxygen positions is entirely dependent on the particular configuration of the hydrogens, so that it does not lead to any further entropy, but it does contribute to the broadening of spectral lines (§3.4) and must have implications for any precise modelling of the properties of ice.

The study of the actual structure of ice is very difficult, and in some respects reliable conclusions have not yet been reached. The subject is discussed in detail by Kuhs and Lehmann (1986). We will consider first the evidence for the geometry of the water molecule in the ice structure and then what can be learnt about the molecular disorder. Subtle differences between H_2O and D_2O will be ignored.

2.5.1 *The molecular geometry*

When a molecule forms hydrogen bonds within the ice lattice its electronic structure will be modified, with consequent small changes in the equilibrium positions of the protons or deuterons that constitute the hydrogen nuclei. Averaging over any effects of disorder, the mean state of the molecule is described by just two parameters: the O—H distance r_{OH} and the H—O—H angle θ_{HOH}. An inter-site O—H distance of 1.000–1.008 Å, as in Table 2.6, is consistently obtained in neutron diffraction experiments, but is too large when compared with hydrogen-bonded water in other chemical situations. The evidence for a shorter actual O—H bond length was reviewed by Whalley (1974).

A powerful probe of local molecular geometry is nuclear magnetic resonance. The resonant frequency of a particular proton in an applied field is modified by the local field contributed by the magnetic moments of the nearby protons (^{16}O has no magnetic moment). As the field of a dipole varies as $1/r^3$ and as the average fields of more distant protons tend to average out, the local field at a given proton is dominated by the effect of the other proton on the same molecule. In experiments at low temperatures, where the protons do not change site on the time scale of an n.m.r. experiment, the width of the resonance, expressed as the second moment of the magnetic field of the resonance at a fixed frequency, yields the distance $r_{HH} = 2r_{OH} \sin\frac{1}{2}\theta_{HOH}$ between the two protons on a given molecule. In practice an equivalent measurement is best made from the free-induction decay in a pulsed n.m.r. experiment (see e.g. Hore 1995). An early experiment of this kind by Barnaal and Lowe (1967) provided important support for the Pauling model, and later measurements by Rabideau *et al.* (1968) and Baianu *et al.* (1978) yielded consistent values of $r_{HH} = 1.58 \pm 0.02$ Å. Figure 2.11 shows the values of r_{OH} and θ_{HOH} which are consistent with this result. The range lies between the marked points corresponding to a free molecule and an average site in the half-hydrogen model.

Ice II and ice IX are high pressure phases of ice (see Chapter 11) with fourfold co-ordination, O—O distances of 2.71–2.80 Å, and lattice energies differing from ice Ih by less than 340 J mol^{-1}. These phases are both ordered so that the hydrogen co-ordinates can be determined by normal neutron crystallography. After averaging over non-equivalent sites and correction for thermal motions these structures have $r_{OH} = 0.985$ Å (Whalley 1974, 1976). As the water molecules in ice Ih are in very similar situations to those in ice II or ice IX, it is reasonable to adopt a similar value for r_{OH} in ice Ih. The similarity between the bonding in these three phases is confirmed by similarities in n.m.r., neutron quadrupole resonance, and the frequency of the OH stretching vibration (§3.4.5 and §11.11).

Floriano *et al.* (1987) have analysed the total angular-dependent neutron scattering from polycrystalline ice to yield a pair distribution function in a similar manner to that used in the study of liquids. From this they obtain a mean value for the actual O—H distance $r_{OH} = 0.985 \pm 0.007$ Å. From all these measurements it seems reasonable to adopt a mean actual O—H distance of 0.985 ± 0.005 Å, for which Fig. 2.11 gives $\theta_{HOH} = 106.6 \pm 1.5°$. This accords well with water molecules in other chemical situations. These values for r_{OH} apply to the mean atomic positions at low temperatures; they are larger than the equilibrium bond length by about 0.01 Å due to zero-point motion in the anharmonic potential well of the proton (Kuhs and Lehmann 1986).

A sophisticated deuteron spin alignment experiment was carried out by Fujara *et al.* (1988) at temperatures where protons jump from one site to another by the motion of Bjerrum defects (see §4.4) in a time scale comparable with the observation. The fact that the molecular H—O—H angle is less than the tetrahedral angle means that, when one of the protons on a molecule jumps onto a different O—O bond, the remaining O—H bond swings through about 3°. The experiment of Fujara *et al.* was interpreted as observing this 3° swing.

2.5.2 *Molecular disorder*

The r.m.s. displacements $\langle u_O^2 \rangle^{1/2}$ of the oxygen atoms given in Table 2.7 represent a combination of the thermal (including zero-point) motion and the extra displacements which are, as explained above, a consequence of the Pauling disorder. As is usual in statistics, the squares of the different contributions are additive. The

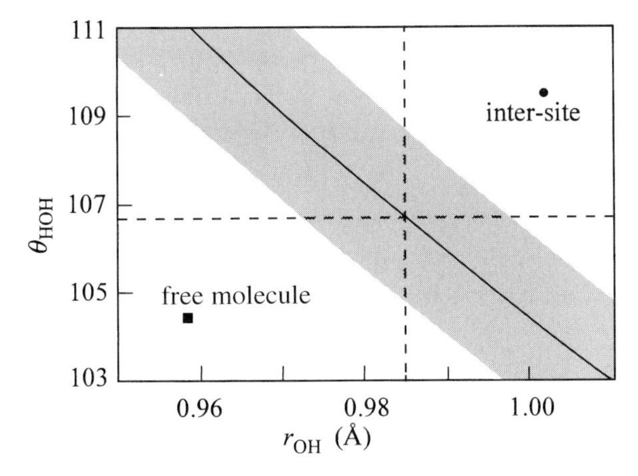

Fig. 2.11 Relationship between the covalent O—H bond length r_{OH} and the bond angle θ_{HOH} for a molecule in ice as required by the evidence of n.m.r. experiments that $r_{HH} = 1.58 \pm 0.02$ Å. The shaded zone corresponds to this limit of error. The points corresponding to a free molecule and to the inter-site separation in the average structure are also shown. The broken lines indicate the currently favoured values for a molecule in ice. (See Whalley 1976.)

zero-point mean-square displacement $\langle u^2 \rangle$ of a harmonic oscillator of mass m and angular frequency ω is easily shown to be $\hbar/m\omega$, and from a knowledge of the lattice frequencies in ice Leadbetter (1965) has calculated the zero-point component of $\langle u_O^2 \rangle$ for H_2O ice to be 0.0085 Å^2. At 15 K the observed $\langle u_O^2 \rangle$ corresponding to the entry in Table 2.7 is 0.0091 Å^2. The observed displacements are therefore dominated by the zero-point motion, or at higher temperatures by thermal motion. However, the small difference between these figures can be attributed to structural disorder, and Kuhs and Lehmann (1987) estimate the r.m.s. displacement from this cause to be 0.036 ± 0.001 Å. Detailed analysis of their neutron diffraction data indicates that these displacements do not have the ellipsoidal symmetry expected for the hexagonal lattice but are directed more toward the axes of symmetry of the molecule (i.e. the bisectors of the H—O—H angles). The disorder will also contribute to the displacements of the hydrogens, but the dominant effect giving large $\langle u_H^2 \rangle$ perpendicular to the bond is due to the lower frequency of librational motion rather than to the fact that θ_{HOH} is less than the tetrahedral angle.

We noted in §2.2.1 that the Pauling disorder will lead to elastic scattering of neutrons into directions that do not satisfy the Bragg condition, and that the distribution of this scattering in reciprocal space is determined by the ice rules. However, this diffuse scattering also contains detailed information about the molecular disorder within the crystal. Figure 2.12(a) from Li *et al.* (1994) shows an example of such scattering in one plane of reciprocal space. The dark spots at integer values of h and l represent the Bragg peaks, and the contours represent the

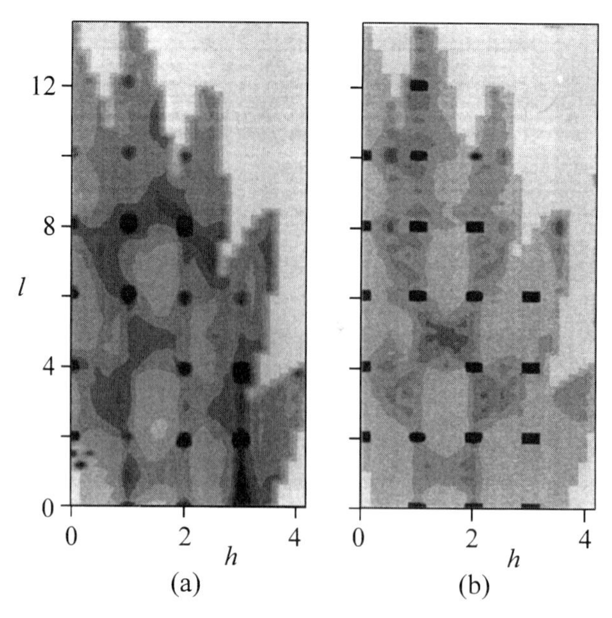

Fig. 2.12 Contour plot of the intensity of diffuse neutron scattering from D_2O ice at 20 K in the *hhl* plane of the hexagonal reciprocal lattice. (a) Experimental observations, and (b) simulation based on the ice rules with all atoms on their average sites. (From Li *et al.* 1994, with permission from Taylor and Francis.)

intensity of the diffuse (mainly elastic) scattering between these peaks. The intensity of the diffuse scattering is orders of magnitude less than that in the Bragg peaks. Figure 2.12(b) shows the scattering in this plane calculated according to the ice rules with all atoms located on their average sites. The similarity provides support for the Pauling model, but there are clear differences, most particularly in this plane a column of extra scattering along the line 33*l* in reciprocal space; this feature was independently observed under different experimental conditions by Schneider and Zeyen (1980). This column of scattering can be understood physically as arising from the coherent displacement of groups of molecules lying in $\{0001\}$ planes by small (≈ 0.05 Å) amounts in different $\langle 2\bar{1}\bar{1}0 \rangle$ directions (Nield and Whitworth 1995). These are static displacements within a given Pauling configuration, not thermal motions. The reason for such correlated displacements is not known, but they suggest some degree of ordering, though with very little effect on the entropy. Nield and Whitworth (1995) describe an attempt to interpret the diffuse scattering in more detail by a reverse Monte Carlo (RMCX) simulation using 1728 molecules. This involved finding actual sets of atomic co-ordinates which are consistent with the observed scattering, and hence determining mean molecular co-ordinates and displacements. However, such methods are not yet sufficiently well developed to be applied successfully in the case of ice (Beverley and Nield 1997).

2.6 Summary

We now summarize the main features of the crystal structure of ice Ih which are important for the remainder of the book.

Ice is built up of H_2O molecules in such a way that each molecule is linked by hydrogen bonds to four others at the corners of a regular tetrahedron, offering protons to two neighbours and accepting hydrogen bonds from two others. The O—O distance is approximately 2.76 Å, and the proton lies about 0.985 Å from one of the oxygen atoms.

The molecules are arranged in a hexagonal lattice as shown in Figs. 2.3 and 2.6. The lattice parameters at a typical temperature of 253 K are $a = 4.519$ Å and $c = 7.357$ Å, and there are four molecules per unit cell, giving a density of 3.074×10^{28} molecules per m^3, which is equivalent to 0.9197 Mg m^{-3}. For convenience these and other well-established parameters of ice Ih are summarized inside the back cover.

Subject to the above constraints there is no long-range order in the orientations of the molecules. For a specified arrangement of oxygen atoms the disorder amongst the hydrogens is described by the 'ice rules' (or 'Bernal–Fowler rules') that there are two hydrogens adjacent to each oxygen and there is only one hydrogen per bond. This disorder leads to the Pauling entropy approximately equal to $k_B \ln(\frac{3}{2})$ per molecule, and many properties of ice are a consequence of this disorder.

The H—O—H angle of 104.52° in the free water molecule does not exactly fit into the O—O—O angle of 109.47° in the tetrahedral structure, and a complication arising from the disorder is that molecules are subject to small displacements off their average sites. These facts lead to difficulties in determining the 'actual' structure or understanding it in detail in terms of the bonding between molecules. For most physical properties of ice these refinements are not significant.

3
Elastic, thermal, and lattice dynamical properties

3.1 Introduction

The last chapter focused on the structure of ice Ih. The unifying theme of this chapter will be those effects which depend on small displacements of atoms or molecules off their equilibrium sites. Later chapters will deal with properties where atoms (often just protons) move from one site to another.

A simple model of a crystalline solid treats the atoms as point masses linked together by springs. If this assembly is subjected to an external stress, the springs are stretched or compressed and the lattice is distorted. Macroscopically this distortion constitutes an elastic strain, so the chapter deals first with the subject of elasticity. When the material is heated the masses vibrate under the restoring forces of the springs, and the model can then account for the thermal properties. For this purpose the coupling of the motions is very significant. Springs obeying Hooke's law can account for heat capacity, but non-linear properties of the inter-atomic forces represented by the springs are essential to explain thermal expansion or conduction. In describing these properties of ice we first present the macroscopic properties, which readers may require for macroscopic applications, and only then do we turn to their interpretation.

More specific spectroscopy of the modes of vibration of the molecules in ice can be achieved using infrared absorption, Raman scattering, or inelastic neutron scattering. These techniques are described in §3.4 and provide information about the properties and bonding of the molecules in ice. They are also relevant to the study of the other solid phases of water to be described in Chapter 11. Finally we consider the modelling of all these properties of ice, first with two empirical force constants describing bond stretching and bond bending, and then by more elaborate theories.

3.2 Elasticity

In the most elementary treatment of the elastic deformation of a solid material the strain ε is related to the stress σ by the equation

$$\sigma = E\varepsilon, \tag{3.1}$$

where E is the appropriate elastic modulus. However, the situation is not quite so simple because a tensile stress, for example, produces not only an extension but a lateral contraction $\nu\varepsilon$, where ν is Poisson's ratio. In the case of single crystals such as ice it is also necessary to take account of the anisotropy of the material, and the standard theory is given by Nye (1957) or other textbooks on elasticity. We use here the matrix rather than the tensor formulation.

The stresses and strains each have six components σ_i and ε_i defined relative to the axes of the crystal. The subscripts $i = 1$ to 3 correspond to extensions parallel to the axes, with $i = 3$ meaning the hexagonal c-axis in the case of ice, and the subscripts $i = 4$ to 6 correspond to shears in the planes perpendicular to the axes 1 to 3 respectively. The elastic properties are then described either by the six equations:

$$\varepsilon_i = \sum_{j=1}^{6} s_{ij}\,\sigma_j \tag{3.2}$$

or by the inverse set of equations:

$$\sigma_i = \sum_{j=1}^{6} c_{ij}\,\varepsilon_j. \tag{3.3}$$

The constants s_{ij} are called the *elastic compliances* and the c_{ij} are the *elastic constants* or *stiffnesses*. For a hexagonal crystal there are five independent compliances:

$$s_{11} = s_{22},$$

$$s_{33},$$

$$s_{12} = s_{21},$$

$$s_{13} = s_{31} = s_{23} = s_{32},$$

and

$$s_{44} = s_{55}.$$

One further quantity is

$$s_{66} = 2(s_{11} - s_{12}),$$

and the remaining s_{ij} are all zero. The properties can be equivalently described by the corresponding set of c_{ij}, and the relationships between these sets of constants are given in Nye (1957). The elastic properties of a hexagonal crystal are isotropic in the (0001) plane.

For an isotropic material, such as polycrystalline ice with randomly oriented grains, there are only two independent elastic constants, which are usually chosen as s_{11} and s_{12} (or c_{11} and c_{12}). There are two distinct physical properties: a resistance to a change in volume and a resistance to a change in shape. These are described by the bulk modulus K (equal to the reciprocal of the compressibility) and the shear modulus G respectively. For an *isotropic* material these are related to the compliances by the equations:

$$K = \frac{1}{3(s_{11} + 2s_{12})}, \tag{3.4}$$

and

$$G = \frac{1}{2(s_{11} - s_{12})}. \tag{3.5}$$

Other familiar quantities *for an isotropic material* are Young's modulus

$$E = \frac{1}{s_{11}}, \tag{3.6}$$

and Poisson's ratio

$$\nu = -\frac{s_{12}}{s_{11}}. \tag{3.7}$$

If we are concerned about the anisotropy within a single crystal, we cannot use these standard isotropic parameters and have to work with the matrices s_{ij} or c_{ij}.

In the case of ice it is technically difficult to make reliable measurements of the elastic constants by the application of a static stress. Except for hydrostatic compression, this is because of the limitations imposed by creep, particularly in the case of polycrystalline ice (Gold 1958; Sinha 1978*a*). Much more success has been obtained by measuring the velocity of sound which is equal to $(c/\rho)^{1/2}$, where c is one of the c_{ij} or a combination of them and ρ is the density. For sound propagating along the c-axis longitudinal and transverse waves have different velocities. However, for waves propagating at an angle to the c-direction the two polarizations of the transverse waves are not equivalent, with the result that there are then three possible velocities. Measurements of these velocities as functions of the direction of propagation give sufficient information to determine all five independent elastic constants. For ice there have been many ultrasonic experiments of this kind, the most detailed being those of Dantl (1968, 1969) who observed velocities that were independent of frequency over the range 5–190 MHz for temperatures between −140 °C and the melting point.

A more recent way of determining the velocities of sound waves in ice is Brillouin spectroscopy, in which a highly monochromatic laser beam is scattered off high-frequency acoustic waves thermally excited within the crystal. The scattered light is Doppler shifted, and when the spectrum is analysed with a Fabry–Perot interferometer there are found to be three components. The frequencies of these components are determined by the velocities of the acoustic modes appropriate to the orientations of the incident and scattered light. There is no need to attach transducers to the crystal, and the scattering is a property of the ice independent of any porosity etc. Detailed measurements by Gammon *et al.* (1983*a,b*) on various types of ice at the single temperature of −16 °C yield the velocities of sound given in Table 3.1 and the elastic parameters in Table 3.2. These data are considered to supersede the values quoted by Dantl (1968, 1969). The degree of anisotropy is apparent from the difference between the pairs s_{11} and s_{33}, s_{12} and s_{13}, and s_{44} and $s_{66} = 2(s_{11} - s_{12})$. The bulk modulus derived from these

Table 3.1 Velocities of sound in single crystals of ice at −16 °C

Direction of propagation	Mode	Velocity (km s^{-1})
Parallel to *c*	longitudinal	4.040
	transverse	1.810
Perpendicular to *c*	longitudinal	3.892
	transverse $\perp c$	1.930
	transverse $\| c$	1.810

From Gammon *et al.* (1983*b*), reproduced by courtesy of the International Glaciological Society from *Journal of Glaciology*, Vol. 29, No. 103, 1983, pp. 433–60, Table VII.

Table 3.2 Elastic parameters of monocrystalline ice at −16 °C

Property and units	Symbol	Adiabatic value	Isothermal value
Compliance (10^{-12} m^2 N^{-1})	s_{11}	103.2 ± 0.5	103.5
	s_{12}	-42.9 ± 0.5	−42.5
	s_{13}	-23.2 ± 0.2	−22.8
	s_{33}	84.4 ± 0.4	84.8
	s_{44}	331.8 ± 0.2	331.8
Elastic constant (10^8 N m^{-2})	c_{11}	139.3 ± 0.4	136.7
	c_{12}	70.8 ± 0.4	68.3
	c_{13}	57.6 ± 0.2	55.1
	c_{33}	150.1 ± 0.5	147.6
	c_{44}	30.1 ± 0.1	30.1
Bulk modulus (10^8 N m^{-2})	K	89.0 ± 0.2	86.5

From Gammon *et al.* (1983*b*), with some recalculations. Reproduced by courtesy of the International Glaciological Society from *Journal of Glaciology*, Vol. 29, No. 103, 1983, pp. 433–60, Table VI.

data is included in Table 3.2, and agrees well with more direct measurements such as those of Gow and Williamson (1972). Appropriately weighted average values of the elastic parameters of isotropic polycrystalline ice are given in Table 3.3.

All elastic parameters derived from the velocity of sound waves apply to dynamic (or adiabatic) deformation. A small correction derived from classical thermodynamics and involving the coefficient of thermal expansion is necessary to obtain the isothermal elastic parameters, and these are given in the final column of Table 3.2. As the temperature falls towards 0 K the compliances s_{ij} decrease and the c_{ij} and moduli increase by amounts of the order of 25%. The detailed temperature dependence has to be taken from the measurements by Dantl (1968, 1969), Gagnon *et al.* (1988), or for lower temperatures by Proctor (1966). Over the commonly used higher part of the temperature range Gammon *et al.* (1983*b*) showed that any of the parameters $X(T)$ is given approximately by

Table 3.3 Averaged adiabatic elastic parameters for isotropic polycrystalline ice at –16 °C

Parameter	Value
Bulk modulus K	89.0×10^{8} N m^{-2}
Young's modulus E	93.3×10^{8} N m^{-2}
Shear modulus G	35.2×10^{8} N m^{-2}
Poisson's ratio ν	0.325

From Gammon *et al.* (1983*b*), reproduced by courtesy of the International Glaciological Society from *Journal of Glaciology*, Vol. 29, No. 103, 1983, pp. 433–60, Table VIII.

the single equation

$$X(T) = X(T_0)[1 \pm a(T - T_0)] \qquad \text{with} \quad a = 1.42 \times 10^{-3}\,\text{K}^{-1}. \tag{3.8}$$

The positive sign is for compliances and the negative sign for elastic constants or moduli.

The pressure dependence of the elastic constants over the range of stability of ice Ih has been studied using Brillouin spectroscopy by Gagnon *et al.* (1988).

3.3 Thermal properties

3.3.1 *Heat capacity*

The temperature dependence of the molar heat capacity of ice at constant pressure C_p is shown in Fig. 3.1. The values for H_2O are those of Haida *et al.* (1974) shown as open circles supplemented by some low-temperature values from Flubacher *et al.* (1960) (solid circles). The values for D_2O are taken from Matsuo *et al.* (1986) on an 0.01 M KOD-doped sample, omitting measurements in the region of the ice Ih–ice XI phase transition at 76 K (see §11.2). The results plotted are very close to earlier measurements by Giauque and Stout (1936) and Long and Kemp (1936) on H_2O and D_2O ice respectively.

At the lowest temperatures the heat capacity varies as T^3 in accordance with the Debye model for the thermal excitation of long-wave acoustic phonons, and the Debye temperature Θ_D appropriate to these measurements is approximately 226 K. However, this parameter has little significance because the low temperature data of Flubacher *et al.* (1960) show significant deviations from the Debye model as low as 5 K. In the higher temperature region there is appreciable excitation of modes involving molecular motions that depend on the mass of the hydrogen atom rather than the whole molecule. For D_2O these modes have lower frequencies than for H_2O, so that they are more easily excited, giving a larger heat capacity. Such modes can be modelled as sets of harmonic oscillators

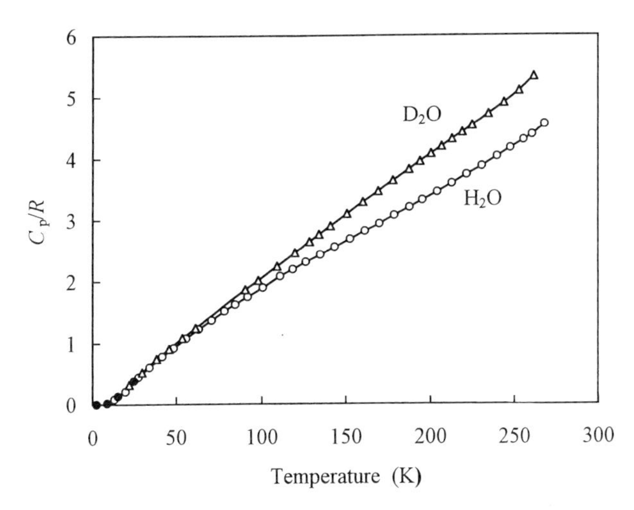

Fig. 3.1 The molar heat capacity of ice at constant pressure C_p expressed in terms of the gas constant R. H_2O data from Haida *et al.* (1974) with low temperature values (solid circles) from Flubacher *et al.* (1960). D_2O data from Matsuo *et al.* (1986).

all of the same frequency, giving Einstein terms in the heat capacity (see Matsuo *et al.* (1986) or Leadbetter (1965)), but serious modelling can only be done in terms of the full spectrum of lattice vibrations to be described in §3.4. This was attempted by Klug *et al.* (1991*b*), who obtained close agreement with the experimental heat capacity up to 80 K, but concluded that at higher temperatures anharmonic effects (even at constant volume) increase the heat capacity above that calculated (by ~12% at 200 K).

3.3.2 *Thermal expansion*

The most precise and wide ranging data on the thermal expansivity of ice are provided by the lattice parameters given in Table 2.5, and these are generally compatible with results from other techniques as reviewed by Röttger *et al.* (1994). The expansion is isotropic to within the accuracy of the measurements (~2%), and mean linear expansion coefficients $\alpha(T)$ derived from the lattice parameters of H_2O and D_2O are plotted in Fig. 3.2. A remarkable feature of this figure is that $\alpha(T)$ is negative below 73 K, but this is in fact a common property of tetrahedrally co-ordinated crystals (Smith and White 1975).

The interpretation of the thermal expansion of a crystalline material depends on the anharmonic nature of the inter-atomic forces, and this is discussed theoretically in terms of the translational modes of lattice vibration using the Grüneisen parameter $\gamma(T)$ (Collins and White 1964; Ashcroft and Mermin 1976, Chap. 25). The lattice modes for ice will be described in §3.4.5. For a specific

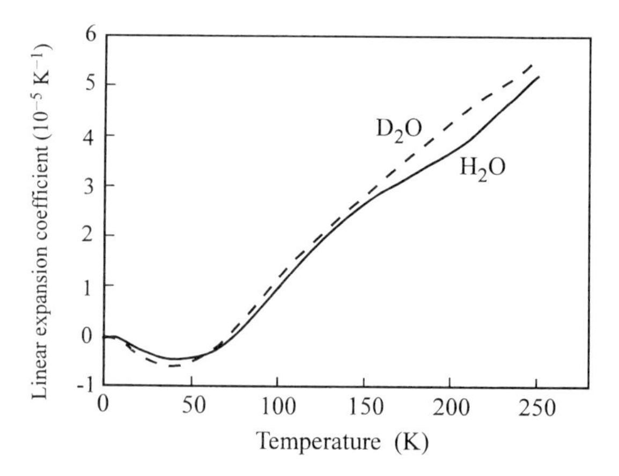

Fig. 3.2 Linear thermal expansion coefficients of H_2O and D_2O ice as calculated from the lattice parameter data in Table 2.5. (From Röttger *et al.* 1994.)

mode i we define the quantity

$$\gamma_i = -\frac{\partial \ln \nu_i}{\partial \ln V}, \tag{3.9}$$

in which ν_i is the mode frequency and V is the volume. For a truly harmonic potential the γ_i will be zero. In an isotropic material general thermodynamic arguments show that the linear expansion coefficient α is related to the heat capacity C_V (per unit volume) and the bulk modulus K by the equation

$$\alpha = \frac{\gamma(T)C_V}{3K}, \tag{3.10}$$

in which $\gamma(T)$ is the thermally weighted average of the γ_i. For most modes in most materials γ_i is positive (i.e. the mode frequencies fall as the crystal expands) and $\gamma(T)$ is approximately constant and of the order of unity. In such cases the crystal expands on heating and the temperature dependence of α is dominated by that of the heat capacity. However, for some transverse acoustic modes it is possible for γ_i to be negative, and if these modes have low energy and dominate the average at low temperatures, $\gamma(T)$ and hence α can be negative. In a lattice dynamical model of the structure of ice Tanaka (1998) has identified hydrogen-bond-bending modes which have this property, and he has derived negative expansion coefficients of approximately the observed magnitude. The significance of negative α is that the anharmonicity of the low energy modes is such that as they become excited the crystal shrinks.

3.3.3 *Thermal conductivity*

There have been many experiments to measure the thermal conductivity of ice and those that seem most reliable are plotted in Fig. 3.3. The solid line represents the 'best' mean values according to the analysis of Slack (1980). There is no established anisotropy, and many of the experiments were performed on polycrystalline material. The estimated accuracy of the mean line was quoted as $\pm 10\%$. The conductivity λ above 60 K is reasonably well fitted by the relation

$$\lambda = \frac{651\,\mathrm{W\,m^{-1}}}{T}, \tag{3.11}$$

except for a significant fall below this line at the higher temperatures. This deviation is clearly established in the measurements of Andersson *et al.* (1980) which were performed under a pressure of 70 MPa. Adjusted values from their experiments were included in calculating the solid line in Fig. 3.3, but actual data points have not been plotted because of the pressure difference.

Measurements on single crystals at low temperatures were made by Klinger (1975) and extended down to 0.6 K by Varrot *et al.* (1978) and Klinger and Rochas (1983). Fig. 3.4 shows typical results, with a maximum in the conductivity at about 7 K. Results at and below the maximum depend on the quality and thermal history of the specimen. This is why it was important to use single crystals, although there was no evidence of anisotropy.

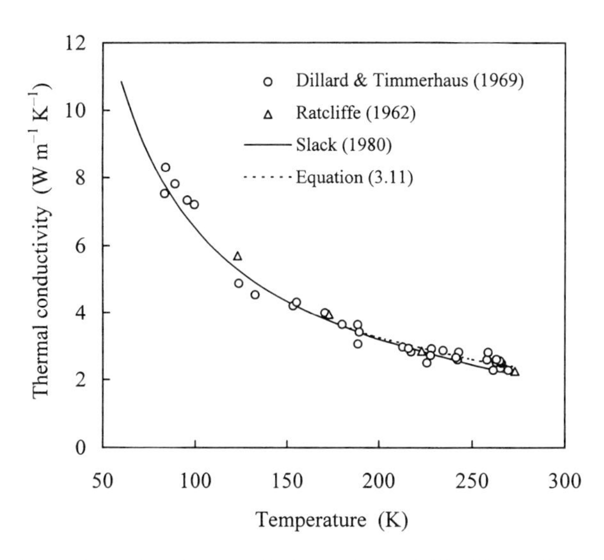

Fig. 3.3 The thermal conductivity of H_2O ice above 60 K. The solid line due to Slack (1980) represents his 'best fit' to all available data, and deviates at higher temperatures from the broken line representing equation (3.11).

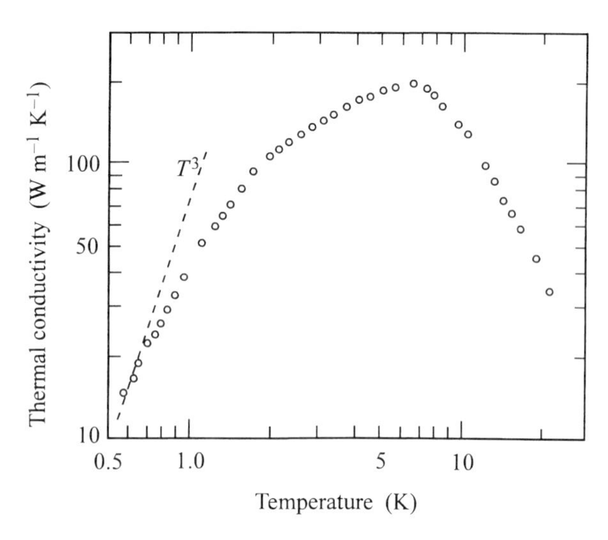

Fig. 3.4 Low-temperature thermal conductivity (on logarithmic scales) of freshly grown monocrystalline H_2O ice (Klinger and Rochas 1983). Reproduced with permission from *Journal of Physical Chemistry*, **87**, 4155–6. Copyright 1983 American Chemical Society.

These properties of ice can be understood quite well according to the standard theory of the thermal conductivity of insulating crystals (see for example Klemens 1958), and their interpretation at temperatures above the maximum of the conductivity has been discussed quantitatively by Slack (1980). If a branch s in the spectrum of lattice vibrations (see §3.4.6) contains phonons of wave vector $\mathbf{k}$ which make a contribution $C_s(\mathbf{k})\,\mathrm{d}^3\mathbf{k}$ to the heat capacity per unit volume, the thermal conductivity λ is given by the equation

$$\lambda = \frac{1}{3}\sum_s \int C_s(\mathbf{k})v_s^2(\mathbf{k})\tau_s(\mathbf{k})\,\mathrm{d}^3\mathbf{k}, \tag{3.12}$$

where $v_s(\mathbf{k})$ is the group velocity of the phonons and $\tau_s(\mathbf{k})$ is their relaxation time, corresponding to a mean free path $v_s(\mathbf{k})\tau_s(\mathbf{k})$. Only the acoustic phonons contribute to the conductivity because the group velocity of other modes is too small, and for the acoustic modes the relaxation time is dominated by three-phonon umklapp processes. These are processes in which a phonon of wave vector $\mathbf{k}_1$ is converted to two phonons of wave vectors $\mathbf{k}_2$ and $\mathbf{k}_3$ or vice versa in such a way that energy is conserved and the wave vectors satisfy the condition that $\mathbf{k}_2+\mathbf{k}_3-\mathbf{k}_1$ is a vector of the reciprocal lattice. The rate of such processes depends on the anharmonicity of the inter-atomic potentials, but they lead to the general result that $\lambda \propto 1/T$, as in equation (3.11). This is a common result for simple materials at temperatures above the Debye temperature Θ_D. Slack (1980) suggests that the deviation from this dependence in ice at higher temperatures is due to additional scattering of the acoustic phonons by optic modes. Andersson and Suga (1994) have shown that when ice Ih transforms to the ordered phase ice XI at 72 K (see §11.2) the thermal conductivity is increased, and therefore at least in the temperature range 50–70 K the phonon scattering is in part related to the proton disorder; this may be indirectly via increased anharmonicity induced by the disorder.

Using the model of equation (3.12) Slack (1980) estimates that the mean free path of phonons near the melting point is about 95 Å, which is quite long compared to the lattice spacing and in part explains why ice is a better thermal conductor than liquid water (by a factor of about 4).

At low temperatures umklapp processes become rare because very few of the phonons have values of **k** large enough for scattering to involve a vector of the reciprocal lattice. The phonon mean free path then becomes limited by static defects in the crystal or even by the boundaries of the crystal itself. For boundary scattering λ is proportional to the heat capacity, which at such low temperatures varies as T^3. Varrot *et al.* (1978) show that the magnitude and temperature dependence of λ below 1 K is consistent with this model, though the T^3 dependence is barely approached in the example in Fig. 3.4. Above 1 K the temperature dependence of λ is always less than T^3, which implies that the phonons are scattered by small internal defects, and the measurements in this range are sensitive to the state of the crystal. As will be explained in Chapters 6 and 7, we now know that condensation of interstitial molecules during cooling can introduce prismatic dislocation loops or modify the state of existing dislocations. It seems probable that the effects of annealing described by Klinger (1975) and Klinger and Rochas (1983) are connected with such processes.

3.4 Spectroscopy of lattice vibrations

For a crystal of N H_2O molecules there are $3N$ atoms which exert forces on one another, and such a coupled system has $3\times3N$ normal modes of vibration. In the case of ice information about the frequencies of these modes can be determined by infrared absorption, Raman spectroscopy, and inelastic neutron scattering. In this section we will present the data obtained by these techniques and discuss the identification of the main features of the spectrum. The modelling of the modes in terms of inter-atomic forces will be described in §3.5.

The spectra obtained by the three techniques are presented together in Fig. 3.5. All of the spectra depend on the density of states of vibrational modes in the material, but the various modes appear with different relative strengths in the three cases. The horizontal axis represents the mode frequency which should logically be in THz, but is more usefully labelled as the corresponding quantum energy. In optical spectroscopy this axis is conventionally labelled with the wave number of an equivalent photon (not the phonon) in cm^{-1}, but in neutron

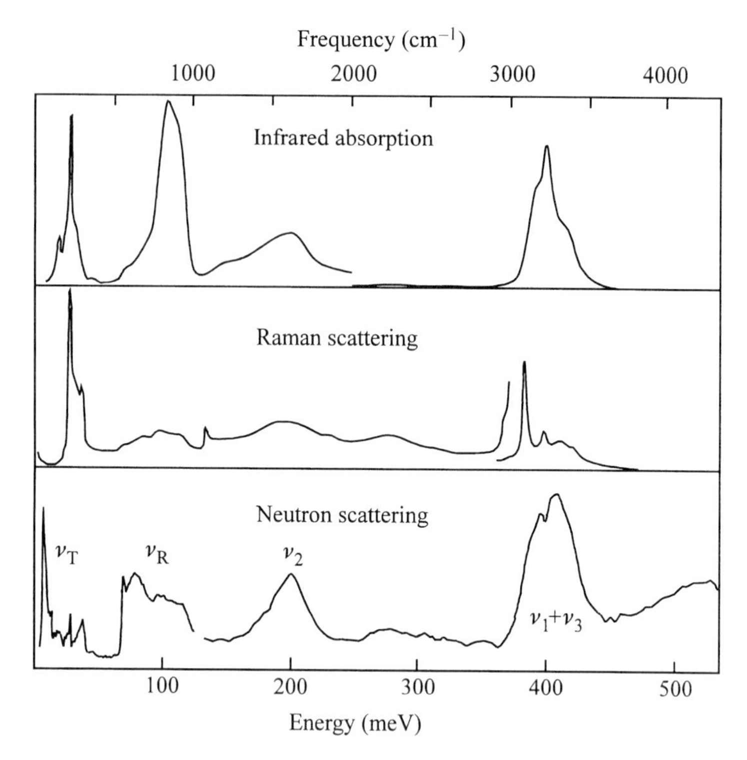

Fig. 3.5 Lattice vibration spectra of H_2O ice as observed by infrared absorption (Bertie *et al.* 1969), Raman spectroscopy (Wong and Whalley 1975; 1976) and inelastic neutron scattering (Li and Ross 1992; Li 1996).

scattering the energy change is that of the neutron expressed in meV. For convenience both scales are marked on Fig. 3.5, but in this chapter we will work primarily in meV. The conversion factor is $1000\,\mathrm{cm}^{-1} \equiv 124.0\,\mathrm{meV}$ and the equivalence with other units can be found in Fig. 9.1 and inside the front cover. The measurements span two decades of frequency, and results from more than one specimen or instrument have been combined in each plot. There are changes to the vertical scale at the breaks in the curves. The results of the three methods clearly have certain mode frequencies in common, but detailed features differ because the techniques do not measure equivalent quantities.

3.4.1 *Infrared absorption*

The transparency of ice to visible light extends into the near infrared (see Fig. 9.3), but absorption rises for wavelengths greater than about 1.3 μm (8000 cm^{-1}) and is very strong over most of the range of Fig. 3.5. The infrared data in the figure are taken from the experiments of Bertie *et al.* (1969) which covered the range 30–4000 cm^{-1}. For measurements at the peak around 400 meV the specimens could only be a few μm thick, but by about 40 meV several mm were required as the ice passes into the range of low absorption characteristic of microwaves and radio waves (see Chapter 9). In passing through the frequency range of the infrared absorption bands the permittivity falls from the value of 3.16 obtained in the high-frequency limit of electrical measurements (§5.3.1) to 1.70, which is the square of the refractive index for visible light (Table 9.2). This change in permittivity represents the contribution to the electrical polarizability of ice that arises from small displacements of atoms from their equilibrium sites in an electric field.

3.4.2 *Raman spectroscopy*

In Raman spectroscopy a beam of monochromatic light, now usually from a laser, is passed through the specimen and the spectrum of the scattered light is analysed. This spectrum contains low-intensity components with frequencies shifted by the energy quanta required to excite appropriate modes of vibration in the lattice. The energy plotted in Fig. 3.5 represents this shift in frequency. In the limit of very small energy changes Raman-type scattering from acoustic phonons is called Brillouin scattering, and it has been described in §3.2 as a way of measuring the velocity of sound. The data for larger energy changes presented in Fig. 3.5 are from the experiments of Wong and Whalley (1975, 1976) on polycrystalline ice. Experiments on single crystals using polarized light are described by Scherer and Snyder (1977) and at low frequencies by Faure and Chosson (1978). Such measurements provide more detailed information about the nature of the modes being excited. Comparison of the infrared and Raman data in Fig. 3.5 shows that some modes are infrared active and others Raman active. In an ordered crystal different selection rules apply to these processes, but with the Pauling disorder in the ice Ih these rules are partially relaxed (Whalley and Bertie 1967; Klug and Whalley 1972).

In both infrared absorption and Raman scattering it is necessary to conserve crystal momentum or wave vector **k**. As the wavelength of infrared or optical radiation is large compared with the lattice spacing, the photon wave vector is small compared with the reciprocal lattice vectors of the crystal, and this means that the only modes that can be observed are ones which will be described in §3.4.6 as being near the Brillouin zone centre. This restriction is not imposed on the inelastic scattering of X-rays, but such experiments require an extremely high-energy resolution. Using synchrotron X-radiation Ruocco *et al.* (1996) were able to observe Brillouin scattering in ice by acoustic phonons of very short wavelength (~10 Å). The interest in such measurements is that they can be extended to the liquid phase, where it is found that the velocity of such short-wavelength sound waves in the liquid is close to that in ice and much higher than the usual velocity in the liquid.

3.4.3 *Inelastic neutron scattering*

Incoherent inelastic neutron scattering experiments are in principle similar to Raman spectroscopy. A mono-energetic beam of neutrons of near-thermal energy passes through the specimen and the energy distribution of the scattered neutrons is determined as a function of the angle of scattering. There are two important differences from optical Raman scattering. The first is that the neutrons interact directly with the nuclei rather than through the coupling of an electromagnetic wave to the charge distribution of the molecule, and this means that there are no selection rules arising from the coupling. The second difference is that the neutron wavelength is comparable with the lattice spacing, and the conservation of both energy and wave vector allow in principle the measurement of the dispersion curves $\varepsilon(\mathbf{k})$ relating the energy (or frequency) ε of a lattice mode to its wave vector $\mathbf{k}$ throughout the whole Brillouin zone. These relations are explained more fully in §3.4.6.

The availability of modern instrumentation on a pulsed neutron source has increased the range and resolution of such measurements in recent years, and their application to ice is described by Li and Ross (1992) and Li (1996). The data in Fig. 3.5 are taken from their experiments using an instrument known as TFXA (Time Focusing X-tal Analyser) for energy transfers below 120 meV and HET (High Energy Transfer) for energies above this. The curves show the amplitude-weighted phonon density of states for polycrystalline ice at 15 K, but measurements on single crystals in different orientations are virtually identical.

3.4.4 *Isotope effects*

Important evidence about the modes of lattice vibration is obtained by comparing the spectra for H_2O and D_2O. Figure 3.6 shows on an expanded scale the low-frequency neutron data of Fig. 3.5 together with the corresponding spectrum for D_2O. The mean ratio of the frequencies of matched features below 45 meV is 1.04 ± 0.01. This is approximately the same as the square root of the ratio of the molecular masses $(20/18)^{1/2}$, showing that this part of the spectrum involves the motion of whole water molecules. On the other hand the small peak which appears for H_2O at 67 meV moves to 48.5 meV in D_2O, giving a ratio of 1.38 which is close to $2^{1/2}$ and shows that features in this part of the spectrum are determined primarily by the motion of hydrogen nuclei. This shift by a factor of $2^{1/2}$ applies to all the higher frequency modes in Fig. 3.5, and matching observations apply to the infrared and Raman spectra.

New peaks appear in the spectra of D_2O doped with H_2O and vice versa. Isotopic exchange means that for small dopings the added isotope will be present in the form of HOD molecules. Figure 3.7 shows Raman peaks at 303.6 meV arising from HOD in H_2O and 410.9 meV for HOD in D_2O. The frequencies of these sharp modes depend significantly on temperature; the value of 410.9 meV given above was obtained at 269 K and it falls to 404.8 meV at 10 K (Sivakumar

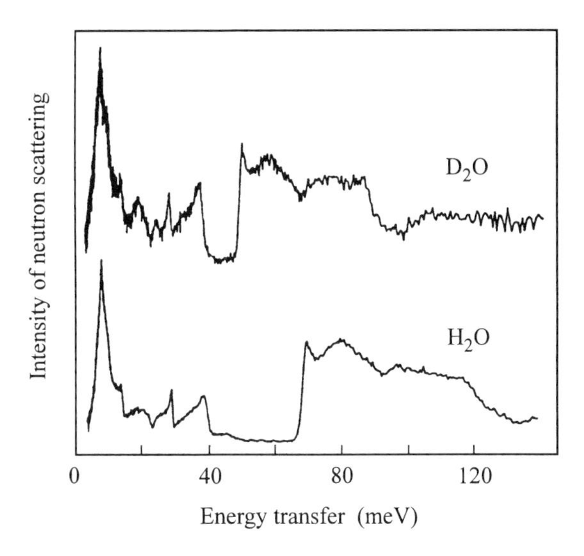

Fig. 3.6 Low-energy incoherent inelastic neutron scattering spectra of polycrystalline H_2O and D_2O ice determined with the TFXA instrument at 15 K. (From Li and Ross 1992; see also Li 1996.)

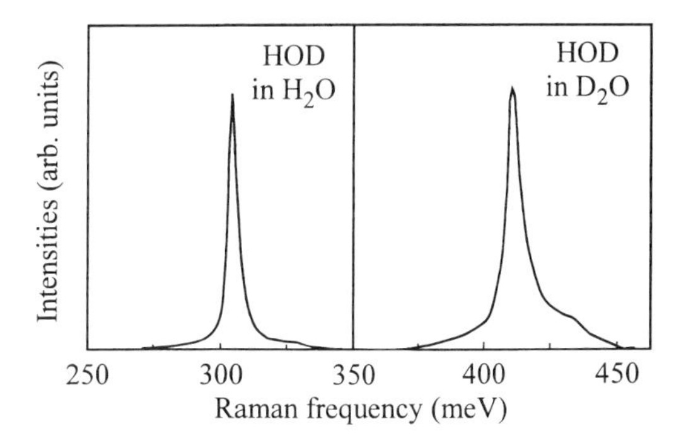

Fig. 3.7 Localized O—D or O—H bond stretching modes of 3 mol% HOD in H_2O and D_2O ice, observed by Raman spectroscopy at −4 °C for the most favourable polarization and after subtraction of scattering by the pure solvent. The vertical scales for the two peaks are not the same. (From Scherer and Snyder 1977.) This is the mode denoted μ_3 in Fig. 3.10.

et al. 1978). Incoherent inelastic neutron scattering is particularly well suited to observing the modes of HOD in D_2O, because the incoherent scattering cross section for H is 20 times larger than that for D. Using this technique Li and Ross (1992, 1994) have observed a line at 408 meV for HOD in D_2O, which is equivalent to the Raman line in Fig. 3.7, and they have also identified two further modes. The first at 104 meV is shown in Fig. 3.8, and the other forms a weaker

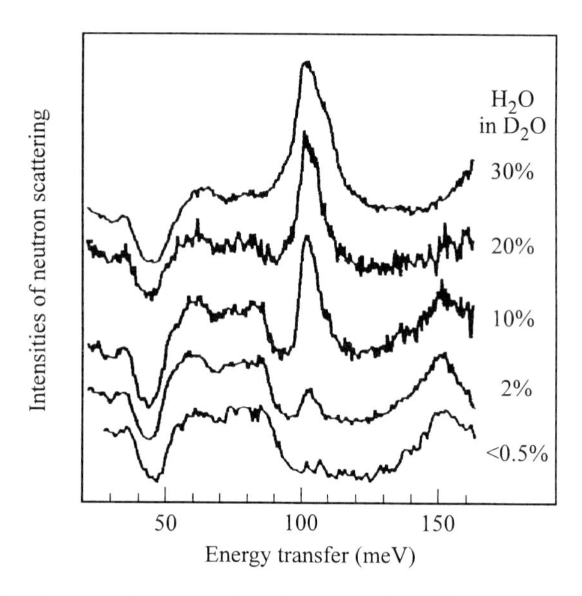

Fig. 3.8 The localized mode (denoted μ_1 in Fig. 3.10) of HOD in D_2O as observed by inelastic neutron scattering on HET for specimens containing the stated percentages of H_2O. (From Li and Ross 1994.)

peak at 185 meV. A related peak at 180 meV was observed in the infrared spectrum of liquid H_2O/D_2O mixtures by Maréchal (1991). Devlin *et al.* (1986) have studied the infrared and Raman spectra of the isotopic defect modes, and Devlin (1990) has reviewed this topic and made elegant use of it to study defects in ice Ic and amorphous ice (see §6.5.3).

3.4.5 *Interpretation of the spectra*

As long ago as 1933 Bernal and Fowler recognized that similarities between the Raman spectra of water vapour, liquid water, and ice implied that all these phases consisted of distinct H_2O molecules, and this was one of the factors leading to the Pauling model of the crystal structure of ice (§2.2.1). The fact that the forces between molecules are weak in comparison with the internal bonding results in a simple division of the lattice modes into three groups involving the internal vibrations, rotations, and translations of the molecules. The frequencies of the first two of these groups depend primarily on the mass of the hydrogen or deuterium nuclei, and the frequencies of the translations depend on the mass of the whole molecule in the manner referred to in the previous section.

A free H_2O molecule has just the three normal modes of vibration illustrated in Fig. 3.9. The comparatively small motions of the oxygen atoms are required to keep the centre of mass stationary, and these motions result in the frequency ν_3

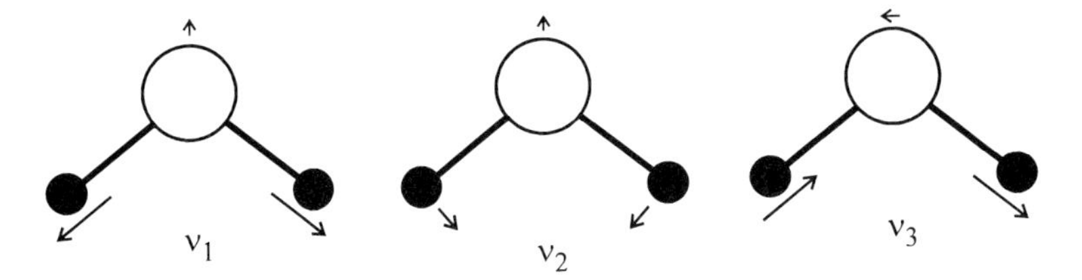

Fig. 3.9 The three normal modes of an isolated water molecule.

being slightly higher than ν_1. These two depend on the force constant for stretching the covalent O—H bond, while the bending mode ν_2 depends on the force constant for changing the bond angle. In the vapour the free molecules have a rich rotation–vibration infrared spectrum (Benedict *et al.* 1956), from which the frequencies of the molecular modes are deduced to be:

$$\nu_1 = 3656.65\,\text{cm}^{-1} \equiv 453.4\,\text{meV}$$
$$\nu_2 = 1594.59\,\text{cm}^{-1} \equiv 197.7\,\text{meV}$$
$$\nu_3 = 3755.79\,\text{cm}^{-1} \equiv 465.7\,\text{meV}.$$

For ice the band around 400 meV is thus identified with the O—H bond stretching modes ν_1 and ν_3. The frequencies are lowered from those of the free molecule by the hydrogen bonding to the neighbouring molecules, but, as a single molecule cannot vibrate independently of its neighbours, this coupling also leads to complex mode structures involving many molecules. By comparing the spectra for ice Ih with those of ices VII and VIII and considering the expected properties of ordered ice Ic, Whalley (1977) proposed a detailed interpretation of the structure of this band in the infrared and Raman spectra. The most characteristic feature of the Raman spectrum is the strong peak at 382.3 meV (3083 cm^{-1}, at 95 K), and this is identified as the in-phase ν_1 mode (i.e. the mode in which all molecules execute a ν_1-type vibration in phase with one another).

The peaks arising from the presence of HOD in D_2O are clearly related to the molecular modes, but they are much narrower than the corresponding modes of the D_2O matrix because they are localized modes which do not couple strongly to the surrounding lattice. Figure 3.10 illustrates the normal modes identified by Li and Ross (1994) for such an isotopic impurity and denoted by the symbols μ_1, μ_2, and μ_3. The mode μ_3 involves O—H stretching and should therefore occur near the frequencies ν_1 and ν_3 for pure H_2O; its frequency is well separated from the ν_1 plus ν_3 band of the D_2O matrix, which is displaced to a lower frequency . The mode μ_3 is therefore assigned to the well-established peak at about 408 meV. This mode has been the subject of detailed study by Raman spectroscopy, where it is referred to as ν_{OH} (or ν_{OD}) or the 'isolated' or 'decoupled' stretch mode (Sivakumar *et al.* 1978; Johari and Chew 1984). The mode μ_2 is related to ν_2 in that it involves bond bending, and it is assigned to the peak at 185 meV, which was also attributed to HOD by Maréchal (1991). The mode μ_1 involves an

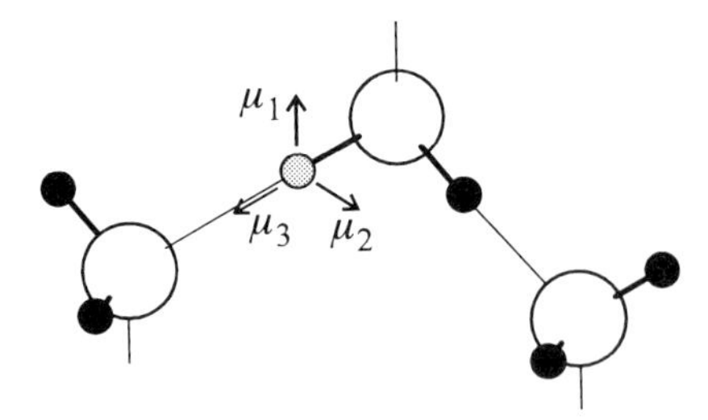

Fig. 3.10 The three localized (or decoupled) modes of an isotopic hydrogen impurity in ice. The motion μ_2 is in the plane of the three oxygen atoms shown, and μ_1 is perpendicular to this plane.

almost rigid rotation of the HOD molecule, so its frequency depends primarily on the weaker forces between neighbouring molecules. It is therefore assigned to the lowest frequency peak at 104 meV.

It might be expected that the frequency of the O—H stretching modes would depend purely on the bond length r_{OO}. However, Johari and Chew (1984) have investigated the dependence on temperature and pressure of the frequencies of both the coupled and decoupled modes in the Raman spectrum, and found that a change of lattice parameter produced by reducing the temperature is 3.5 times more effective than an equivalent change produced by compression. This difference reveals the significance of anharmonicity in the inter-atomic potentials. The frequency of the O—H stretch mode is well correlated with the O—H bond length in other hydrogen-bonded materials (e.g. Lutz and Jung 1997), and can be used in the study of other phases of ice (§11.11).

In Fig. 3.6 the band with complex structure between 68 and 125 meV for H_2O (or 49 and 88 meV for D_2O) shifts according to the mass of the hydrogen atoms and is in the same region as the μ_1 mode. This band must therefore be assigned to the rotations of the molecules under the constraints of their neighbours. Such motions will be strongly coupled to form a band of modes. This band is called the rotation or 'libration' band, and is labelled ν_R in Fig. 3.5. Ordering of the protons by transformation to ice XI (see §11.2) produces a sharpening of the features in the neutron scattering spectrum in this region (Li *et al.* 1995), showing that the spectrum of these modes is in part dependent on the proton disorder in ice Ih.

The small hump at 280 meV in Fig. 3.5 has to be interpreted as a combination band of ν_2 with the libration modes ν_R, and its properties in neutron scattering experiments support this. Further combination bands are observed in the neutron scattering spectrum above 400 meV, the first of these being visible at the end of the range in Fig. 3.5 (Li 1996). Combination bands also account for features in the higher frequency range of the infrared spectrum to be seen in Fig. 9.2.

As already explained on the basis of the frequency shift in changing from H_2O to D_2O, the portion of the spectrum below 50 meV (for H_2O) must be assigned to the translational motion of whole molecules, and this region is labelled ν_T in Fig. 3.5. The simplest of these modes are sound waves or acoustic phonons. Further discussion of these modes requires the use of dispersion curves which will be explained in the following section.

3.4.6 *Dispersion curves*

The theory of phonon dispersion curves in crystalline solids is developed in many books on solid state theory, but the treatment is usually limited to simple illustrative cases or otherwise presented in a very generalized way. We here describe the basic concepts as they apply to the translational modes of an idealized model of ice Ih, in which the four molecules in the primitive unit cell (Fig. 2.6) are treated as point masses and the forces between them depend linearly on the relative molecular displacements.

We consider a crystal of N molecules placed at their equilibrium sites $\mathbf{r}_j$ ($j = 1$ to N). Each molecule will be at one of the four positions i (= 1 to 4) in its unit cell. The displacement of the molecule j from this equilibrium position is denoted by the vector $\mathbf{u}_j$ and the potential energy of the crystal due to these displacements is a quadratic function of all the $\mathbf{u}_j$. This is quite general in the harmonic approximation, and the energy can be represented as a sum over all pairs of molecules which are considered as interacting with one another. The $3N$ linear equations of motion of the N molecules then have solutions which describe $3N$ normal modes. For each mode, identified by its wave vector $\mathbf{k}$ and an index s, the displacement of the atom of type i at site $\mathbf{r}_j$ is the real part of

$$\mathbf{u}_j = \mathbf{a}_{si} \exp[\mathrm{i}(\omega_s(\mathbf{k})t - \mathbf{k} \cdot \mathbf{r}_j)]. \tag{3.13}$$

The wave vectors $\mathbf{k}$ form a lattice of N points distributed over the first Brillouin zone of the reciprocal lattice, and for each value of $\mathbf{k}$ there are 12 modes, arising from the 4 molecules per unit cell times 3 spatial co-ordinates. The index $s = 1$ to 12 identifies the mode, and its eigenfrequency is $\omega_s(\mathbf{k})$. The energies $\varepsilon_s(\mathbf{k})$ are of course equal to $\hbar\omega_s(\mathbf{k})$. The vectors $\mathbf{a}_{si}$ represent the amplitudes of oscillation of the ith molecule in each unit cell for the corresponding values of s. These amplitudes may be complex, indicating that the motions of molecules with different i may differ in phase as well as amplitude and polarization. A plot of $\varepsilon_s(\mathbf{k})$ against $\mathbf{k}$ for a particular direction in reciprocal space is known as the dispersion curve for that value of s, and the curves for different s are referred to as 'branches' of the dispersion relation.

The first three branches ($s = 1$ to 3) are the acoustic modes, which have the property that $\mathbf{a}_{si}$ are the same for all i. Equation (3.13) then describes a sound wave, and these waves have the property that $\varepsilon_s(\mathbf{k})$ is directly proportional to $|\mathbf{k}|$ for small $|\mathbf{k}|$. In one branch $\mathbf{a}_{si}$ is parallel to $\mathbf{k}$ (at least along directions of symmetry), so that these waves are longitudinal, forming the 'longitudinal acoustic' (LA) branch. The two other acoustic branches contain the 'transverse acoustic'

(TA) modes in which $\mathbf{a}_{si}$ are perpendicular to $\mathbf{k}$ and to each other. In the remaining nine branches the four molecules within each unit cell oscillate in different directions relative to one another, with symmetries characterizing the mode. These modes are given the general name of 'optic modes' because in ionic materials they couple strongly to electromagnetic radiation. Their frequencies are usually less strongly dependent on $\mathbf{k}$.

The first Brillouin zone of the reciprocal lattice is shown in Fig. 3.11. Due to the high degree of symmetry it is sufficient to specify $\varepsilon_s(\mathbf{k})$ within the region indicated by the bold lines, which represents 1/24 of the volume of the zone. Key points in the zone are denoted by Γ at the centre and A, M, and K at the boundaries. Dispersion curves are commonly drawn along the lines of symmetry ΓA, ΓM, and ΓK. Faure (1969) has calculated the dispersion relations for the simplest possible model of ice Ih with four point masses in the unit cell and two force constants representing the stretching of the O—O bond length and the changing of the O—O—O bond angle. With the ratio of these constants chosen to fit the elastic constants of ice, he obtained the dispersion curves shown in Fig. 3.12. Note that there are 12 branches, though in some places there is degeneracy and two branches coincide. Subject to the selection rules, the infrared and Raman spectra should have peaks at frequencies corresponding to vertical transitions at the point Γ in Fig. 3.12. The phonon density of states, as determined by neutron scattering, is an appropriately weighted sum of the number of modes of energy ε within the Brillouin zone. As we will see in §3.5, Faure's model does not provide a satisfactory description of the observed density of states in the translational band, and no model of this kind gives a satisfactory explanation of the two highly reproducible sawtooth-shaped peaks at 28 and 38 meV for H_2O (Fig. 3.6).

A model treating the molecules as point masses cannot, of course, describe the librational and molecular bands in the spectrum of lattice modes, but the general principles of describing the coupled modes by functions $\varepsilon_s(\mathbf{k})$ should be

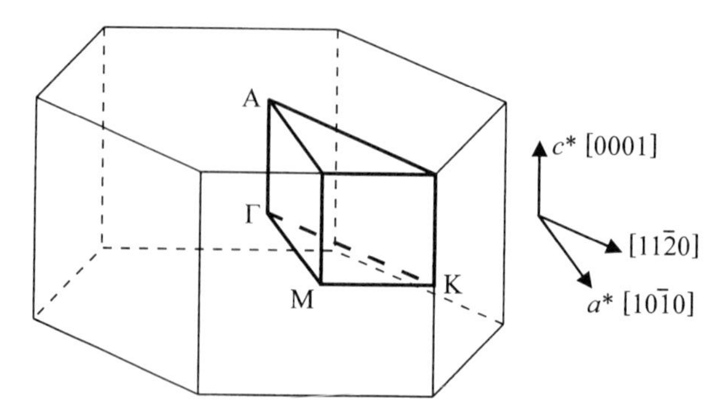

Fig. 3.11 The first Brillouin zone for the structure of ice Ih with the origin at the point Γ. $\Gamma A = \frac{1}{2}c^*$ and $\Gamma M = \frac{1}{2}a^*$, where a^* and c^* are the vectors of the reciprocal lattice.

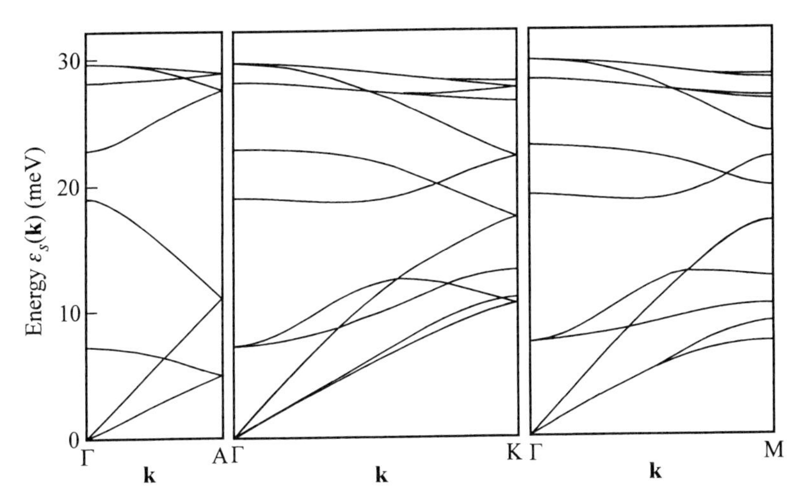

Fig. 3.12 The dispersion curves $\varepsilon_s(\mathbf{k})$ along the directions ΓA, ΓK, and ΓM for the translational modes of ice Ih according to the two-force-constant model. (From Faure 1969.)

applicable to models in which the oxygen and hydrogen atoms are treated as separate interacting masses. The number of branches will then increase from 12 to 36, accounting for $9N$ modes, where N is the number of molecules. However, for ice Ih the proton disorder presents a serious difficulty, because it is no longer possible to describe the structure with just one primitive unit cell.

A partial set of dispersion curves for D_2O ice was measured by Renker (1973), who was able to resolve about eight branches in some directions, though less in others. Li (1996) reports a smaller number of similar dispersion curves obtained with different instrumentation, but the experimental detail available about dispersion curves is not sufficient for much analysis. The interpretation of experiments at the present time is focused on the density of states and features determined optically at the zone centre.

3.5 Modelling

There has been intensive theoretical work over many years on the interpretation of the properties of ice described in this chapter in terms of molecules and the interactions between them, but some questions are still unresolved and the work continues. It is beyond the scope of this book to discuss this subject in detail, so we will confine our treatment to some simple matters which must be appreciated before embarking on further reading. All models start with some assumptions about the molecules and the interactions between them. These assumptions may be purely empirical or based on a model of the molecule itself. For the simplest assumptions results can be derived analytically and parameters fitted to the data, but in more recent modelling numerical solutions or computer simulations are essential.

3.5.1 *Elasticity and translational modes*

The simplest possible model has already been introduced in deriving the dispersion relations in Fig. 3.12. It treats the molecules as point masses in the ice lattice with perfect tetrahedral symmetry at each site. Displacements of molecules from their equilibrium positions are governed by force constants K_f and G_f which relate to changes in bond lengths Δl and bond angles $\Delta\theta$ respectively. The corresponding increase in the potential energy is then

$$\Delta E = \tfrac{1}{2}\sum K_f \Delta l^2 + \tfrac{1}{2}\sum G_f r_{OO}^2 \Delta\theta^2, \tag{3.14}$$

where the summations are over all bonds and all O—O—O angles in the crystal; r_{OO} is the O—O bond length. For a given bond the force resulting from a change in length Δl is then $-K_f\Delta l$, and similar relations apply to the angles.

The elastic properties can be derived from this model, and equations for the c_{ij} were obtained by A. Kahane (unpublished thesis 1962, quoted by Mitzdorf and Helmreich 1971). With only two parameters K_f and G_f and five independent elastic constants there clearly have to be three relations between the c_{ij} (or the s_{ij}), and Penny (1948) had shown earlier that these relations are:

$$c_{11} + c_{12} = c_{33} + c_{13}$$

$$(c_{33} - c_{13})^2 = 2c_{44}(c_{33} - 5c_{13} + 4c_{12}) \tag{3.15}$$

$$12c_{13}^2 + 5c_{13}c_{33} - c_{33}^2 + 2c_{44}(5c_{13} + 2c_{33}) = c_{12}(15c_{13} + c_{33} + 14c_{44}).$$

With the experimental data in Table 3.2 the two sides of these equations are equal within 1, 2, and 6% respectively, which is not far outside experimental error. This fact implies that the elastic anisotropy of ice arises from the topology of the structure and not from differences between non-equivalent bonds. Measurements of the elastic properties of ice cannot be expected to yield significant information about the intermolecular forces other than K_f and G_f. By considering the elastic energy stored in the bonds, with no change of bond angles, it is simple to show that the bulk modulus K is given by

$$K = \frac{K_f}{4\sqrt{3}r_{OO}}. \tag{3.16}$$

Equation (3.16) is consistent with Kahane's equations, and using the dynamic elastic constants at $-16\,°C$ from Table 3.2 his equations give the force constants $K_f = 17.1\ N\,m^{-1}$ and $G_f = 0.476\ N\,m^{-1}$. With these values Young's modulus is determined primarily by G_f and is comparatively insensitive to K_f, which means that under uniaxial stress ice deforms elastically by changing the bond angles rather than the bond lengths.

On the basis of this model the full set of translational dispersion curves can be calculated as in Fig. 3.12. Bertie and Whalley (1967) used these relations to discuss the interpretation of the infrared absorption spectrum, and showed that the model could not account for any modes above 28 meV, although absorption above this limit was observed experimentally. Later neutron scattering experiments showed that there is a peak in the density of states at 38 meV as well as one at 28 meV, and additional features will have to be included in the model to account for these two peaks. An early suggestion by Renker (1973) was that the bonds parallel and oblique to the *c*-axis had different force constants. This would be inconsistent with the conclusion drawn above from the anisotropy of the elastic constants, and Li and Ross (1992) have shown that it could not be the explanation because, in experiments on single crystals, the peaks in the neutron scattering were observed to be independent of the crystal orientation. Corresponding peaks have been observed in several high-pressure phases of ice (Fig. 11.19), and an explanation of this feature has become an important aim of modelling the intermolecular interactions in ice.

The introduction of more elaborate models, possibly including the effect of proton disorder, requires computer modelling, either using lattice dynamics or molecular dynamical simulation. In lattice dynamics the eigenvalues of the linear equations of motion of an assembly of atoms or molecules with periodic boundary conditions are solved by diagonalizing the dynamical matrix (Dolling 1976). Standard programs exist for this purpose, and in the case of ice local disorder is incorporated by using an extra large unit cell. In molecular dynamics the actual motion of the molecules is followed in the simulation as a function of time, and the spectrum can be derived from a Fourier transform of the correlation function of their motions (Sciortino and Corongiu 1993). Examples of molecular dynamic calculations on phases of ice using empirical pair potentials (Table 1.1) or *ab initio* methods are Tse *et al.* (1984), Lee *et al.* (1993), and Burnham *et al.* (1997).

Using the lattice dynamics program PHONON, Li and Ross (1993) and Li (1996, 1997) have shown that merely introducing separate oxygen and hydrogen atoms into the model with force constants derived from various inter-atomic potentials will not generate two distinct peaks at the upper end of the translational band. However, if force constants differing by a factor of the order of 2 are assigned to the hydrogen bonds at random, two peaks are produced in the density of states. On this basis Li and Ross (1993) proposed that there are 'two kinds of hydrogen bond', depending on the four possible relative orientations of a pair of adjacent molecules in the structure, but no reason has yet been found to explain why the force constants should differ by such a large amount. Electrostatic dipole interactions are far too small. In support of their proposal Li and Ross (1994) point to an apparent splitting of the isotopically decoupled mode μ_1, which should be particularly sensitive to the H-bond stretching force constant. The transformation of ice Ih to the proton-ordered phase ice XI (§11.2) will change the proportions of the two types of hydrogen bonds proposed by Li and Ross, but experiments to date on the inelastic scattering from partially transformed specimens have not shown any corresponding change in the ratio of the intensities of the 28 and 38 meV peaks (Li *et al.* 1995; Fukazawa *et al.* 1998).

Other workers concerned with theoretical modelling believe that the two-peak feature should emerge naturally from the inclusion of appropriate long-range, many-body, or anharmonic interaction terms in the model. Lattice dynamical calculations have produced a splitting at the point Γ of the highest frequency (transverse and longitudinal optic) modes, but the effect is much too small (Marchi *et al.* 1986; Klug *et al.* 1991*a*). Of a number of papers using increasingly sophisticated models only a molecular dynamical simulation using polarizable *ab initio* potentials (Clementi *et al.* 1993) shows signs of generating the two peaks, but the physical feature of the model responsible for them has not been identified. For a convincing interpretation the actual molecular motions responsible for the peaks need to be identified.

3.5.2 *Molecular and librational modes*

As already explained, the molecular modes ν_1, ν_2, and ν_3 are primarily determined by the properties of the isolated molecule but modified by interactions with its neighbours, most particularly by the change in the potential well for the protons as a result of hydrogen bonding. The librational modes are determined by the forces acting on a molecule when it is rotated away from its equilibrium position in the lattice. In this case the restoring forces acting are the same as those coupling the motion of one molecule to its neighbours, and this leads to a broad band of eigenfrequencies, in a manner similar to that already described in connection with the translational modes. For the molecular modes, however, the coupling is relatively weak compared to the forces acting within the molecule.

Shawyer and Dean (1972) performed lattice dynamical calculations on up to 500 atoms using three intramolecular and five intermolecular force constants chosen to fit the spectra and required to be compatible with the elastic constants. Their analysis gave a librational band about as broad as that shown for the neutron scattering results in Fig. 3.5, but their molecular bands were too sharp. Sceats and Rice (1980) discuss the force constants applicable to a pair of molecules in more detail, including significant anharmonic terms and fitting their parameters to the librational spectra for ice Ih, the liquid, and other phases. Itoh *et al.* (1996*b*) report molecular dynamic modelling of the librational band, and show that in the proton-ordered structure of ice XI the band breaks up into several more distinct peaks, in line with the experimental observations of Li *et al.* (1995) already mentioned.

The O—H bond stretching modes ν_1 and ν_3 have been modelled in some detail by Rice *et al.* (1983), who were able to account satisfactorily for the shapes of both the infrared and the Raman peaks. A full set of lattice modes obtained using the lattice dynamics PHONON program with eight force constants, including their two hydrogen bond stretching constants, is given by Li and Ross (1994). We make no attempt to summarize the parameters adopted in these various models because they must be understood in the context of the specific model to which they apply. The relevance of spectroscopic models to the crystallographic properties has been reviewed by Kuhs and Lehmann (1986), and the application of these methods to other phases of ice is discussed in §11.11.

In all these studies it has been found that the harmonic approximation to the potentials (i.e. $V(u)=\frac{1}{2}Ku^2$) is inadequate for satisfactory modelling, and anharmonic terms in u^3 or u^4 must be included. Such terms arise in precise modelling of the mean square displacements in crystallography as well as in the spectra. They are essential to produce thermal expansion and they make a contribution to the heat capacity of about 12% by 200 K. They are essential to the umklapp process of phonon scattering and hence to limiting the thermal conductivity at all temperatures above about 10 K.

4
Electrical properties—theory

4.1 Basics

Although at first sight ice does not appear to be an 'electrical material', its dielectric and conductive properties have been extensively studied both experimentally and theoretically. Crystallography tells us about the arrangement of molecules, lattice vibrations depend on the intermolecular forces, but the electrical properties show how molecules turn round or protons flow through the lattice. Ice is the model material for studying such effects, and provides understanding applicable to many hydrogen-bonded solids. Unique types of protonic point defects have to be introduced to account for the electrical behaviour, and these are relevant to other properties of ice as well. An understanding of electrical properties and point defects is essential for interpreting many other properties and applications of ice physics.

The theory of the electrical properties of ice has now become well established, but in this chapter it is presented as a whole for the first time. The following chapter describes the experimental results and interprets them in the light of this theoretical understanding. The specific properties of the relevant point defects as determined from these and other experiments are treated in Chapter 6.

When an electric field is applied to a specimen of ice three distinct processes occur:

1. The individual molecules are polarized by the field. This involves displacements of the electrons relative to the nuclei and small distortions of the molecules under the restoring forces considered in Chapter 3. These are processes which occur in any material. The response to a change in field is very rapid, so that the effects are independent of frequency up to microwave frequencies.
2. The ice is polarized by the reorientation of molecules or bonds. Put in another way, the energies of some of the proton configurations that are compatible with the ice rules are lowered relative to others, so that in thermal equilibrium there is a net polarization of the ice. The achievement of this equilibrium state is a comparatively slow process, requiring thermal activation and local violations of the ice rules.
3. With suitable electrodes a steady current flows in accordance with Ohm's law. There is no detectable electronic conduction in ice and the observed current arises from a flow of protons. Because the conduction process has features in

common with electronic conduction in semiconductors ice is sometimes referred to as a 'protonic semiconductor', but the protons do not share the quantum mechanical properties that are an essential feature of the band theory of electrons in solids.

In the processes 1 and 2 above the electric polarization (i.e. the dipole moment per unit volume) P is related to the electric field E by the equation

$$P = \varepsilon_0 \chi E, \tag{4.1}$$

where χ is the electric susceptibility and ε_0 is the 'permittivity of free space'. In accordance with standard theory of dielectrics the relative permittivity (or dielectric constant) is $\varepsilon = 1 + \chi$. We will always include ε_0 explicitly in equations, making χ and ε dimensionless, and we will drop the adjective 'relative' when referring to the permittivity ε. At high frequencies where only process 1 is operative, the permittivity is denoted ε_∞. In the low-frequency limit the 'static' permittivity is ε_s, and the susceptibility due to process 2 is $\chi_s = \varepsilon_s - \varepsilon_\infty$. In the ohmic conduction process (3 above) the current density J is given by

$$J = \sigma_s E, \tag{4.2}$$

in which σ_s is the steady-state conductivity. In the hexagonal symmetry of ice Ih these properties are in principle anisotropic, but in this chapter we will not find it necessary to include this complication.

The polarization process 2 in ice is an almost perfect example of a Debye relaxation process, in which the polarization approaches its equilibrium value $P_s = \varepsilon_0 \chi_s E$ in accordance with the equation

$$\frac{dP}{dt} = \frac{1}{\tau_D}(P_s - P). \tag{4.3}$$

τ_D is called the Debye relaxation time; it depends on the temperature and purity of the ice. At the highest frequencies the polarization is unable to change with the field, and for low frequencies it follows P_s exactly. However, for changes of field on a time scale comparable with τ_D the polarization lags behind the field, and this results in energy dissipation and an apparent resistive component of the current flowing to the specimen. The equations describing Debye relaxation will be derived in §4.2.

Taking all three processes into account, the electrical behaviour of ice can be represented by the equivalent circuit in Fig. 4.1(a). For a plate of area A and thickness L the capacitance is $C = \varepsilon\varepsilon_0 A/L$ and the conductance is $G = \sigma A/L$. We can therefore label capacitors and resistors with the appropriate permittivities ε and conductivities σ. At high frequencies the circuit reduces to that in Fig. 4.1(b); the capacitance is determined by ε_∞ with a parallel conductance σ_∞ that becomes less significant compared with $\omega\varepsilon_\infty$ as the (angular) frequency ω increases. At low

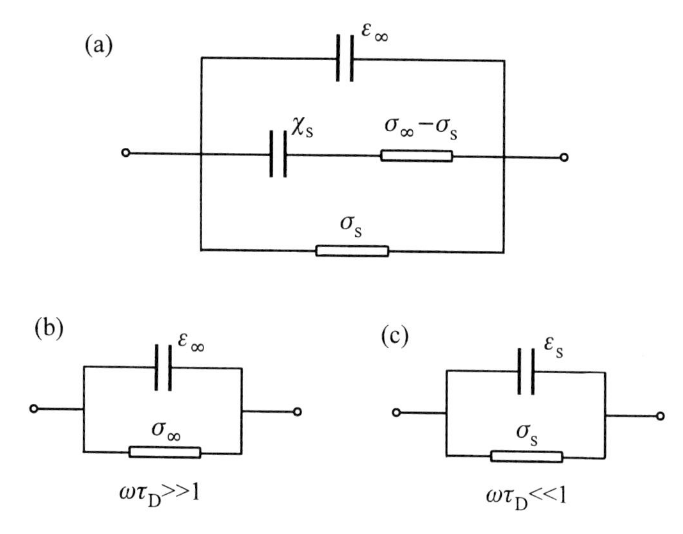

Fig. 4.1 (a) The equivalent circuit for a specimen of ice with ideal ohmic electrodes. (b) High-frequency limit. (c) Low-frequency limit.

frequencies the circuit reduces to Fig. 4.1(c), with the full capacitance determined by $\varepsilon_s = \varepsilon_\infty + \chi_s$ in parallel with the steady-state conductance σ_s. The middle arm of Fig. 4.1(a), consisting of χ_s and $(\sigma_\infty - \sigma_s)$ in series, has exactly the properties of the Debye relaxation process described by equation (4.3). The charge Q flowing into the whole circuit in response to a voltage step of height V at time t_0 is easily seen to be that illustrated in Fig. 4.2.

Because an electric current is carried in ice by the motion of protons rather than electrons there will normally be electrolytic processes occurring at the interface between ice and metallic electrodes. Petrenko and Chesnakov (1990*b*) have shown that the appropriate quantities of hydrogen and oxygen are released just as in the electrolysis of liquid water. This constitutes the direct evidence for protonic rather than electronic charge carriers in ice; an earlier experiment by Decroly *et al.* (1957) aimed at proving this point does not in fact do so because their current would be flowing largely through an HF-contaminated surface layer. The electrolytic process leads to non-ohmic contacts between ice and metallic electrodes and to the build-up of space charges close to the surface of the ice, but such problems can be avoided in special cases (see §5.5). In the main part of this chapter we will assume that the electrodes are ideal so that interfacial problems can be ignored.

The three processes discussed above also occur in liquid water. Figure 4.3 shows the static permittivity ε_s for ice and for water at temperatures near the melting point. The large values of ε_s arise from the dipole moment of the water molecules which are present in both phases, but the close similarity between the

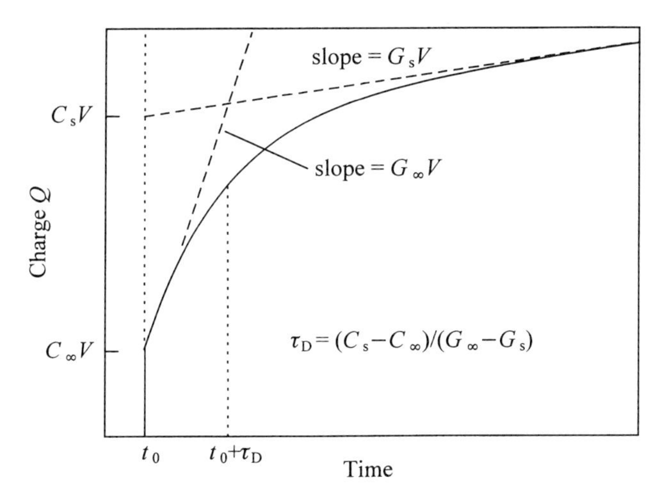

Fig. 4.2 The response of the circuit in Fig. 4.1(a) to a step voltage V applied at time t_0. Q is the integrated charge flowing into the circuit, $G_s = \sigma_s A/L$, $C_s = \varepsilon_s \varepsilon_0 A/L$, etc.

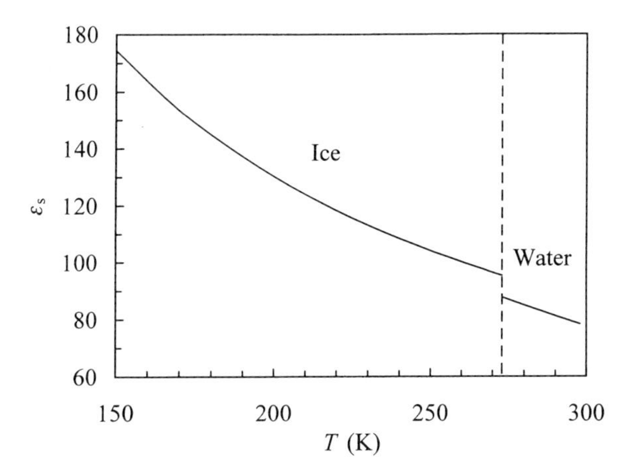

Fig. 4.3 Steady-state permittivity ε_s of pure polycrystalline ice and pure water as functions of temperature T. Data for ice from Johari and Whalley (1981), for water from Malmberg and Maryott (1956).

values is rather a coincidence. The real difference between the solid and liquid phases is in the time scales for polarization to occur. In pure ice at $-10\,°C$ $\tau_D \approx 5 \times 10^{-5}$ s, whereas in water at $+10\,°C$ $\tau_D \approx 1.2 \times 10^{-11}$ s. Because the molecules in ice are bound in the lattice and subject to the ice rules, they are much more difficult to reorient than those in the liquid. This has the practical consequence that microwaves in a typical microwave cooker working at about 2.5 GHz are strongly absorbed by water but not at all by ice. A similar difference in rates is observed in the static conductivities. That for pure ice at $-10\,^{\circ}\mathrm{C}$ is not more than $6 \times 10^{-10}\,\mathrm{S\,m^{-1}}$ compared with about $5 \times 10^{-6}\,\mathrm{S\,m^{-1}}$ for water, but in this case the mechanisms are rather more different than for the permittivity.

Different aspects of the theory of the electrical properties of ice have been developed in parallel largely during the period 1950 to 1980, with especially significant contributions by Bjerrum, Onsager, Nagle, Gränicher, and Jaccard. Later work has brought a fuller understanding of the relationship between these approaches, and in this chapter the subject is presented in a fully integrated way. After a phenomenological description of the Debye relaxation process (§4.2) we discuss in some detail the equilibrium thermodynamics of the polarization of ice treated as an assembly of water molecules (§4.3). The basic topology of the point defects which permit the equilibration of these molecules and the process of protonic conduction are then described (§4.4), and this leads to the Jaccard theory which accounts for the properties of ice in terms of the motion of these defects (§4.5). Section 4.6 deals with the fact that specimens have to have boundaries, and §4.7 aims to give some insight into the significance of the various time constants that can be observed. The chapter ends with a summary of the essential equations arising in the theory of the electrical properties of ice (§4.8).

4.2 Frequency dependence of the Debye relaxation

The electrical properties of ice are most commonly studied by determining the alternating current admittance (reciprocal of impedance) of a specimen of ice as a function of frequency. This reveals the Debye and sometimes other relaxation processes. Before considering the mechanisms of polarization in ice it is appropriate to establish the theory and notation used to describe the a.c. properties of any system obeying the equation

$$\frac{\mathrm{d}P}{\mathrm{d}t} = \frac{1}{\tau_{\mathrm{D}}}(\varepsilon_0 \chi_{\mathrm{s}} E - P), \tag{4.4}$$

which is equation (4.3) with $P_{\mathrm{s}} = \varepsilon_0 \chi_{\mathrm{s}} E$. The relaxation process is characterized by two parameters: its strength χ_{s}, which determines the maximum in-phase response in the low-frequency limit, and its time constant τ_{D}. The theory was developed by Debye (1929), who applied it to the case of molecules in ice even before Pauling's model had been devised.

We represent the sinusoidal alternating field of angular frequency ω by the complex quantity $E = E_0 \mathrm{e}^{\mathrm{i}\omega t}$, where the amplitude E_0 is real and $\mathrm{i} = \sqrt{-1}$. The response is $P = P_0 \mathrm{e}^{\mathrm{i}\omega t}$, where P_0 is complex. Simple mathematics then leads to a complex frequency-dependent susceptibility

$$\chi(\omega) = \frac{P_0}{\varepsilon_0 E_0} = \frac{\chi_s}{1 + i\omega\tau_D}. \tag{4.5}$$

Dividing $\chi(\omega)$ into real and imaginary parts

$$\chi(\omega) = \chi'(\omega) - i\chi''(\omega), \tag{4.6}$$

we find

$$\chi'(\omega) = \frac{\chi_s}{1 + \omega^2\tau_D^2} \tag{4.7a}$$

$$\chi''(\omega) = \frac{\omega\tau_D\chi_s}{1 + \omega^2\tau_D^2}. \tag{4.7b}$$

The negative sign in equation (4.6) indicates that with $\chi''(\omega)$ positive the polarization lags behind the field. The functions $\chi'(\omega)$ and $\chi''(\omega)$ are shown in Fig. 4.4.

In expressing the response as a susceptibility we are considering the ice as a capacitor of complex admittance

$$\begin{aligned} Y(\omega) &= i\omega\frac{A\varepsilon_0(\varepsilon_\infty + \chi(\omega))}{L} \\ &= \frac{A\varepsilon_0\omega\chi''(\omega)}{L} + i\omega\frac{A\varepsilon_0(\varepsilon_\infty + \chi'(\omega))}{L} \\ &= G(\omega) + i\omega C(\omega). \end{aligned} \tag{4.8}$$

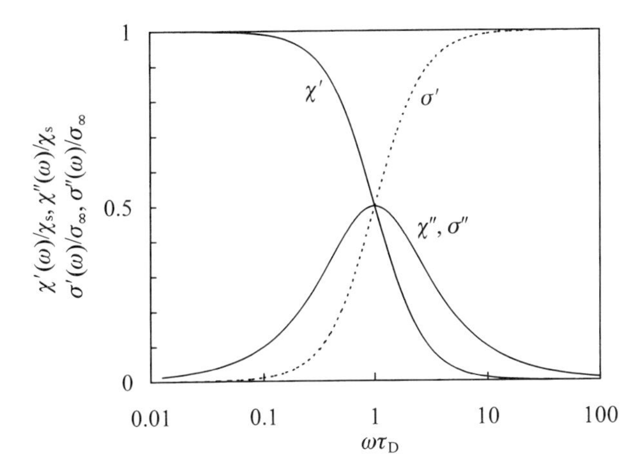

Fig. 4.4 Normalized real and imaginary parts of the susceptibility $\chi(\omega)$ and conductivity $\sigma(\omega)$ in a Debye relaxation process, according to equations (4.7) and (4.12) with $\sigma_s = 0$.

The equivalent circuit thus behaves as a frequency-dependent capacitance $C(\omega)$ and conductance $G(\omega)$ in parallel. Alternatively we may think of the ice as a resistor, in which case we focus on the current density

$$J = \frac{\mathrm{d}P}{\mathrm{d}t} = \mathrm{i}\omega P_0 \mathrm{e}^{\mathrm{i}\omega t}. \tag{4.9}$$

The complex conductivity is then

$$\sigma(\omega) = \frac{J}{E} = \mathrm{i}\omega\varepsilon_0\chi(\omega). \tag{4.10}$$

Dividing this into real and imaginary parts

$$\sigma(\omega) = \sigma'(\omega) + \mathrm{i}\sigma''(\omega), \tag{4.11}$$

and, allowing for the parallel conductance σ_s in Fig. 4.1, we find

$$\sigma'(\omega) = \sigma_s + \frac{\omega^2\tau_D^2(\sigma_\infty - \sigma_s)}{1 + \omega^2\tau_D^2} \tag{4.12a}$$

$$\sigma''(\omega) = \frac{\omega\tau_D(\sigma_\infty - \sigma_s)}{1 + \omega^2\tau_D^2}, \tag{4.12b}$$

in which

$$\sigma_\infty - \sigma_s = \frac{\varepsilon_0\chi_s}{\tau_D}. \tag{4.13}$$

The apparent conductance $G(\omega)$ in equation (4.8) is seen to be $A\sigma'(\omega)/L$. Results are often presented simply as the permittivity $\varepsilon(\omega) = \varepsilon_\infty + \chi'(\omega)$ and the conductivity $\sigma(\omega) = \sigma'(\omega)$ as functions of ω. This gives graphs of the form shown in Fig. 4.5, which includes experimental points for a specimen of nominally pure ice which exhibits all the features described.

Equations (4.7a,b) have the property that

$$\chi'(\omega) = \chi_s - \tau_D\,\omega\,\chi''(\omega), \tag{4.14}$$

and equations (4.6), (4.10), and (4.11) imply that $\omega\chi''(\omega) = \sigma'(\omega)/\varepsilon_0$. These results lead to a useful linear plot (Gränicher 1969) of $C(\omega)$ against $G(\omega)$ according to the equation

$$C(\omega) - C_\infty = C_s - \tau_D(G(\omega) - G_s). \tag{4.15}$$

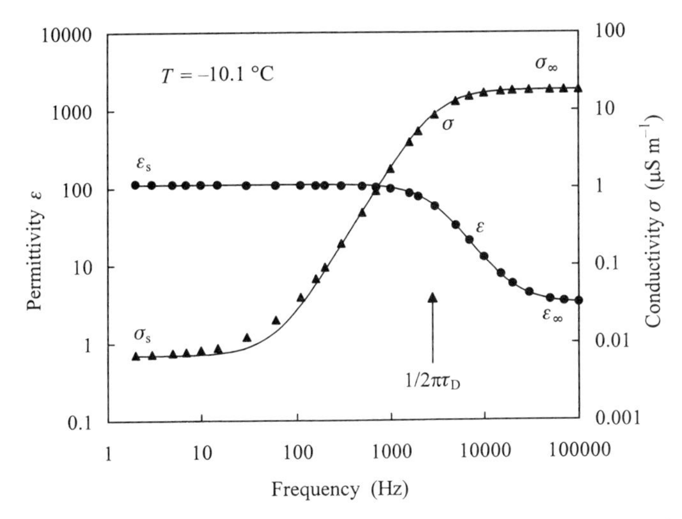

Fig. 4.5 Logarithmic plots of permittivity ε and conductivity σ as functions of frequency for ice described by the circuit in Fig. 4.1. Real data for 'pure' ice at $-10.1\,^{\circ}\mathrm{C}$ are included from Takei and Maeno (1997).

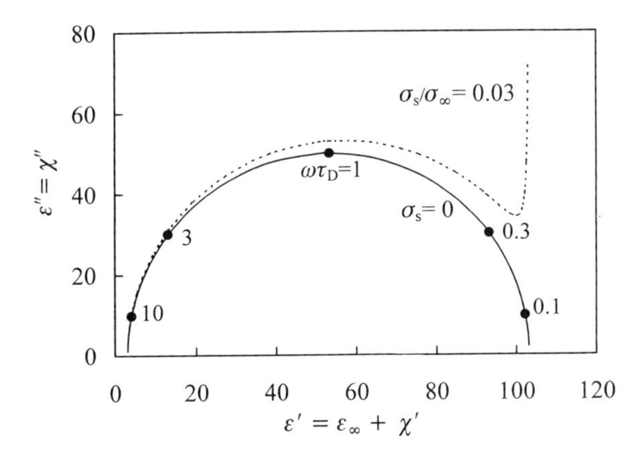

Fig. 4.6 Typical Cole–Cole plots according to equation (4.16) for ice with $\chi_s = 100$, both for $\sigma_s = 0$ and for $\sigma_s/\sigma_\infty = 0.03$.

Another way of presenting data on a relaxation process is the Cole–Cole plot (Cole and Cole 1941). Eliminating $\omega\tau_D$ between equations (4.7a and b) gives

$$(\chi'(\omega) - \chi_s/2)^2 + \chi''(\omega)^2 = (\chi_s/2)^2. \tag{4.16}$$

Thus as ω varies, the locus of the point $\chi'(\omega)$, $\chi''(\omega)$ is a semicircle of radius $\chi_s/2$. Incorporating the term ε_∞, this plot is shown in Fig. 4.6. If a steady state conductivity σ_s is present the Cole–Cole plot will break away from the semicircle at low frequencies as shown by the broken line. The extent of this deviation depends solely on the ratio σ_s/σ_∞. If the dispersion cannot be described by a single relaxation time the Cole–Cole plot will be spread out along the χ' axis, and, for widely separated relaxation times, it will break into more than one semicircle. A similar type of Cole–Cole plot can be used to represent the conductivities $\sigma'(\omega)$ and $\sigma''(\omega)$.

4.3 The static susceptibility χ_s

For an assembly of molecules the static susceptibility is an equilibrium property; its value does not depend on the mechanism by which molecules can be oriented to form the polarized state. In an electric field the different Pauling configurations of the water molecules in ice do not all have the same energy, and it is only necessary to minimize the free energy of this system to find the resulting polarization and hence χ_s.

The next two sections (§§4.3.1 and 4.3.2) deal with the theoretical calculation of χ_s and yield a result which is essential to the full treatment of the electrical properties of ice. It is logically placed here, but it is more difficult than the rest of the chapter, and first time readers or others more interested in the interpretation of experiments may prefer to proceed directly to §4.4, referring back to the result when it is needed at equation (4.43).

4.3.1 *Electric field parallel to the c-axis*

We will start by developing the theory for the simplest possible case of ice Ih with the electric field parallel to the c-axis. In this case the argument is essentially the same as that for the susceptibility of a spin-1/2 paramagnet treated in many undergraduate texts on statistical mechanics, but modified to take account of the correlations imposed by the ice rules. The theory is based on that used by Slater (1941) for KH_2PO_4 and applied to ice Ic by Hollins (1964). It is quoted for ice by Onsager and Dupuis (1962).

Each water molecule has a dipole moment $\mathbf{p}$ which must have one of the six orientations shown in Fig. 2.9. (We use $\mathbf{p}$ rather than the conventional $\boldsymbol{\mu}$ to avoid confusion with mobility in the following sections.) For a field E along the c-axis the component of the dipole moment parallel to the field is $\pm p_z$, where $p_z = p/\sqrt{3}$ and $p = |\mathbf{p}|$. In an assembly of N molecules per unit volume, a configuration with n_+ molecules with the component parallel to the field and n_- molecules opposed to the field will have the net polarization

$$P = (n_+ - n_-)p_z = np_z, \tag{4.17}$$

where $n = (n_+ - n_-)$, and of course $(n_+ + n_-) = N$. Relative to an unpolarized state in the same field the energy of this configuration is $-PE$, and the free energy of the assembly is

$$F = -np_z E - TS(n), \tag{4.18}$$

where $S(n)$ is the entropy $k_B \ln W(n)$ of those configurations which have this value of n, and T is the temperature. For very large N the total number of configurations is effectively the same as its maximum value so that $S(0)$ is equal to the Pauling entropy $Nk_B \ln(3/2)$. The equilibrium value of n is determined by minimizing F, so that

$$\left(\frac{\partial F}{\partial n}\right)_{T,E} = -p_z E - T\frac{dS}{dn} = 0. \tag{4.19}$$

From symmetry $dS/dn = 0$ at $n = 0$, but we find it for non-zero n by expanding $S(n)$ as a Taylor series to second order and differentiating with respect to n. Equation (4.19) then gives

$$Tn\left[\frac{d^2S(n)}{dn^2}\right]_{n=0} = -Ep_z. \tag{4.20}$$

The susceptibility is then

$$\chi_s = \frac{P}{\varepsilon_0 E} = -\frac{p_z^2}{\varepsilon_0 T(d^2S(n)/dn^2)} = -\frac{p^2}{3\varepsilon_0 T(d^2S(n)/dn^2)}, \tag{4.21}$$

and from now on we will assume that the second derivative is evaluated as in equation (4.20) at $n = 0$. Following Nagle (1974, 1978) we introduce a dimensionless constant G defined such that

$$\chi_s = \frac{GNp^2}{3\varepsilon_0 k_B T}. \tag{4.22}$$

This constant is of great importance in the electrical properties of ice; it can in principle be anisotropic. In the present case with the field along the c-axis equation (4.21) shows that

$$G^{-1} = -N\frac{d^2 \ln W(n)}{dn^2}. \tag{4.23}$$

We can evaluate G in this case using an argument similar to the first of the methods for calculating the Pauling entropy in §2.3.1, but now considering *N pairs* of molecules linked by bonds parallel to the c-axis. The two members of each pair will always have the z-components of their dipole moments in the same direction. Let there be n_+ pairs with a positive component and n_- $(= N - n_+)$

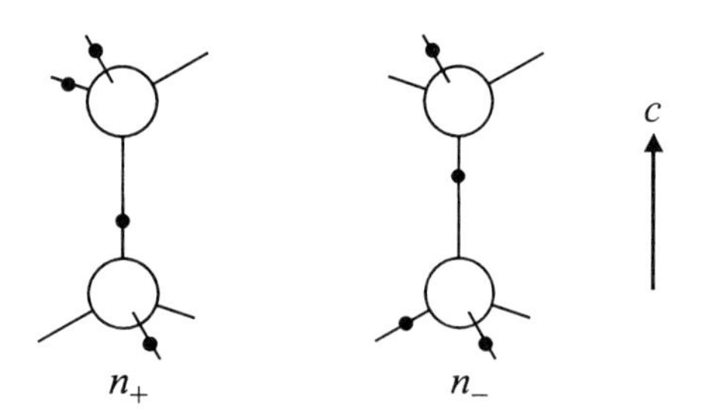

Fig. 4.7 The two kinds of molecular pair considered in the elementary calculation of G.

pairs with a negative component. These pairs are shown in Fig. 4.7, and as before we define $n = n_+ - n_-$. Each of the molecules in a pair has three possible arrangements of the one or two protons which lie on the three bonds that are oblique to the c-axis. Ignoring the second ice rule that there is only one proton per bond, the total number of configurations of the N pairs of molecules with these values of n_+ and n_- is

$$W_1 = \frac{N!}{n_+!\,n_-!}3^{2N}. \tag{4.24}$$

There are a total of $3N$ oblique bonds, of which $2n_+ + n_- = (3N + n)/2$ have a proton at the lower end and $n_+ + 2n_- = (3N - n)/2$ have a proton at the top end as indicated in Fig. 4.7. Within the above W_1 configurations, the probability that an oblique bond with a proton at the bottom also has no proton at the top is just the fraction of oblique bonds $\frac{1}{2}(1 + n/3N)$ for which this is the case. For bonds with no proton at the bottom the probability is $\frac{1}{2}(1 - n/3N)$. Thus, of the W_1 configurations, the number which satisfy the second ice rule is

$$W(n) = W_1\left[\frac{1}{2}\left(1 + \frac{n}{3N}\right)\right]^{(3N+n)/2}\left[\frac{1}{2}\left(1 - \frac{n}{3N}\right)\right]^{(3N-n)/2}. \tag{4.25}$$

Using Stirling's approximation and expanding the logarithms as power series, this gives

$$\ln W(n) = 2N\left[\ln\left(\frac{3}{2}\right) - \frac{1}{6}\left(\frac{n}{N}\right)^2 + \mathrm{O}\left(\frac{n}{N}\right)^4\right]. \tag{4.26}$$

Realizing that there are $2N$ *molecules*, this equation yields the Pauling entropy for $n = 0$. Equation (4.23), which applies to N molecules, then shows that in this approximation $G = 3$.

4.3.2 *The general case*

For other orientations of the field the theory becomes more complicated, but calculations of the same general character as those given above show that at this level of approximation G is isotropic with the value 3 (Kawada 1981; Minagawa 1981, 1990).

Considerable theoretical effort has been devoted to calculating G to a higher degree of approximation, taking account of correlations arising from closed loops within the structure. Such studies have included cubic and two-dimensional ice as well as the more difficult case of ice Ih. Nagle (1974, 1978) developed a series expansion which showed that G is isotropic and closely equal to 3.0, and this has been confirmed by Monte Carlo calculations of correlation functions (Adams 1984). The effects of long-range correlations on G appear to be even less significant than in the case of the entropy itself.

A major problem in the theory of dielectrics is to take account of the effects on one molecule of the internal fields arising from the polarizations of its neighbours. In the case of a gas this leads to the well-known Clausius–Mossotti correction, but for densely packed highly polar molecules the problem is both more serious and more difficult. There has been much controversy over the relative merits of the Onsager–Slater treatment of ice which we have followed above and the alternative approach due to Kirkwood (1939) and Fröhlich (1949); see the review by Stillinger (1982). A critical step is to justify the inclusion of the energy as $-np_zE$ in equation (4.18), with E being the electric field as normally defined from the potential difference divided by the thickness. This was done by Nagle (1979) using his 'unit model of ideal ice', which shows how this apparently simple result is a special feature of the structure of ice and leads our thinking forward to the introduction of protonic point defects.

The aim of the unit model is to build up the ice not from polar molecules but using units that have no net charge or dipole moment. Following Nagle we illustrate this idea in Fig. 4.8 using two-dimensional square ice. The portion of ice shown in (a) satisfies periodic boundary conditions and has a net polarization in the vertical (z) direction. In (b) the molecules are replaced by units consisting of exactly one oxygen with effective charge $-2q$ and parts of the charges of the four surrounding protons as shown in (c). The effective charge $+q$ of each proton is divided into two parts $\frac{1}{2}e_A$ and $\frac{1}{2}e_B$ such that $\frac{1}{2}e_A r_{OH} = \frac{1}{2}e_B r_{O\cdots H}$, where r_{OH} and $r_{O\cdots H}$ are the shorter and longer distances to the neighbouring oxygens. Each unit is then neutral and has no dipole moment, provided we can neglect the small effects of residual electronic charge in the interstices between the units. The huge electrostatic interactions at the points where the units link up are constant (i.e. do not depend on the polarization) and so can be ignored. Apart from this, the only electrostatic interactions between the units are of quadrupole or higher order.

At the top and bottom surfaces there are unbalanced charges $\frac{1}{2}e_B$ marked '+' lying outside units, and there are places marked '−' where such charges would

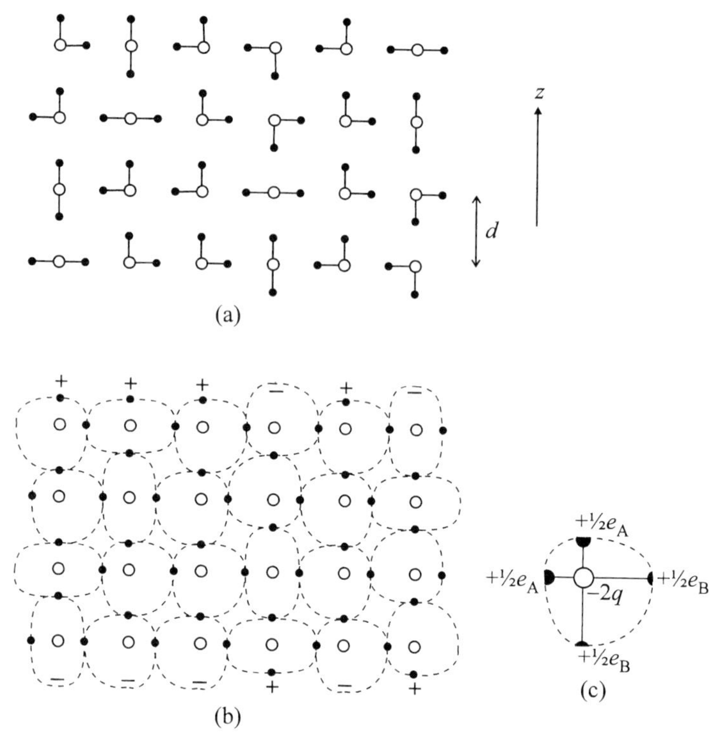

Fig. 4.8 Nagle's 'unit model' illustrated for two-dimensional square ice. (a) and (b) are reproduced from *Chemical Physics*, **43**, J.F. Nagle, Theory of the dielectric constant of ice, pp. 317–28, Copyright 1979, with permission from Elsevier Science.

have to be added to complete units. For unpolarized ice these charges cancel, but if the ice is polarized there will be a net number n_s of excess positive charges per unit area on the top face and opposite charges on the bottom face; these constitute the surface polarization charge $\pm P$ in the usual theory of dielectrics. If, in the notation of the previous section, there are $n = n_+ - n_-$ excess dipoles per unit volume oriented in the z-direction, $n_s = nd$, where d is the thickness of one layer of molecules lying perpendicular to z. In the case of square ice $d = r_{OO} = r_{OH} + r_{O\cdots H}$. For a specimen of area A and thickness L the electrostatic energy U of the surface polarization charge is the potential difference times the charge giving $U = -EL \times \frac{1}{2}n_s e_B A$, but in equation (4.18) we wrote this as $-np_z E$ per unit volume. These results are equivalent if $p_z = \frac{1}{2}e_B d$. The Onsager–Slater–Nagle approach is therefore valid, without any correction for dipole–dipole interactions, provided that p is chosen appropriately. The value of G can be taken from theory, and experiment will then yield a value for p.

In ice Ih the geometry is more complicated, but for polarization along the c-axis d is just $c/4 = 2r_{OO}/3$ and $p = p_z\sqrt{3}$, giving

$$p = e_B r_{OO}/\sqrt{3}. \tag{4.27}$$

The next section will provide an interpretation of this equation by identifying e_B as the charge on a Bjerrum defect.

It is very important to realize that equation (4.21) and the calculations arising from it relate to an assembly of molecules whose moments **p** can be oriented by the field. This is not necessarily the case, because polarization can in some circumstances occur by the hopping of protons from one molecule to the next along a hydrogen bond, and the theory then has to be quite different. The case described above is appropriate for pure ice at normal temperatures, but it is just one limiting case of the Jaccard theory, which will be described in §4.5.

4.4 Protonic point defects

The above theory concerns the thermal equilibrium polarization of an assembly of molecules in an applied field, but in a crystal in which the ice rules are strictly obeyed it is impossible to attain such equilibrium. The orientation of every molecule is determined by its neighbours. The simplest change that can be envisaged is that, in a ring in which all bonds point in a particular direction around the ring, the directions of all these bonds are reversed simultaneously, but this would involve a large activation energy and would not change the overall polarization anyway. Bjerrum (1951) recognized that for polarization or electrical conduction to be possible in ice there must be a number of places in the lattice at which the ice rules are locally violated. Once such defects are present, even in small numbers, their motion has the effect of reorienting the molecules along their path and so changing one Pauling configuration into another. The long Debye relaxation times for ice as compared with water are determined by the concentration and mobility of these defects. The defects proposed by Bjerrum are specific to ice-like structures and are referred to as *protonic point defects* to distinguish them from more conventional point defects such as vacancies and interstitials.

Figure 4.9 shows how the four types of protonic point defect can be formed from the perfect structure of Fig. 2.4. If the single molecule A is turned to a new orientation, it produces one bond with no protons and another in which two protons are pointing towards one another. This is clearly a very unstable situation but further turns of neighbouring molecules can separate the defective bonds. Bjerrum called these incorrectly formed bonds 'orientation faults', but, following Gränicher (1958), they are now known as *Bjerrum defects*. The empty bond is an *L-defect* (from the German *leere* = empty) and the bond with two protons is a *D-defect* (from *doppeltbesetzte* = doubly occupied). The other two defects in Fig. 4.9 are the H_3O^+ and OH^- *ionic defects*, formed by the transfer of a proton

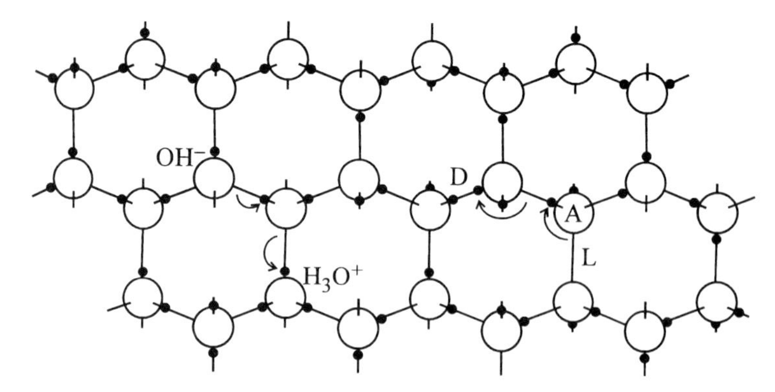

Fig. 4.9 Ionic and Bjerrum defects introduced into the structure of Fig. 2.4.

from one molecule to a neighbouring molecule, and separated by successive jumps of protons from one end of a hydrogen bond to the other. This ionization reaction

$$2H_2O \rightleftharpoons H_3O^+ + OH^-$$

is a familiar process in liquid water, but in ice the ions do not move as complete entities. The oxygens cannot move from one site to another, and the motion of a proton along a bond transfers the state of ionization from one molecule to another.

The motion of an ion or Bjerrum defect through the crystal will follow a zig-zag path along appropriately oriented bonds, reorienting both molecules and bonds along its track. To analyse the electrical consequences of such defect motion we will imagine the actual path followed to be straightened out as in Fig. 4.10. In (a) a proton is introduced at the left-hand end to form an H_3O^+ ion, and then in (b) protons jump along successive bonds as shown by the arrows until the ion is at the right-hand end of the chain. Note that the polarization of the chain has been reversed so that a second H_3O^+ ion cannot pass along this path. If we now create a D-defect at the left-hand end of the chain by rotating the end molecule as in (c), we can then turn successive molecules so that the D-defect travels to the right leaving the chain in state (d), which is the same as (a). A second D-defect cannot pass along the chain, but it is now 'open' for the passage of another H_3O^+ ion.

If we focus on the ends of the chain, we see that a proton was added at the left in (a) and can be removed from the right in (b), so that the net effect of passing both the ion and the Bjerrum defect along the chain has been to transfer a proton from one end to the other and leave the chain in its original state. The sum of the proton displacements in (b) and (d) amounts to a continuous track. Steady-state conduction requires the presence of both types of defect. The motion of the D-defect in steps (c) and (d) consists purely of the reorientation of *molecules* in the chain,

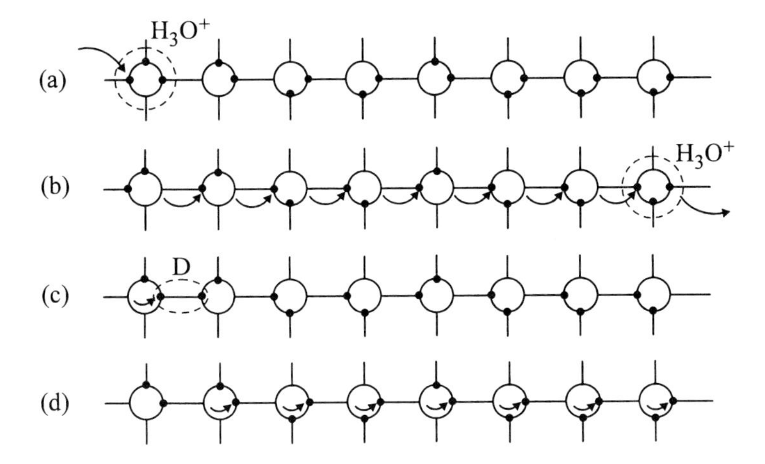

Fig. 4.10 The propagation of an H_3O^+ ion followed by a D-defect along a straightened-out chain of water molecules.

generating the kind of polarization discussed in §4.3. On the other hand passage of an H_3O^+ ion produces a reorientation of *bonds* in the direction of its motion, and would leave the molecules oriented against the field. The total effect, however, includes the transfer of one proton along the chain.

Ionic and Bjerrum defects carry effective charges of magnitudes $e_{\pm}$ and e_{DL} respectively, which must be defined in terms of the effects of their motion. As the net effect of moving an H_3O^+ ion and a D-defect along a path is to move a proton of charge e the effective charges must satisfy the relation

$$e_{\pm} + e_{DL} = e. \tag{4.28}$$

Another way of understanding this is to insert a single proton into an otherwise perfect network of bonds as in Fig. 4.11. This generates an adjacent H_3O^+ ion and D-defect, which can then move apart.

The OH^- ions and L-defects have properties equivalent to the H_3O^+ ions and D-defects, but with opposite sign. In Fig. 4.10 an OH^- ion moving from right to left would convert the chain in (a) to that in (b), blocking it for further motion of OH^- from right to left or of H_3O^+ from left to right. Likewise an L-defect moving from right to left will have the same effect as the D-defect moving from left to right in (c) and (d). In summary, oppositely charged defects of the same kind move in opposite directions under an applied field but produce the same effects in polarizing the ice. A steady current requires the motion of both ionic and Bjerrum defects, though if, for example, the ionic defects involved are positive it does not matter whether the Bjerrum defects are positive or negative. At this point the

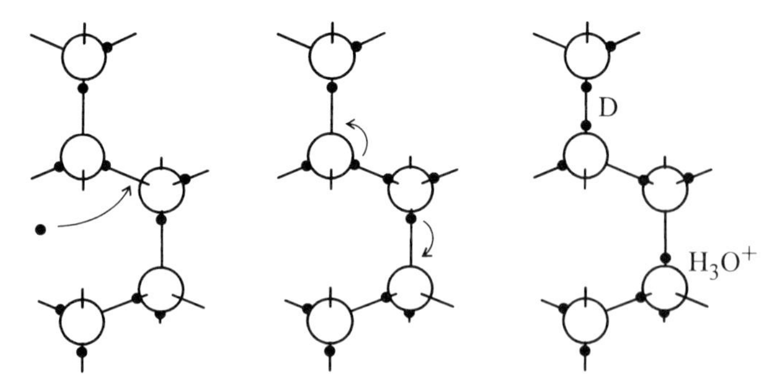

Fig. 4.11 Insertion of a proton into a perfect ice lattice to create both an H_3O^+ ion and a D-defect.

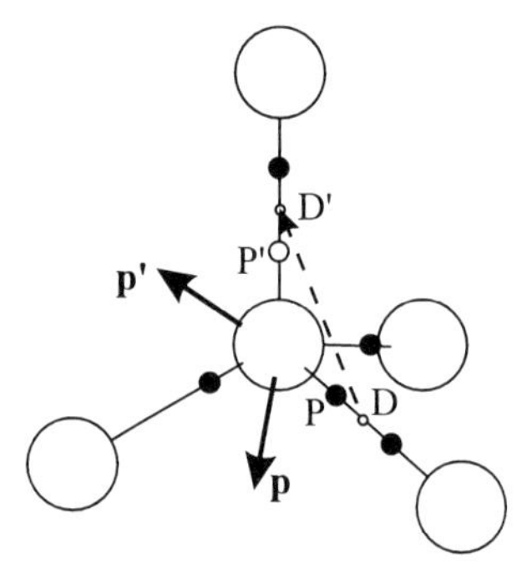

Fig. 4.12 The jump of a D-defect from D to D′ turns the molecular dipole moment **p** through 90° to **p**′.

reader may be surprised that defects which were introduced as pairs in Fig. 4.9 are now being thought of independently of one another. This is in fact essential; one of a pair may be mobile and the other not, and we will find later that extrinsic defects can be introduced by doping (for example with HF or NH_3) or single kinds of defects may be swept out by a field.

The partition of the protonic charge e into the components $e_{\pm}$ and e_{DL} in equation (4.28) is ultimately a matter for experimental determination or theoretical modelling. The simple picture of moving protons through specified distances in Fig. 4.10 is too crude, because the distribution of electronic charge within the molecules responds to displacements of the protons. A fundamental relation due to Onsager and Dupuis (1962) is illustrated in Fig. 4.12. A reorientation of the central molecule, involving the displacement of the proton P to P′, occurs by the motion of a D-defect located at the mid-point of the bond D to the point D′ through a distance $r_{OO}\sqrt{(2/3)}$, where r_{OO} is the distance between adjacent oxygen sites. As a result the dipole moment vector **p** turns through 90° to **p**′. The vector change in **p** is parallel to DD′ and of magnitude $p\sqrt{2}$. Equating the change in dipole moment to the change in polarization resulting from moving the charge e_{DL} on the defect gives

$$p = e_{\mathrm{DL}} r_{\mathrm{OO}} / \sqrt{3}. \tag{4.29}$$

The partition of the charge therefore depends on the size of the molecular dipole moment appropriate for the theory of static polarization. This result will be recognized as the same as equation (4.27), if the charge on a Bjerrum defect e_{DL} is identified with the charge e_{B} introduced in Nagle's unit model. Figure 4.13 illustrates a D- and an L-defect in the unit model, and identifies e_{B} as the extra charge at the D-defect which is not included in the neutral units. In the case of an L-defect e_{B} is the charge missing where two units overlap.

As we will show in §5.4.1, the experimental data indicate that in practice $e_{\mathrm{DL}} \approx 0.38e$ and $e_{\pm} \approx 0.62e$. When an ion with charge $0.62e$ moves from one site to another through a distance $r_{\mathrm{OO}} = 2.76$ Å, the proton of charge e moves only 0.76 Å, but the resulting product of defect charge times displacement ($0.62e \times 2.76 = 1.71e$) includes the displacement of electronic charge in the opposite direction. Further issues concerning the partitioning of the charge and application of these concepts to other hydrogen bonded systems are discussed by Nagle (1992).

It is intuitively improbable that a D-defect has the linear O—H—H—O structure drawn in the diagrams used here, because the energy could be lowered by the protons moving off the straight line in opposite directions. The whole question of the structure of this and the other protonic point defects will be considered in Chapter 6, but for the general theory of the electrical properties of ice these details do not matter. The mere existence of mobile defects with the properties we have described is sufficient to allow polarization and conduction, and the theory depends only on the charges, concentrations and mobilities of the defects.

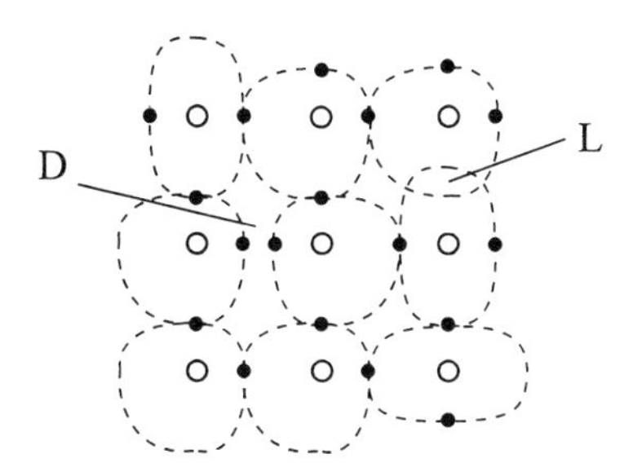

Fig. 4.13 D- and L-defects in the unit model of Fig. 4.8.

In pure ice the intrinsic thermal equilibrium concentrations $n_{\pm}$ and n_{DL} (per unit volume) of pairs of ionic or Bjerrum defects depend on temperature according to the equations

$$n_{\pm} = \frac{2}{3} N \exp\left(-\frac{E_{\pm}}{2k_B T}\right), \tag{4.30}$$

and

$$n_{DL} = N \exp\left(-\frac{E_{DL}}{2k_B T}\right), \tag{4.31}$$

in which $E_{\pm}$ and E_{DL} are the energies (or strictly free energies) of formation of *pairs* of ionic and of Bjerrum defects respectively and N is the number of molecules per unit volume. These equations will be justified in §6.2.

The mobility μ_i of a defect i is defined by its drift velocity in unit electric field, and these mobilities are normally expected to vary with temperature according to an Arrhenius relation of the form

$$\mu_i \propto \frac{1}{T} \exp\left(-\frac{E_{im}}{k_B T}\right), \tag{4.32}$$

where E_{im} is the (free) energy of activation for motion.

In this last paragraph it was convenient to denote a species of defect by the suffix i, and in writing equations this is a very useful practice which we will adopt throughout the next sections of this chapter. The standard convention for numbering defects is

1. H_3O^+ ion
2. OH^- ion
3. D-defect
4. L-defect.

In terms of our previous notation the charges would then be

$$e_1 = -e_2 = e_{\pm}$$
$$e_3 = -e_4 = e_{DL}.$$

4.5 Jaccard theory

In most materials the processes of dielectric polarization and electrical conduction are totally distinct and can be analysed independently. This is not the case for ice, where both are properties of the protonic subsystem and arise from motions of the two pairs of protonic point defects. The quantitative theory of these properties of ice was developed by Jaccard (1959, 1964). It is based on the protonic point defects introduced by Bjerrum and subsequently discussed by Gränicher (1958) (see also Gränicher *et al.* 1957). A parallel development of very similar concepts is due to Onsager and Dupuis (1960, 1962). Jaccard's first approach in 1959 used kinetic arguments to describe the fluxes of the four protonic point defects. The principles are very simple, but the calculation of the geometrical and statistical factors involved is difficult, and this led to some incorrect numerical factors which made the equations unnecessarily cumbersome. In his 1964 paper he adopted a more general approach via the thermodynamics of irreversible processes, and this enforces some simplicity on the final results. The latter treatment was further developed by Hubmann (1979). In this chapter we start with a kinetic approach, but using the understanding and terminology that emerges from the thermodynamic treatment; the main features of that model will be described subsequently.

Because polarization and conduction are linked in ice it is necessary to distinguish carefully between these two effects. Labelling the protonic point defects with the indices $i = 1$ to 4 as in the previous section, the fluxes of these defects (i.e. the number crossing unit area per unit time) are denoted by the vectors $\mathbf{j}_i$. Starting at time $t = 0$ with totally unpolarized ice, the cumulative charge transfer through the ice per unit area is

$$\int_0^t (e_1\mathbf{j}_1 + e_2\mathbf{j}_2 + e_3\mathbf{j}_3 + e_4\mathbf{j}_4)\,\mathrm{d}t' = \int_0^t [e_{\pm}(\mathbf{j}_1 - \mathbf{j}_2) + e_{\mathrm{DL}}(\mathbf{j}_3 - \mathbf{j}_4)]\,\mathrm{d}t'. \quad (4.33)$$

This charge transfer includes both the polarization $\mathbf{P}$ and the integrated conduction current, which remains as a surface charge unless removed by an electrode. In discussing Fig. 4.10 we noted how the passage of an ion or Bjerrum defect blocked a chain of bonds for subsequent passage of a defect of the same type. An important parameter in describing the state of the ice is the number of chains of bonds crossing a given plane that are oriented in one direction or the other. This quantity is described by the vector

$$\boldsymbol{\Omega}(t) = \int_0^t (\mathbf{j}_1 - \mathbf{j}_2 - \mathbf{j}_3 + \mathbf{j}_4)\,\mathrm{d}t'. \quad (4.34)$$

This is called the 'configuration vector'. The significant difference between it and the expression (4.33) is not in the absence of the e_i but in the signs of the fluxes. A flow of H_3O^+ ions transfers charge and increases $\boldsymbol{\Omega}$; a subsequent flow of D-defects in the same direction adds to the transfer of charge but tends to cancel the effect on $\boldsymbol{\Omega}$, opening the chains of bonds for the passage of further ions.

4.5.1 *Kinetic approach—one kind of defect*

We will now develop the basic ideas of the kinetic approach by considering just one kind of defect, the positive ion ($i = 1$), in a geometrically simplified structure consisting of planar layers of molecules lying perpendicular to the direction of polarization, the separation of the layers being a. Each layer is linked to the next by N_p bonds per unit area, and the number of these bonds oriented to accept an ion travelling in the positive direction is $\frac{1}{2}N_p - \Omega$, where Ω is the component of $\mathbf{\Omega}$. The number of bonds blocked for such ions is $\frac{1}{2}N_p + \Omega$. The ions will move around at random, and the frequency at which an ion makes jumps along a bond which is open to it is ν_1. If the number of ions per unit volume is n_1, the number per unit area in a single layer is $n_1 a$ and the net rate at which such ions make jumps between adjacent layers is

$$\mathbf{j}_1 = -2\nu_1(n_1 a)\frac{\mathbf{\Omega}}{N_p} = -\xi\nu_1 n_1 a^3\mathbf{\Omega}, \tag{4.35}$$

in which $\xi/2$ is the geometrical factor $1/(N_p a^2)$.

For such a structure the diffusion coefficient of the ions $D_1 = \nu_1 a^2$, and by the Einstein relation (see §6.3) the mobility $\mu_1 = e_1\nu_1 a^2/k_B T$. In the presence of an electric field the ions also have a drift velocity $\mu_1\mathbf{E}$, and the total flux can be written

$$\mathbf{j}_1 = n_1\mu_1\mathbf{E} - \xi\nu_1 n_1 a^3\mathbf{\Omega}. \tag{4.36}$$

Changing to the ice Ih structure, we replace a by the oxygen–oxygen distance r_{OO}, write ν_1 in terms of μ_1 and absorb all geometrical and statistical factors into a revised dimensionless constant ξ. Then

$$\mathbf{j}_1 = \frac{n_1\mu_1}{e_1}(e_1\mathbf{E} - \xi r_{OO}k_B T\mathbf{\Omega}). \tag{4.37}$$

The numerical factor ξ is extremely difficult to calculate by kinetic arguments (this is where Jaccard's 1959 theory ran into trouble), but, anticipating future developments, we introduce a new factor $\Phi = \xi r_{OO}k_B T$, and the final equation for the flux becomes

$$\mathbf{j}_1 = \frac{n_1\mu_1}{e_1}(e_1\mathbf{E} - \Phi\mathbf{\Omega}). \tag{4.38}$$

This means that the configuration vector gives rise to an effective force on the ions $-\Phi\mathbf{\Omega}$ which is added to the electric force $e_1\mathbf{E}$. We will show later that Φ is independent of the type of defect, so it is not given a subscript.

For only one type of defect there can be no steady-state conductivity, because for that to be possible there must be a second type of defect to re-open blocked chains of bonds. The current density is $\mathbf{J} = e_1\mathbf{j}_1$, and the polarization $\mathbf{P} = \int\mathbf{J}\,dt'$ is in this case equal to $e_1\mathbf{\Omega}$ (eqn (4.34)). In terms of the macroscopic parameters $\mathbf{P}$ and $\mathbf{E}$, equation (4.38) becomes

$$\frac{d\mathbf{P}}{dt} = n_1\mu_1\left(e_1\mathbf{E} - \Phi\frac{\mathbf{P}}{e_1}\right) = \frac{n_1\mu_1\Phi}{e_1}\left(\frac{e_1^2}{\Phi}\mathbf{E} - \mathbf{P}\right). \tag{4.39}$$

This has the same form as equation (4.4) with

$$\chi_s = \frac{e_1^2}{\varepsilon_0 \Phi} = \frac{e_\pm^2}{\varepsilon_0 \Phi}, \tag{4.40}$$

and

$$\tau_D = \frac{e_1}{n_1 \mu_1 \Phi} = \frac{\varepsilon_0 \chi_s}{\sigma_\infty}, \tag{4.41}$$

in which $\sigma_\infty = n_1 \mu_1 e_1$. This latter expression is the same as (4.13), with the naturally expected value for the conductivity σ_∞ on a time scale for which there is no build-up of polarization or $\mathbf{\Omega}$.

If we had performed the calculation for Bjerrum L-defects, which have the effect of reorienting molecules, equation (4.40) would be

$$\chi_s = \frac{e_4^2}{\varepsilon_0 \Phi} = \frac{e_{DL}^2}{\varepsilon_0 \Phi}. \tag{4.42}$$

This is the case discussed in §4.3, for which χ_s is given by equation (4.22), and, using the relations (2.1) and (4.29), we find

$$\Phi = \frac{8\sqrt{3}}{G} r_{OO} k_B T. \tag{4.43}$$

The determination of the numerical factor in Φ, which we have denoted as ξ in (4.35), is therefore equivalent to the calculation of G. Apart from the partitioning of the charges $e_\pm$ and e_{DL}, this quantity is the one numerical factor involved in the Jaccard theory. We will return to it later at the end of §4.5.3.

4.5.2 *More than one kind of defect*

In the general case there are four equations like (4.38) describing the fluxes of the four kinds of defect. To account for the signs in a convenient way equation (4.34) is written as

$$\mathbf{\Omega}(t) = \int_0^t \sum_{i=1}^{4} \eta_i \mathbf{J}_i \, dt', \tag{4.44}$$

where $\eta_i = +1, -1, -1$ and $+1$ for $i = 1$ to 4. At this stage we will add for completeness a diffusion term $-D_i \nabla n_i$ that will be needed later, and the equations for the four fluxes finally become

$$\mathbf{j}_i = \frac{n_i \mu_i}{|e_i|}(e_i \mathbf{E} - \eta_i \Phi \mathbf{\Omega}) - D_i \nabla n_i. \tag{4.45}$$

The current density is

$$\mathbf{J} = \sum_{i=1}^{4} e_i \mathbf{j}_i. \tag{4.46}$$

In these equations the fluxes represent the actual directions of motion of the defects, and for the negatively charged defects 2 and 4, e_i is a negative quantity. Equation (4.45) has been written in such a way that all the μ_i are positive quantities.

In cases where $\nabla n_i = 0$ the set of coupled equations (4.44)–(4.46) may be solved for alternating fields of the form $E = E_0 e^{i\omega t}$. The result represents a *single* Debye relaxation in parallel with a steady-state conductivity, just as discussed in §4.2 with the macroscopic parameters χ_s, σ_∞, σ_s and τ_D (see Fig. 4.1). To write equations for these quantities in the most informative way we introduce the partial conductivities

$$\sigma_i = n_i \mu_i |e_i|, \tag{4.47}$$

and group them into an ionic conductivity

$$\sigma_\pm = \sigma_1 + \sigma_2 \tag{4.48}$$

and a Bjerrum conductivity

$$\sigma_{DL} = \sigma_3 + \sigma_4. \tag{4.49}$$

The equations for the parameters of the ice are then

$$\sigma_\infty = \sigma_\pm + \sigma_{DL}, \tag{4.50}$$

$$\frac{e^2}{\sigma_s} = \frac{e_\pm^2}{\sigma_\pm} + \frac{e_{DL}^2}{\sigma_{DL}}, \tag{4.51}$$

$$\frac{1}{\tau_D} = \Phi\left(\frac{\sigma_\pm}{e_\pm^2} + \frac{\sigma_{DL}}{e_{DL}^2}\right), \tag{4.52}$$

and

$$\chi_s = \frac{(\sigma_\pm/e_\pm - \sigma_{DL}/e_{DL})^2}{\varepsilon_0 \Phi(\sigma_+/e_\pm^2 + \sigma_{DL}/e_{DL}^2)^2}. \tag{4.53}$$

These four equations satisfy the relation (4.13).

The high- and low-frequency limits are of special significance. At high frequencies all four defects move independently and their contributions to the total current are summed. If one component is dominant its effect will dominate in σ_∞ and in χ_s. When Bjerrum defects dominate χ_s is given by equation (4.42), which is the equilibrium value for the reorientation of molecules. However, if ionic defects are dominant equation (4.53) reduces to (4.40), and χ_s is larger by a factor $(e_\pm/e_{DL})^2$.

The steady-state current is limited by the *lower* of the components $\sigma_\pm$ and σ_{DL}, because this determines the rate at which chains of bonds are unblocked to allow the majority species to flow. Another special case arises if $\sigma_\pm/e_\pm = \sigma_{DL}/e_{DL}$, so that the numerator of equation (4.53) vanishes. In this case the fluxes of the two types of defect are balanced and a current flows without any build-up of polarization; $\mathbf{\Omega} = 0$ and $\chi_s = 0$.

4.5.3 *Thermodynamic approach*

The model of Jaccard (1964) and Hubmann (1979), which is based on the theory of irreversible thermodynamics (see de Groot 1951), is developed in a very general form to include gradients of defect concentrations, temperature and pressure as well as anisotropy. We use the approach here to give meaning to the constant Φ, and we confine the treatment to isotropic ice with no gradients of temperature, pressure or concentration.

The configurational entropy S_c of the ice is a function of state depending on the configuration vector $\mathbf{\Omega}$, and as it must be a symmetrical function, it can be expanded for small $\Omega = |\mathbf{\Omega}|$ as

$$S_c(\Omega) = S_c(0) + \frac{1}{2}\frac{d^2 S_c}{d\Omega^2}\Omega^2 + O(\Omega^4). \tag{4.54}$$

The starting point of the theory of irreversible thermodynamic processes is an expression for the rate of increase of entropy. This arises from the Joule heating which for unit volume is $\mathbf{J} \cdot \mathbf{E}$, and to this must be added any increase in the configurational entropy S_c. We may write this rate as

$$\dot{S} = \frac{\mathbf{J} \cdot \mathbf{E}}{T} + \dot{S}_c, \tag{4.55}$$

or

$$T\dot{S} = \mathbf{J} \cdot \mathbf{E} + T\frac{d^2 S_c}{d\Omega^2}\mathbf{\Omega} \cdot \dot{\mathbf{\Omega}}. \tag{4.56}$$

Using equations (4.44) and (4.46), this becomes

$$T\dot{S} = \sum_{i=1}^{4}\left(e_i\mathbf{E} + \eta_i T\frac{d^2 S_c}{d\Omega^2}\mathbf{\Omega}\right) \cdot \mathbf{j}_i. \tag{4.57}$$

In the theory this must be expressed as a sum of products of generalized forces $\mathbf{x}_i$ and fluxes $\mathbf{j}_i$

$$T\dot{S} = \sum_{i=1}^{4} \mathbf{x}_i \cdot \mathbf{j}_i, \tag{4.58}$$

and we assume that each flux is linearly dependent on the corresponding force according to the equations

$$\mathbf{j}_i = \lambda_i \mathbf{x}_i, \tag{4.59}$$

where the coefficients λ_i are constants. Writing

$$T\frac{\mathrm{d}^2 S_\mathrm{c}}{\mathrm{d}\Omega^2} = -\Phi, \tag{4.60}$$

we identify the forces $\mathbf{x}_i$ with the expressions $(e_i\mathbf{E}-\eta_i\Phi\boldsymbol{\Omega})$ appearing in equations (4.45), and equations (4.59) become equivalent to (4.45). The important feature of this approach is that it establishes that there is only one quantity Φ, which appears in the four equations (4.45) without the need for a subscript i. This quantity is related to the configurational entropy in a manner that closely resembles the special case treated in §4.3.1.

Ryzhkin and Whitworth (1997) have formulated the theory of $S_\mathrm{c}(\Omega)$ in such a way that Φ can be calculated using a mean-field cluster approximation at the level of the Pauling approximation (i.e. ignoring closed rings). This gives

$$\Phi = \frac{8}{\sqrt{3}} r_\mathrm{OO} k_\mathrm{B} T. \tag{4.61}$$

The result depends only on the tetrahedral symmetry around a single molecular site and is therefore isotropic and would be the same for ice Ic. The Jaccard model is linked to theories of the static permittivity via equation (4.43), and this calculation shows that both approaches are consistent, with the parameter G being very closely equal to 3.0. In his formulation of the theory Hubmann (1979) used a different dimensionless parameter Γ to describe the numerical factor in Φ, and he determined the value of this parameter from experiment (see §5.4.1). It is now clear that his theory is equivalent to that presented here if Γ is given the value $G/4$. We recommend the use only of Φ or of G.

4.6 Ice with blocking electrodes

It is in practice difficult to prepare specimens with the ideal electrodes assumed so far. At such electrodes there would be free exchange between electronic and protonic charge carriers. A different situation, which is easier to realize in practice, is that of blocking electrodes for which there is no transfer of charge across the surface of the ice. This can be achieved with an extremely thin insulating layer on the surface of the metal (e.g. Al_2O_3 on aluminium). No steady current can then flow, but there is a build-up of space charge within the ice adjacent to the electrodes. The results of a.c. measurements then depend on the specimen thickness, and additional low-frequency relaxations appear in plots like Figs 4.5 or 4.6.

The theory of the a.c. characteristics of ice with blocking electrodes has been developed by Petrenko and Ryzhkin (1984*a*) for plate-like specimens of thickness L with the field in the x-direction perpendicular to the plate. The polarization charge is assumed to consist of point defects within the region close to the surface, and no account is taken of charges on dangling bonds at the surface or of other special surface states. The space charges are thus represented by the deviations $\Delta n_i(x)$ of the point defect concentrations $n_i(x)$ from their bulk values n_{i0}, and the electric field E, configuration vector Ω and fluxes j_i all become functions of x. To the definition of Ω (eqn (4.44)) and the four equations for the fluxes (4.45), including the diffusion terms, must be added Poisson's equation

$$\varepsilon_0 \varepsilon_\infty \frac{\partial E}{\partial x} = \sum_{i=1}^{4} e_i \Delta n_i, \qquad (4.62)$$

and four equations of continuity

$$\frac{\partial \Delta n_i}{\partial t} + \frac{\partial j_i}{\partial x} = 0. \qquad (4.63)$$

This gives a set of ten equations for the ten variables n_i, j_i, E and Ω. Assuming a sinusoidal variation of the charge on the electrodes these simultaneous equations must be solved to yield the p.d. between the electrodes and hence the real and imaginary parts of the permittivity. For potential differences $\Delta V \ll k_B T/e$, analytical solutions have been obtained which demonstrate physically significant features of experiments on ice. We show here the results for some important simple cases.

In the low-frequency limit the space charge which forms at the electrodes is characterized by two Debye–Hückel type screening lengths κ_1^{-1} and κ_2^{-1}. In the case where Bjerrum defects are the majority carriers

$$\kappa_1^{-1} = \left[\frac{\varepsilon_0 \varepsilon_\infty k_B T}{e_{DL}^2 (n_{30} + n_{40})}\right]^{1/2} \quad \text{and} \quad \kappa_2^{-1} = \left[\frac{\chi_s \varepsilon_0 \varepsilon_\infty k_B T}{e^2 (n_{10} + n_{20})}\right]^{1/2}, \qquad (4.64)$$

in which, in this approximation, $\chi_s = e_{DL}^2/\varepsilon_0 \Phi$. The quantities n_{i0} are the bulk concentrations of the four types of defect. For $\kappa_1^{-1} \ll L \ll \kappa_2^{-1}$ the field distribution is as in Fig. 4.14(a). The Bjerrum defects have moved to form the space charge layers and the central part of the ice carries a uniform field $E_2 = \sigma_{sc}/\varepsilon_0 \varepsilon_s$, where σ_{sc} is the surface charge density on the electrode. However, if both κ_1^{-1} and

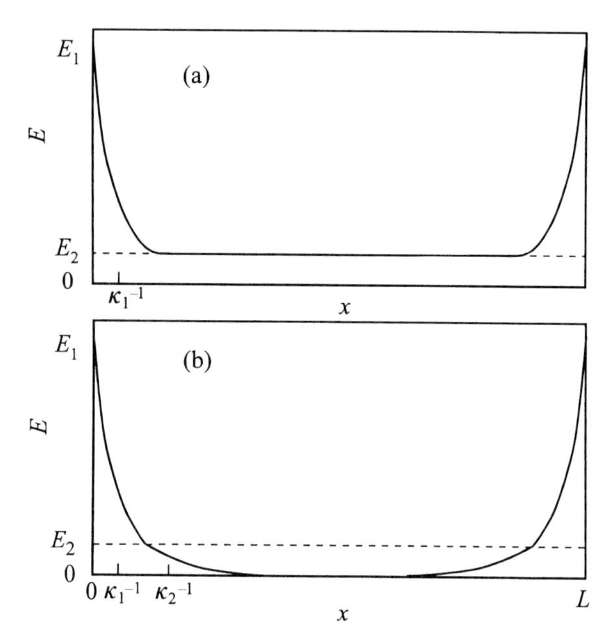

Fig. 4.14 Distribution of the electric field E in an ice specimen with blocking electrodes (a) with only Bjerrum defects, and (b) with both Bjerrum and ionic defects. (From Petrenko and Ryzhkin 1984*a*.)

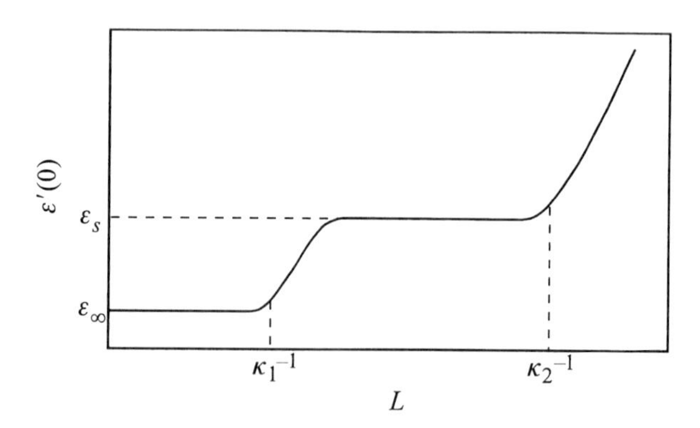

Fig. 4.15 Schematic dependence of the apparent permittivity $\varepsilon'(0)$ in the low-frequency limit on the thickness L (on a logarithmic scale) for a specimen with blocking electrodes. (From Petrenko and Ryzhkin 1984*a*.)

$\kappa_2^{-1} \ll L$, the field in the centre falls to zero as in Fig. 4.14(b). This is because the bulk ice behaves as a conductor, and the second, wider, region of space charge is formed by minority carriers. The apparent low-frequency permittivity $\varepsilon'(0)$ will depend on the specimen thickness as shown schematically in Fig. 4.15.

The establishment of the space charges involves new time constants giving relaxation processes in the low-frequency part of the spectrum. In the simplest case with the active charge carriers being H_3O^+ ions and L-defects, as in pure ice at normal temperatures, Petrenko and Ryzhkin predicted the behaviour shown schematically in Fig. 4.16. The Debye relaxation frequency is $\omega_D = \tau_D^{-1}$ as given by equation (4.52), but there is a single lower-frequency relaxation at $\omega_1 = D_4\kappa_2^2$, where D_4 is the diffusion coefficient of the L-defects. The plateaux of $\varepsilon'(0)$ are given by

$$\varepsilon_D = \varepsilon_\infty \left[\frac{\varepsilon_\infty}{\varepsilon_s} + \frac{1 - \exp(-\kappa_1 L)}{\kappa_1 L} \right]^{-1}, \tag{4.65a}$$

and

$$\varepsilon_1 = \varepsilon_\infty \left[\frac{1 - \exp(-\kappa_1 L)}{\kappa_1 L} + \frac{\varepsilon_\infty}{\varepsilon_s} \frac{1 - \exp(-\kappa_2 L)}{\kappa_2 L} \right]^{-1}. \tag{4.65b}$$

In general there can be up to three relaxations in addition to the Debye relaxation. Their significance depends on L and they may of course merge into one another.

This analysis provides some warnings over the interpretation of complicated observations made on ice specimens which may have ill-defined electrode properties. As the screening lengths are strongly dependent on temperature through n_i, Fig. 4.15 can have serious implications for experiments on the temperature dependence of ε_s in specimens for which L may not remain big enough at low temperatures.

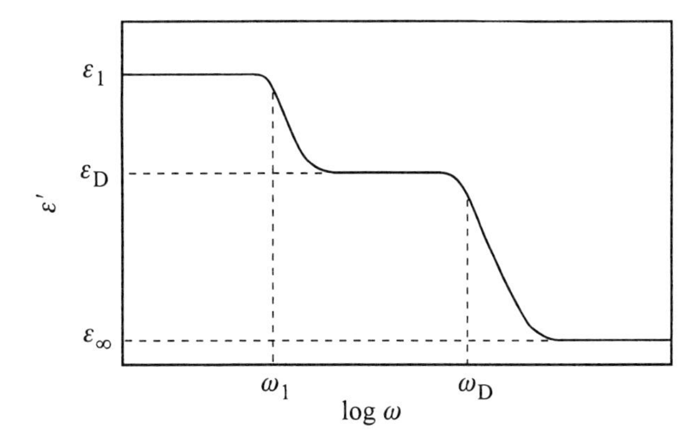

Fig. 4.16 Schematic dependence of the apparent permittivity ε' on frequency ω for an ice crystal with blocking electrodes and both H_3O^+ and L-defects active. (From Petrenko and Ryzhkin 1984*a*.)

4.7 Time constants

It is useful to review briefly the physical significance of the time constants that arise in the foregoing theories. The most significant of these is the Debye relaxation time τ_D. In the simplest case of molecular reorientations produced by the motion of L-defects (as in pure ice at normal temperatures) equation (4.52) reduces to

$$\frac{1}{\tau_D} = \frac{\Phi \sigma_4}{e_4^2}. \tag{4.66}$$

Using equations (2.1), (4.47), (4.61), the Einstein relation (6.16), and the standard expression for a diffusion coefficient $D_i = r_{OO}^2 \nu_i/6$ this becomes

$$\frac{1}{\tau_D} = \frac{n_4 \nu_4}{3N}. \tag{4.67}$$

$n_4 \nu_4$ is the total number of molecular reorientations occurring in unit volume per unit time, and thus, apart from a numerical factor, τ_D represents the average interval between successive reorientations of a single molecule. Such relaxation processes and associated correlation effects have been considered in some detail by Onsager and Runnels (1969).

The Debye relaxation time was introduced in equation (4.4) as the time constant for the polarization of the ice in an applied electric field, and the equation is also applicable to the depolarization if $\mathbf{E} = 0$. This will be the case if the p.d. across the ice is set to zero by shorting the electrodes. With $\mathbf{E} = 0$ the equations describe the decay of $\mathbf{\Omega}$ by the random motion of defects, as considered in deriving equation (4.35). However, this is not the only situation that should be considered.

The free decay of a polarized specimen of ice is what would happen if the polarizing electrodes were suddenly removed (not disconnected or shorted). In this case there will be an internal field $-\mathbf{P}/\varepsilon_\infty \varepsilon_0$ inside the ice. Inclusion of this field in equation (4.45) for the case of a single charge carrier i being dominant leads to a shorter relaxation time τ given by

$$\frac{1}{\tau} = \frac{1}{\tau_M} + \frac{1}{\tau_D}, \tag{4.68}$$

where τ_M is the Maxwell relaxation time $\varepsilon_\infty \varepsilon_0/\sigma_i$ which is familiar for a normal dielectric. In ice τ_M is less than τ_D by the factor $\varepsilon_\infty/\chi_s \approx \frac{1}{30}$. The flux of defects is driven by the internal electric field. If electrodes are left on but shorted, this internal field is cancelled and the depolarization is slower.

When space charges are formed as in the case of blocking electrodes, longer time constants are involved which depend on the time required to establish the appropriate space charge of majority or minority defects by diffusion. Thus we see in connection with Fig. 4.16 that the screening length κ_1^{-1} is the diffusion length $\sqrt{(D_4\tau)}$ corresponding to the relaxation time $\tau = \omega_1^{-1}$.

4.8 Summary

The electrical properties of a specimen of ice with ideal non-blocking electrodes are described by three components represented by parallel equivalent circuit elements in Fig. 4.1(a). The high-frequency permittivity ε_∞ is that common to any kind of material, almost independent of frequency and temperature. The Debye relaxation is described by the susceptibility $\chi_s = \varepsilon_s - \varepsilon_\infty$ in series with the conductivity $\sigma_\infty - \sigma_s$, and the Debye relaxation time τ_D is given by

$$\sigma_\infty - \sigma_s = \frac{\varepsilon_0 \chi_s}{\tau_D}. \tag{4.13}$$

In the low-frequency limit the ice has a steady-state conductivity σ_s.

The Debye dispersion and conductivity are described by the Jaccard theory in terms of the concentrations n_i, mobilities μ_i, and effective charges e_i of the ionic and Bjerrum defects. The results of this theory are summarized by the equations:

$$\sigma_i = n_i \mu_i |e_i| \tag{4.47}$$

$$\sigma_\pm = \sigma_1 + \sigma_2 \tag{4.48}$$

$$\sigma_{DL} = \sigma_3 + \sigma_4 \tag{4.49}$$

$$\sigma_\infty = \sigma_\pm + \sigma_{DL} \tag{4.50}$$

$$\frac{e^2}{\sigma_s} = \frac{e_\pm^2}{\sigma_\pm} + \frac{e_{DL}^2}{\sigma_{DL}} \tag{4.51}$$

$$\frac{1}{\tau_D} = \Phi\left(\frac{\sigma_\pm}{e_\pm^2} + \frac{\sigma_{DL}}{e_{DL}^2}\right) \tag{4.52}$$

$$\chi_s = \frac{(\sigma_\pm/e_\pm - \sigma_{DL}/e_{DL})^2}{\varepsilon_0 \Phi (\sigma_\pm/e_\pm^2 + \sigma_{DL}/e_{DL}^2)^2}. \tag{4.53}$$

The quantity Φ is given by

$$\Phi = \frac{8\sqrt{3}}{G} r_{OO} k_B T, \tag{4.43}$$

with $G = 3.0$.

There are two important limiting cases. For pure ice at normal temperatures the Debye relaxation arises from the reorientation of molecules by Bjerrum defects and $\sigma_{DL} \gg \sigma_{\pm}$. In this case

$$\sigma_{\infty} = \sigma_{DL},$$

$$\sigma_{s} = \frac{e^2}{e_{\pm}^2}\sigma_{\pm},$$

and

$$\chi_{s} = \frac{e_{DL}^2}{\varepsilon_0 \Phi} = \frac{GNp^2}{3\varepsilon_0 k_B T}. \qquad (4.42 \text{ and } 4.22)$$

Equation (4.22) follows from purely equilibrium thermodynamics. It involves the dipole moment p of a water molecule in ice, which is related to the effective charges on the defects by the equation

$$p = e_{DL} r_{OO}/\sqrt{3}. \qquad (4.29)$$

In the other limiting case $\sigma_{\pm} \gg \sigma_{DL}$, and the polarization occurs by the reorientation by ionic defects not of whole molecules but of hydrogen bonds. In this case

$$\chi_{s} = \frac{e_{\pm}^2}{\varepsilon_0 \Phi}. \qquad (4.40)$$

For intrinsic defects in pure ice equations (4.31) and (4.32) combine to give

$$\sigma_{\infty} - \sigma_{s} \propto \frac{1}{T} \exp\left(-\frac{E_{\sigma_{\infty}}}{k_B T}\right), \qquad (4.69)$$

and for conduction dominated by, for example, L-defects

$$E_{\sigma_{\infty}} = \tfrac{1}{2}E_{DL} + E_{Lm}. \qquad (4.70)$$

Using equations (4.13) and (4.22), the temperature dependence of the Debye relaxation time is given by

$$\tau_{D} \propto \exp\left(\frac{E_{\tau}}{k_B T}\right), \qquad (4.71)$$

with $E_{\tau} = E_{\sigma_{\infty}}$.

The results of experiments on pure and doped ice and the further interpretation of these equations are described in Chapter 5. A further summary giving the properties of the protonic point defects will be found at the end of Chapter 6.

5 Electrical properties—experimental

5.1 Introduction

When metal electrodes are placed on a specimen of ice and a p.d. is applied, a current will flow which dies away over a period typically of several minutes. If the electrodes are subsequently shorted a stored charge is released. Such experiments were described by Ayrton and Perry (1877) and gave a very high apparent permittivity for ice. We now understand that the initial current represents the d.c. conductivity of the ice and that there is partial blocking of charge transfer at the electrodes leading to the build-up of space charge layers.

To observe the true dielectric polarization of ice it is necessary to make measurements on a shorter time scale, and it was known by the 1890s that at radio frequencies ice had a much smaller 'refractive index' than water. Dewar and Fleming (1897; see also Fleming and Dewar 1897) using a vibrator to charge and discharge an ice-filled capacitor at 120 Hz showed that the permittivity rose from about 2.4 at −200 °C to 60 at −20 °C. Later measurements by Errera (1924) revealed the frequency dependence of the permittivity between about 0.5 and 200 kHz, and these observations were identified by Debye (1929) as characteristic of a relaxation process.

There have been innumerable experiments over more than a century applying the latest techniques of instrumentation or specimen preparation to investigate the electrical properties of ice. These experiments include the investigation of the properties of electrodes, different kinds of ice, ageing effects, and other anomalies. In this chapter we focus on a selection of experiments which provide the most reliable information about good quality specimens. After a brief account of techniques we consider experiments on pure ice and the effects of doping. We incorporate in our account the interpretation of these observations in terms of the protonic point defects introduced in Chapter 4. The later parts of the chapter describe a range of other electrical experiments which provide further information about these defects.

5.2 Techniques

5.2.1 *Electrical instrumentation*

For alternating current measurements it is necessary to determine as a function of frequency the complex admittance $Y(\omega)$ given by equation (4.8). Because of the

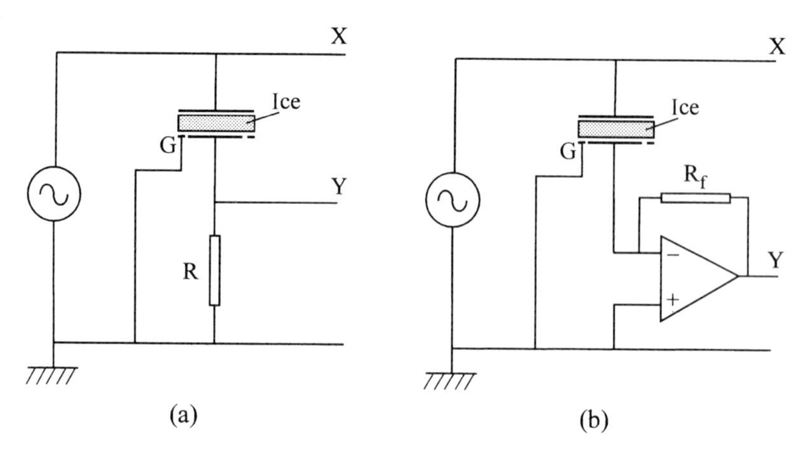

Fig. 5.1 (a) Simple circuit for measurement of the complex admittance of an ice specimen with a guard ring G. (b) Modification of (a) using an operational amplifier to measure the current without introducing the impedance of R into the circuit.

large surface conductivity of ice it is essential for measurements above about −30 °C to use a guard ring electrode, and the simplest configuration for such measurements is shown in Fig. 5.1(a). For small values of the resistor R, the lower electrode on the ice is approximately at the same potential as the guard ring G, and the voltage at Y gives the current while that at X gives the p.d. across the specimen. The complex admittance $Y(\omega)$ is the ratio of current to p.d. taking account of phase. In a more modern system the resistor is replaced by an operational amplifier with negative feedback (Fig. 5.1(b)) to measure the current with zero input impedance, and the real and imaginary parts of $Y(\omega)$ are determined with a lock-in amplifier (e.g. Loria *et al.* 1978). For low frequencies an electrometer-quality amplifier is required. Without a lock-in amplifier the voltages at X and Y produce an ellipse on an X–Y plotter and measurements are made from this; the system is then referred to as the 'loop method', and is particularly suited to measurements down to 10^{-4} Hz (Ruepp 1973; Kawada 1978).

For frequencies between 10 and 10^6 Hz the majority of experiments have used an a.c. bridge, usually of the transformer ratio-arm type illustrated in Fig. 5.2. This three-terminal arrangement with the guard ring and screen connected to ground is insensitive to stray capacitances, and so well suited to measurements on ice specimens in a cryostat.

An experimentally very different technique is to record the charge flow in response to a step voltage as illustrated in Fig. 4.2 (Noll 1978; Zaretskii *et al.* 1991). The appropriate circuit is the same as Fig. 5.1(b) but with a step voltage applied and the resistor R_f replaced by a capacitor. For a single relaxation in the ice this technique provides full information from the application of a single step of voltage, instead of having to make deductions from a.c. measurements over a range of frequencies. In the case of more than one relaxation, several exponentials

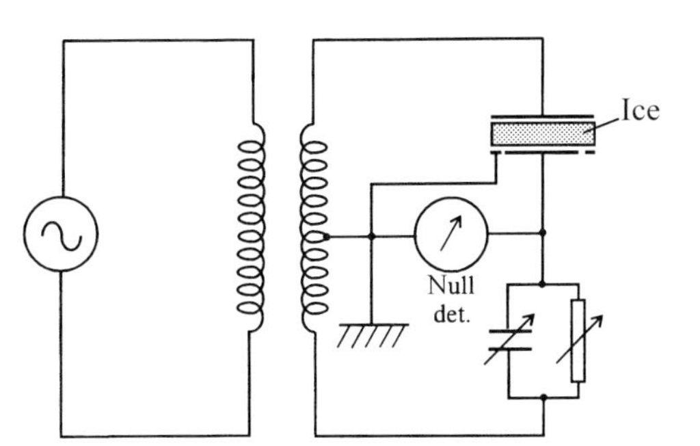

Fig. 5.2 Transformer ratio-arm bridge for determination of the complex admittance of an ice specimen with a guard ring.

can be fitted to the observed data. The method is especially suited to long relaxation times, but with fast digital data acquisition techniques it can be used over the whole range of relaxation times observed in ice.

5.2.2 *Electrodes*

Many types of metallic electrodes have been used by different workers in experiments on ice, such as plates pressed against the surface, gold leaf, gold evaporated onto the surface or indium amalgam which adheres to ice and remains liquid down to -38 °C. All such electrodes allow only partial exchange of charge between electrons in the metal and protons in the ice, resulting in the build-up of space charge layers with associated relaxations and reduced d.c. conductivity. In some cases there may also be air gaps over part of the contact area.

One way of producing a more precisely defined situation is deliberately to include insulating layers between the electrodes and the ice (Mounier and Sixou 1969; Gross and Johnson 1983). This introduces two capacitors in series with the equivalent circuit of Fig. 4.1(a), and adds an extra relaxation process which has to be fitted in the data analysis. However, it may be better to do this than to have the ill-defined space charge relaxation of a plain metallic electrode. For best results the blocking layer must have as large a capacitance as possible, and good quality anodized layers on aluminium or SiO_2 on silicon may be better than mica or Teflon.

Hydrogen-loaded palladium electrodes can exchange protons directly with ice and this makes possible special kinds of experiment which are described in §5.6.3. A new way of preparing metallic electrodes which make non-blocking contact to ice is described in §5.5. Such electrodes have not been used in the main experiments on the properties of ice described in §§5.3 and 5.4, but they open up possibilities of better measurements in the future.

Another way of avoiding the build-up of space charge at the electrodes is to move them over the surface of the ice. This is the basis of the 'electrical conductivity method' (ECM) devised by Hammer (1980) to determine the variation of the steady-state conductivity of polar ice cores along their length (see §5.4.4 and §12.5.1). A pair of brass electrodes typically 1 mm in diameter and 10 mm apart is dragged across the ice at a speed of about 100 mm s^{-1}, so that the time scale of the measurement on a particular region of the ice is of the order of 0.1 s. The p.d. used is at least 1 kV, which is three orders of magnitude greater than that typical of laboratory experiments on small specimens. The method has been shown to give results characteristic of the steady-state conductivity σ_S of the bulk ice, without appreciable contribution from the surface (Schwander *et al.* 1983). It has been proved successful for showing variations in conductivity from point to point along an ice core, but it cannot yield absolute values.

5.2.3 *Typical results*

Under favourable conditions pure crystals of good quality can give results which correspond closely to the ideal Debye relaxation described by the equivalent circuit of Fig. 4.1(a). Such data are shown in Fig. 4.5, and yield Cole–Cole plots like those in Fig. 4.6. Ice can give some of the most perfect Debye relaxations of any material. More commonly, however, the results deviate from ideal in a number of ways.

The Debye dispersion may have a spread of relaxation times, giving a Cole–Cole plot which is spread out along the ε' axis so that it appears to be part of a circle with the centre below the axis (von Hippel *et al.* 1971). This spread can be attributed to inhomogeneity within the specimen, but consideration of the equivalent circuit shows that a sandwich of layers parallel to the electrodes will give only a single mean relaxation time. To produce the spread of relaxation times there must be inhomogeneity over the area rather than through the thickness of the specimen (Petrenko 1993*b*).

As already explained, experiments with metallic electrodes produce a space charge dispersion at frequencies lower than that corresponding to τ_D. This can often be resolved into two or more dispersions (e.g. von Hippel *et al.* 1971; Noll 1978), and there is sometimes a further dispersion at higher frequencies. This high-frequency dispersion is not well understood. It is sometimes attributed to roughness at the electrodes; it could also arise from surface conduction. In the next sections we will concentrate on the Debye dispersion and static conductivity after allowance has been made for additional effects. We will return in §§5.6–7 to what can be learnt from the space charge effects in appropriately designed experiments.

5.3 Pure ice

5.3.1 *The high-frequency permittivity* ε_∞

The limiting permittivity at frequencies above the Debye relaxation has been determined in the course of normal dielectric measurements over the temperature range 200–230 K by Johari and Jones (1978), but for higher temperatures it was necessary to use special measurements in the range 1–100 MHz (Johari 1976). A typical value is $\varepsilon_\infty = 3.16 \pm 0.02$ at 253 K, falling to 3.14 at 200 K. Measurements by Gough (1972) show a continual decrease down to 3.093 ± 0.003 at 2 K. For D_2O ε_∞ is slightly smaller, varying from 3.06 at 253 K to 2.96 at 80 K (Johari and Jones 1976). Using a microwave resonator at 39 GHz Matsuoka *et al.* (1997) found that ε_∞ is anisotropic, which is equivalent to the ice being birefringent. At 252 K they obtained values of 3.1734 ± 0.0055 and 3.1396 ± 0.0049 for electric fields parallel and perpendicular to the c-axis respectively. They also made measurements with a parallel plate capacitor at 1 MHz. The anisotropy was independent of frequency and temperature, but their 1 MHz values were higher (by about 0.044) than their values at 39 GHz or those by other workers quoted above.

The origin of the permittivity ε_∞ is the polarization of the molecules by displacements of the electronic charge distribution and distortion of the molecules in the electric field. For the purposes of the present chapter ε_∞ is just a constant that has to be subtracted when studying the Debye relaxation.

5.3.2 *The Debye relaxation*

A good set of dielectric measurements on a pure crystal is shown in Fig. 4.5, and such results can be fitted well to equations (4.7) and (4.12–16), yielding the susceptibility $\chi_s (= \varepsilon_s - \varepsilon_\infty)$ and relaxation time τ_D. There is always some degree of approximation arising from the effect of steady-state conduction or space charge relaxation at the low-frequency end of the range. Results which are believed to represent the intrinsic properties of pure ice are obtainable at temperatures down to about −40 °C. Values of χ_s measured perpendicular and parallel to the c-axis at 253 K and of the corresponding τ_D are given for selected experiments in Table 5.1. (In the case of polycrystalline samples the single value of χ_s is listed as $\chi_{s\perp}$ because this orientation is dominant in a polycrystal.) There is an unresolved problem about the anisotropy of χ_s. Careful measurements by Johari and Jones (1978) specifically to look for anisotropy found none, but several other experiments show a definite effect always with $\chi_{s\parallel}$ greater than $\chi_{s\perp}$. The theory of §4.3 predicts that χ_s is isotropic and varies with temperature as $1/T$ (eqn (4.22)), but observations with $\chi_{s\parallel} > \chi_{s\perp}$ are always associated with a temperature dependence of $\chi_{s\parallel}$ as $1/(T - \Delta)$. This is particularly clear in the measurements over a wide temperature range by Kawada (1978) and Takei and Maeno (1987) with values of $\Delta \approx 47$ K. Such Curie–Weiss behaviour is characteristic of interactions not included in the theory which would lead to a ferroelectric type of ordering at low temperatures. The values of χ_s which correspond most closely to the theory of Chapter 4 are those perpendicular to the c-axis, for which the experimental results are consistent at $\chi_{s\perp} = 96 \pm 2$ at 253 K. There is less close agreement between the two experiments reported on D_2O.

The relaxation times in Table 5.1 are reasonably consistent at $(1.4 \pm 0.2) \times 10^{-4}$ s at 253 K. The two results for D_2O are less consistent, but the remarkable fact is that these relaxation times are close to those for H_2O even though the

Table 5.1 Debye relaxation in pure ice. The symbols $\|$ and $\perp$ refer to measurements with **E** parallel and perpendicular to the c-axis

Reference	Measurements at 253 K			E_τ (eV)
	$\chi_{s\perp} = \varepsilon_{s\perp} - \varepsilon_\infty$	$\chi_{s\|} = \varepsilon_{s\|} - \varepsilon_\infty$	$10^4\tau_D$ (s)	
H_2O				
Humbel *et al.* (1953)	95	115	1.5	$\approx$0.57
Wörz and Cole (1969)	97		1.5	0.575
von Hippel *et al.* (1971)	96	112	1.4	
Johari and Jones (1978)	98	98	1.2	0.55
Kawada (1978)	94	110	1.7	0.575
Johari and Whalley (1981)†	99 (polycrystalline)			
Takei and Maeno (1987)	96	113		
D_2O				
Johari and Jones (1976)	104 (polycrystalline)		1.8	0.54
Kawada (1979)	92	121	4.3*	*

*Already becoming impurity dominated, but allowing for this E_τ is compatible with H_2O values.
†Data shown in Fig. 4.3.

process involves the motion of the much heavier deuteron. Figure 5.3 illustrates a typical temperature dependence of τ_D in 'pure' ice. The steep slope at higher temperatures corresponds to equation (4.71), with the activation energy E_τ given in Table 5.1. The values of E_τ are consistent at about 0.55–0.575 eV (53–55 kJ mol^{-1}), and represent the intrinsic behaviour of pure ice. The smaller slope in Fig. 5.3 at lower temperatures corresponds to an activation energy of 0.23 eV. It extrapolates directly into the range of relaxation times observed calorimetrically for partial ordering of pure ice at around 100 K (Haida *et al.* 1974). This region shows more variation between experiments and is thought to depend on trace impurities in the ice. At first sight one expects that the susceptibility, being an equilibrium property of the ice, should not be sensitive to the impurities which primarily affect the rate at which equilibrium polarization can be established, but in certain cases the Jaccard theory of §4.5 predicts major departures from this rule; these effects will be described in §5.4.

The effect of hydrostatic pressure is to produce a small increase in the permittivity ε_s that is roughly proportional to the density, and to raise τ_D by 28% for 189 MPa at $-23.4\,°$C (Chan *et al.* 1965). This increase corresponds to a positive activation volume of $(2.9 \pm 0.3) \times 10^{-6}\,\mathrm{m^3\,mol^{-1}}$ (see §6.5.2).

5.3.3 *The steady-state conductivity* σ_s

The d.c. conductivity of ice is difficult to determine because of polarization at the electrodes, which inhibits the flow of current and generates the extra relaxations

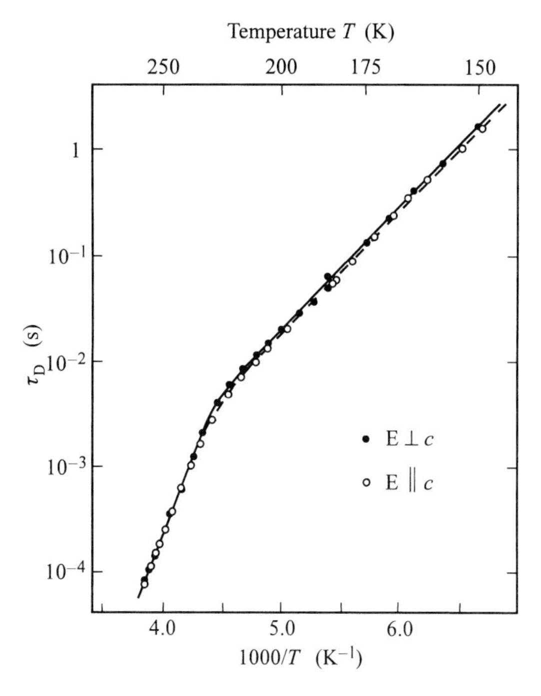

Fig. 5.3 Temperature dependence of the Debye relaxation time τ_D in 'pure' ice. (From Kawada 1978.) The low-temperature portion is probably determined by residual impurities.

referred to in §5.2.3. It is more usual to determine σ_s as the low-frequency limit of the conductivity below the Debye dispersion as seen in Fig. 4.5, but in 'pure' ice σ_s is so small that this limit is often not observed before space charge relaxations become important. Unless great care is taken about the use of guard rings the bulk conductivity at temperatures above about $-30\,°C$ is masked by surface conduction (Bullemer *et al.* 1969; Camp *et al.* 1969). Even taking account of these factors the above workers and others have found that the value of σ_s varies between specimens over several orders of magnitude. The conclusion is that this property is very sensitive to residual impurities, and there is no evidence that a truly intrinsic conductivity of pure ice has ever been measured. The lowest conductivity reported to date was obtained by Petrenko *et al.* (1983) for a d.c. measurement using ohmic electrodes (see §5.6.3). They found $\sigma_s = 6.4 \times 10^{-10}\,S\,m^{-1}$ at $-10\,°C$ with an activation energy of $0.70 \pm 0.07\,eV$ down to $-56\,°C$. (The siemens, denoted by the symbol S, is the SI unit of conductance equivalent to ohm^{-1}, and we will use it throughout.) This measurement represents an upper limit on the intrinsic value of σ_s. As we can learn much more from measurements on ice with known dopants, we will not give further consideration to the numerous experiments on 'pure' ice with unknown trace impurities.

In experiments on semiconductors measurements of conductivity would be coupled with measurements of the Hall effect to separate the roles of charge carrier concentration and mobility. However, if protonic defects move in discrete jumps no Hall effect will be observed in ice, and this appears to be true even more generally (Sokoloff 1973; Gosar 1974).

5.4 Doped ice

5.4.1 *General properties*

Although water is an excellent solvent for many substances, almost all of them are virtually insoluble in ice, and when present they are incorporated as inclusions or clusters. The electrical properties of ice are, however, very sensitive to small concentrations of certain impurities that can be incorporated in the hydrogen-bonded network to generate protonic point defects. The classic example is HF, but other acids and hydroxides also have effects, and we will also pay particular attention in this section to HCl, NH_3 and KOH.

In Fig. 5.4 an HF molecule is substituted for an H_2O molecule in ice, leaving one bond which lacks a proton. If the molecule below it in the figure is rotated as shown to put a proton on this bond, an L-defect is released into the ice. This could happen spontaneously, or some activation energy may be required to separate the L-defect from what will become an oppositely charged fluorine site. In water HF is known to be a weak acid, partially dissociating according to the equation

$$H_2O + HF \rightleftharpoons H_3O^+ + F^-.$$

In ice the same thing may happen releasing H_3O^+ ions, and such dissociation is included in Fig. 5.4. Note that the chemical properties of HF dominate over a

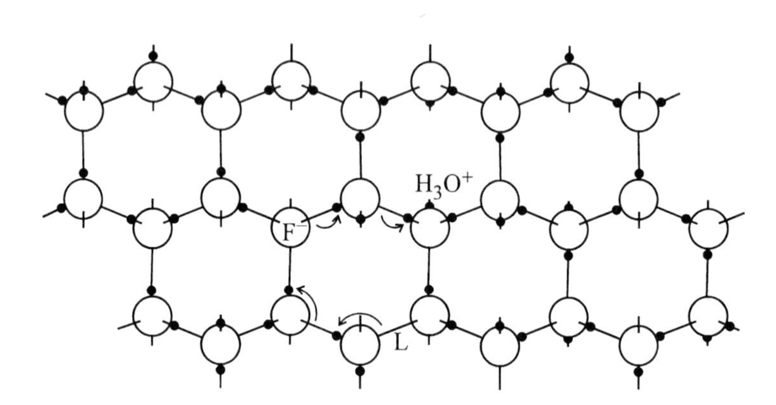

Fig. 5.4 The introduction of an HF molecule in place of one of the H_2O molecules in Fig. 2.4, with the subsequent release of an L-defect and an H_3O^+ ion.

purely geometrical inclination to create the species H_2F, which is energetically unfavourable.

A converse process can be envisaged for the substitutional incorporation of NH_3, with the possible release of an OH^- ion and a D-defect, leaving an NH_4^+ ion hydrogen bonded on an H_2O site.

The creation of ions and Bjerrum defects by such doping has a direct effect on the electrical properties. The defect concentrations n_i determine the partial conductivities σ_i, and hence the measurable electrical properties σ_∞, σ_s and χ_s through the equations (4.50)–(4.53) of the Jaccard theory. This is similar to the effect of doping a semiconductor with donor or acceptor atoms to release electrons or holes which then participate in the conduction process. The models proposed for the incorporation of HF and NH_3 are justified or not by their electrical consequences. There is as yet no independent evidence such as n.m.r. for the actual location of these impurity atoms in the lattice.

Much of our understanding of the electrical properties of ice grew out of the study of HF-doped ice by the Zurich group in the 1950s (A. Steinemann 1957; Jaccard 1959; Gränicher 1963), but we illustrate the effects of this doping in Fig. 5.5 using more recent results selected from Camplin *et al.* (1978). The results are shown in terms of the high-frequency and steady-state conductivities as these are most readily interpreted in terms of the added defects. Note that σ_s for really pure ice would be well below the bottom of this figure. There are four significant features in these graphs:

1. For higher temperatures and low doping σ_s is almost independent of temperature and increases with the concentration of HF.
2. At these higher temperatures small doping which has a large effect on σ_s produces only a small fractional increase in σ_∞.
3. For curves in the middle of the range there is a 'cross-over' in which σ_∞ at high temperatures extrapolates into σ_s at low temperatures and vice versa.
4. At high concentrations of HF (though still only a few p.p.m.) σ_∞ and σ_s are similar in magnitude.

Using the Jaccard theory (equations (4.47)–(4.53)) these features can all be explained in terms of the introduction of H_3O^+ ions and L-defects by the HF, with the defect mobilities remaining constant. For clarity of explanation we will now assume that at higher temperatures and relatively low HF concentrations the majority charge carriers are the Bjerrum defects, so that $\sigma_{DL} \gg \sigma_\pm$; this assignment will be justified later. The Bjerrum defects then determine σ_∞ and the ions determine the much smaller conductivity σ_s. The fact that σ_s is independent of temperature in this range shows both that the ion mobility μ_1 is independent of temperature and that the HF is fully ionized. The L-defects contributed by the HF add to those already present as the majority carriers in pure ice. At low HF concentrations they produce only a small fractional increase in σ_∞. The fact that σ_∞ rises smoothly with HF concentration indicates that D-defects do not make

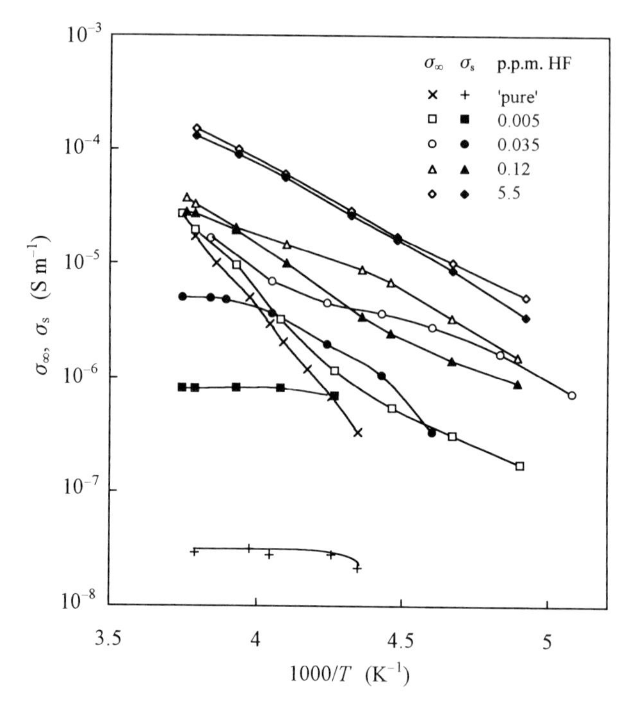

Fig. 5.5 Temperature dependence of the conductivities σ_∞ and σ_s above and below the Debye relaxation for selected specimens of HF-doped and nominally pure ice (Camplin *et al.* 1978). Note how the two curves for a given specimen, denoted by points of the same shape, appear to cross one another, but the upper curve is always that corresponding to σ_∞. Reproduced with changes by courtesy of the International Glaciological Society from *Journal of Glaciology*, **21**(85), pp. 123–41, 1978, Fig. 1.

the dominant contribution to σ_∞ in pure ice, because an increase in L-defects will decrease the concentration of D-defects and no consequent decrease in σ_∞ is observed.

At the higher temperatures the results show that σ_∞, which is approximately equal to σ_{DL}, falls rapidly with decreasing temperature while $\sigma_s \approx \sigma_\pm$ remains almost constant. Extrapolating to lower temperatures, the lines for σ_{DL} and $\sigma_\pm$ will cross. In accordance with equations (4.50) and (4.51) the ionic defects then become the majority carriers, so that $\sigma_\infty \approx \sigma_\pm$ and $\sigma_s \approx \sigma_{DL}$. This accounts for the 'cross-over' which we have noted in Fig. 5.5. Equation (4.53) predicts that at the same time the susceptibility χ_s will pass through a minimum. The precise condition is that $\sigma_\pm/\sigma_{DL} = e_\pm/e_{DL}$, at which point $\chi_s = 0$ and the ice behaves as a pure resistor in parallel with the capacitance due to ε_∞. This drop in χ_s was indeed observed in the results of Camplin *et al.* (1978), and will be demonstrated even more clearly in Figs 5.6, 5.7 and 5.9.

The fourth feature of Fig. 5.5 was that at high doping levels both σ_s and σ_∞ are comparable in magnitude. This happens when the doping has raised σ_s above the intrinsic value of σ_∞, and both conductivities are completely dominated by the doping. A fuller consideration of these conditions will follow.

Hubmann (1979) has shown that the equations of the Jaccard theory imply a universal relationship between χ_s and the ratio σ_s/σ_∞. This enables the theory to be tested in terms of observable quantities without knowledge of the concentration or state of dissociation of the dopants. He uses the experimentally determinable dimensionless quantities

$$s = \sigma_s/\sigma_\infty, \tag{5.1}$$

and

$$\Lambda = \frac{2\sqrt{3}r_{OO}\varepsilon_0 k_B T}{e^2}\chi_s. \tag{5.2}$$

According to equations (4.50), (4.51), and (4.53) these quantities can both be expressed as functions of the ratio $\rho = \sigma_\pm/\sigma_{DL}$ by the equations

$$\Lambda(\rho) = \frac{G}{4}\left(\frac{e_\pm}{e}\right)^2 \frac{(\rho - e_\pm/e_{DL})^2}{(\rho + e_\pm^2/e_{DL}^2)^2}, \tag{5.3}$$

and

$$s(\rho) = \left(1 + \frac{e_\pm}{e_{DL}}\right)^2 \frac{\rho}{(1+\rho)(\rho + e_\pm^2/e_{DL}^2)}. \tag{5.4}$$

G is the parameter introduced in equations (4.22) and (4.43), which has been shown to be closely equal to 3.0. For a given s, Λ is a double valued function depending on whether ionic or Bjerrum defects are the majority carriers (or more precisely whether ρ is greater or less than $e_\pm/e_{DL}$). The data from Hubmann's experiments on HF- and NH_3-doped ice over a wide range of temperature are plotted in this form in Fig. 5.6, and the line shows the function $\Lambda(s)$ fitted to them. The ratio of the limiting values of $\Lambda(s)$ as $s \to 0$ is $(e_\pm/e_{DL})^2$, corresponding to the change of equation (4.40) into (4.42). Hubmann's data give $e_\pm = (0.62 \pm 0.01)e$ and $e_{DL} = (0.38 \pm 0.01)e$, and when substituted into equation (5.3) they yield $G = 3.6 \pm 0.2$. However, his data were obtained for fields parallel to the c-axis, and in this orientation we have seen in §5.3.2 that the theory needs to take account of the Curie–Weiss parameter Δ. To compare experimental results with the theory of Chapter 4 it is more appropriate to use measurements of χ_s with the field perpendicular to the c-axis. If we take Hubmann's value of $e_\pm$ together with the mean value of $\chi_{s\perp} = 96 \pm 2$ for pure ice from Table 5.1, we obtain $G = 3.1 \pm 0.2$, in complete agreement with the theoretical value of 3.0 (the

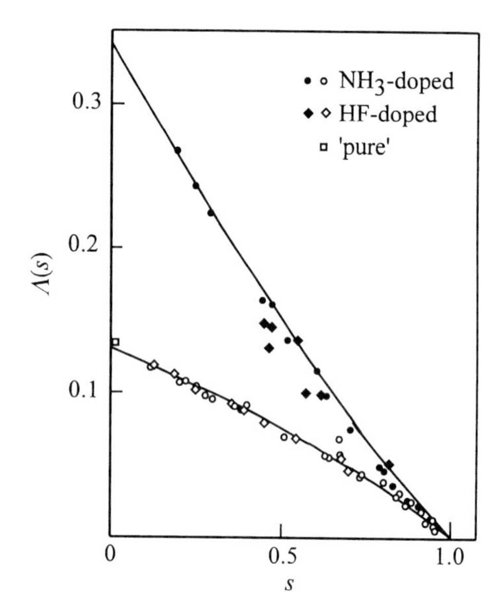

Fig. 5.6 Dielectric measurements on HF- and NH_3-doped ice over a wide range of concentrations and temperatures, analysed according to equations (5.1) and (5.2) and fitted to the universal function $\Lambda(s)$. The upper line corresponds to ionic defects being the majority charge carriers, and the solid points are identified as fitting this line, while the open points fit the lower line with Bjerrum defects being the majority carriers. Reproduced with permission from *Zeitschrift für Physik*, **B32**, M. Hubmann, Polarization processes in the ice lattice, pp. 127–39, Fig. 6, Copyright 1979 Springer-Verlag.

comparison with experiment is discussed by Ryzhkin and Whitworth 1997). According to equation (4.29) Hubmann's experimentally determined value of e_{DL} corresponds to a dipole moment of the water molecule in ice of 9.69×10^{-30} C m. This should be compared with the value for the free molecule of 6.186×10^{-30} C m quoted in §1.3. The increase is to be expected from the theory of Coulson and Eisenberg (1966) referred to in §2.4.

According to equation (5.3) Λ and thus $\chi_s = 0$ for $s = 1$, and this is the condition for the cross-over of σ_s and σ_∞ discussed above. The effect of hydrostatic pressure on doped ice is to increase $\sigma_\pm$ and reduce σ_{DL} (Hubmann 1978). For a doped specimen close to the cross-over conditions χ_s is very sensitive to these conductivities, and Fig. 5.7 shows that a single specimen can be taken reversibly through the minimum in χ_s by the application of such a pressure.

The justification for the assignment of σ_∞ to Bjerrum defects and σ_s to ionic defects in pure ice, and hence the labelling of components throughout the above discussion, comes from this analysis. In Fig. 5.6 pure ice lies at small s on the lower branch of the curve, and this is the branch for which the dominant partial conductivity is that of the species with the lower effective charge. This can be

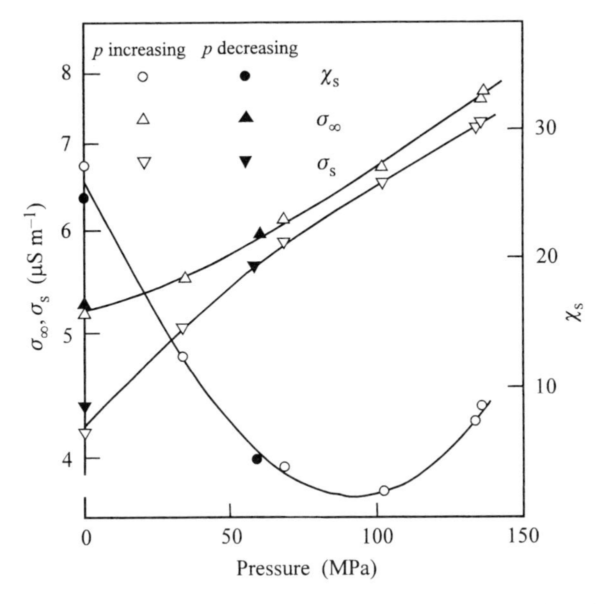

Fig. 5.7 The effect of hydrostatic pressure on the dielectric properties of ice doped with 5×10^{-5} M NH_3 at −24.3 °C. This specimen is very close to the cross-over condition and is taken through the cross-over by the application of pressure. Reproduced with permission from *Zeitschrift für Physik*, **B32**, M. Hubmann, Polarization processes in the ice lattice, pp. 127–39, Fig. 5, Copyright 1979 Springer-Verlag.

understood from equations (4.40) and (4.42) in which the lower charge gives the lower susceptibility. From theoretical modelling (e.g. Scheiner and Nagle 1983) it is clear that the smaller charge has to be that of the Bjerrum defect, and this assignment agrees with many earlier arguments about the likely effects of HF in the ice.

5.4.2 *Interpretation in terms of point defects*

Knowing the values of $e_\pm$ and e_{DL}, equations (4.50) and (4.51) can be inverted to calculate $\sigma_\pm$ and σ_{DL} from the experimental data. In a doped crystal each of these quantities will usually be dominated by the partial conductivity for a single defect. In the case of HF doping, for example, the majority defects are the L-defects, for which equation (4.47) can be written as

$$\sigma_{DL} = \sigma_L = n_L \mu_L e_{DL}. \tag{5.5}$$

The minority charge carriers are the H_3O^+ ions for which

$$\sigma_\pm = \sigma_+ = n_+ \mu_+ e_\pm. \tag{5.6}$$

The temperature dependencies of $\sigma_\pm$ and σ_{DL} involve those of both the concentrations and the mobilities. The mobilities are expected to follow the Arrhenius relation

$$\mu_i \propto \frac{1}{T}\exp\left(-\frac{E_{im}}{k_B T}\right), \tag{5.7}$$

though for H_3O^+ ions the results indicate no significant temperature dependence at high temperatures. In a doped specimen the concentrations n_i depend on the chemistry of the incorporation and dissociation of the specific dopant. In the case of HF illustrated in Fig. 5.4 the fluorine atom can be in one of four states

1. HF together with its L-defect—concentration n_{HFL};
2. HF without the L-defect—concentration n_{HF};
3. F with the L-defect but without the H—concentration n_{FL};
4. F^- without the L-defect or H, and carrying a full negative charge—concentration n_{F^-}.

There are chemical equilibria between all these species and the L and H_3O^+ defects in the lattice. For a simple analysis some approximations are essential, and it is usual in this case to assume that all the L-defects are dissociated from the fluorine site so that the one controlling equilibrium is the ionization of the species HF. For this equilibrium

$$\frac{n_{F^-}\cdot n_+}{n_{HF}} = K_{HF} \propto \exp\left(-\frac{E_{aHF}}{k_B T}\right), \tag{5.8}$$

where E_{aHF} is the energy required to dissociate the HF. Assuming n_+ is much greater than the intrinsic concentration, we must have $n_+ = n_{F^-}$, and the total concentration of HF present in any form $N_{HF} = n_{HF} + n_{F^-}$. These equations give

$$n_+ = \frac{1}{2}\left(-K_{HF} \pm \sqrt{K_{HF}^2 + 4K_{HF}N_{HF}}\right), \tag{5.9}$$

where the positive sign must be chosen. At high temperatures and low concentrations ($N_{HF} \ll K_{HF}$) this equation predicts almost complete dissociation, with $n_+ \approx N_{HF}$ and independent of temperature. However, for higher concentrations or lower temperatures, where $N_{HF} \gg K_{HF}$,

$$n_+ \approx (K_{HF}N_{HF})^{1/2} \propto N_{HF}^{1/2}e^{-E_{aHF}/2k_B T}. \tag{5.10}$$

Kröger (1974) has surveyed the point defect equilibria that arise in pure and doped ice and the range of approximations that are possible (though he misses the relevant case of complete ionization given above). Further complications are possible if charge carriers can associate with one another or with trapping centres such as vacancies or dislocations, and there are reasons to believe that trapping does occur (see §6.11). The equilibria for the formation of intrinsic ionic or Bjerrum defects in pairs require that the products $n_+ n_-$ and $n_D n_L$ are constant at their intrinsic values. This means that doping to increase n_+ will automatically suppress n_- and *vice versa*.

Early experiments on doped ice were interpreted by concentrating on specific features, such as linear segments of the Arrhenius plots, which give the activation energies for particular processes. A better procedure is to fit the whole data set to a model based on specific approximations, though such treatments do not necessarily discriminate well between models (see Camplin *et al.* 1978). In the following section we tabulate results derived by various authors for specific dopants, but all interpretations have to be treated with some caution.

5.4.3 *Specific dopants*

5.4.3.1 Hydrogen fluoride

Hydrogen fluoride is one of the most soluble and active dopants for ice, though doped samples are unstable with respect to loss of the dopant by out-diffusion. There are two sets of experiments which have been analysed in terms of the properties of the point defects involved. In the classic analysis by Jaccard (1959), which includes data from A. Steinemann (1957), parameters are extracted from the slopes of Arrhenius plots in limiting cases. In the analysis by Camplin *et al.* (1978) of experiments by Camplin and Glen (1973) various models are fitted to the whole data set, and this is important where clearly separated regions are not identifiable. The raw data from these experiments are far from ideal, not giving such well-defined single Debye relaxations as can be obtained for pure ice, and there are clear differences between the sets of observations before the analysis begins. The general features do, however, support the model described above. In particular σ_∞ rises linearly with the total HF concentration N_{HF} in the range where L-defects are the majority carriers; this was the basis of the simplifying assumption in §5.4.2 that the L-defects are fully dissociated from the fluorine sites. At temperatures above the cross-over σ_s is proportional to N_{HF} at low concentrations, but it deviates towards $N_{HF}^{1/2}$ at higher concentrations and lower temperatures in agreement with equation (5.9).

Table 5.2 gives some activation energies and mobilities deduced from these analyses. In the case of Camplin *et al.* these are for the simplest of their models, which corresponds to the assumptions described above, but the effective charges have been allowed to adopt values which are rather far from those which have been subsequently well established (see §5.4.1). For the higher concentrations the analysis is difficult because σ_s and σ_∞ are not very different in magnitude. In this region Steinemann (1957) reports a famous double cross-over, resulting in two minima in the dependence of χ_s on the HF concentration at $-10\,^\circ$C.

Table 5.2 Defect parameters for HF- and HCl-doped ice

	HF		HCl
	Jaccard (1959)	Camplin *et al.* (1978)	Takei and Maeno (1987)
Experimental activation energies			
Pure ice E_{σ_∞} (eV)	0.575	0.624	0.585 ± 0.024
Heavy doping			
E_{σ_∞} (eV)	0.235	0.24	
E_{σ_s} (eV)	0.325	0.32	
Derived quantities			
Activation energy for L-defect mobility E_{Lm} (eV)	0.235	0.315	0.190 ± 0.017
Activation energy of formation of Bjerrum defect pair E_{DL} (eV)	0.68	0.664	0.790 ± 0.082
Dissociation energy of H_3O^+ from acid E_{aHF} or E_{aHCl} (eV)	0.65	0.59	0.650 ± 0.018
Dissociation energy of L-defect from trap $E_{aL,trap}$ (eV)			0.152 ± 0.072
L-defect mobility (218 K) ($m^2 V^{-1} s^{-1}$)	2.3×10^{-9}	2.5×10^{-9}	
H_3O^+ mobility (273 K) ($m^2 V^{-1} s^{-1}$)		2.8×10^{-8}	

The mobilities quoted in Table 5.2 depend on the assumed concentrations of HF actually in solid solution in the specimens.

5.4.3.2 Hydrogen chloride

The experiments of Gross *et al.* (1978) and Takei and Maeno (1984, 1987) show that HCl behaves similarly to HF. At higher temperatures the L-defects are fully dissociated and the HCl ionizes as a weak acid, even though it is normally considered a strong acid in water. At low temperatures σ_s and σ_∞ are less well separated than for HF, and Takei and Maeno conclude that the L-defects are being trapped either to some state of the Cl site or elsewhere.

Figure 5.8 shows the temperature dependence of the quantities σ_{DL}/e_{DL} and $\sigma_\pm/e_\pm$ derived from σ_s and σ_∞ according to equations (4.50) and (4.51) for one crystal grown from a 4×10^{-5} M HCl solution; the concentration actually incorporated in the ice is much less but not directly measurable. The two curves cross twice, at -39 and $-84\,°C$, and Fig. 5.9 shows the corresponding minima in χ_s. In this crystal the majority carriers between these temperatures will be the H_3O^+ ions, and the L-defects will be the majority carriers elsewhere. Notice from equation (4.53) that it is the plots of σ_{DL}/e_{DL} and $\sigma_\pm/e_\pm$ which must cross to produce this effect. For much smaller levels of doping the curves will not cross.

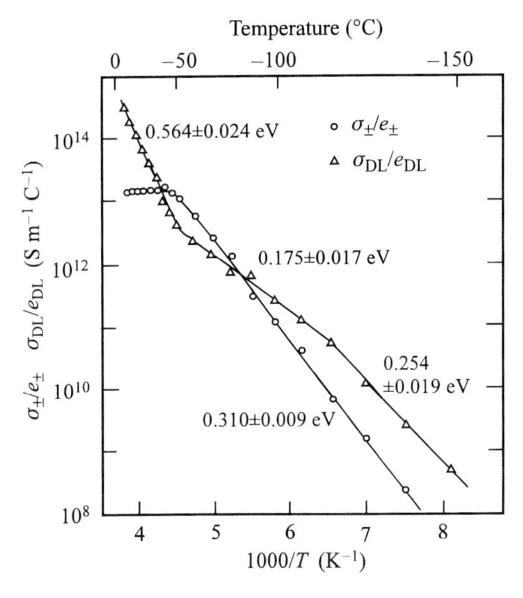

Fig. 5.8 Temperature dependence of $\sigma_\pm/e_\pm$ and σ_{DL}/e_{DL} for an HCl-doped specimen (grown from 4×10^{-5} M solution), showing the double cross-over of these scaled conductivities. Activation energies are given for the straight-line portions of these Arrhenius plots. (From Takei and Maeno 1987.)

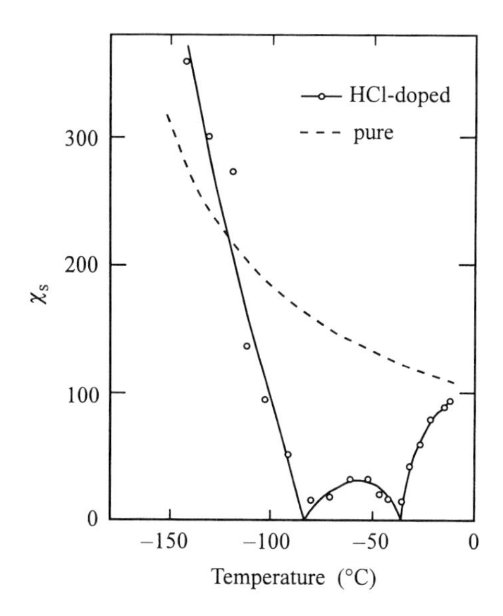

Fig. 5.9 Dielectric susceptibility χ_s of the HCl-doped specimen shown in Fig. 5.8, showing minima of χ_s at the cross-over points. Results for 'pure' ice are included for comparison. (From Takei and Maeno 1987.)

In comparing these Figures with Fig. 5.5 which shows only one cross-over, note the much wider temperature range of the more recent HCl data.

Activation energies deduced from these experiments are compared with those for HF-doped ice in Table 5.2. There are clear differences which arise from the way the data have been interpreted. If the L-defects are taken to be totally dissociated throughout the temperature range, the activation energy for the extrinsic conductivity is attributed entirely to their motion, but if there is evidence for the trapping of L-defects, as in the case of HCl, the activation energy attributed to motion E_{Lm} will be reduced. The activation energy E_{DL} for the creation of a Bjerrum defect pair is deduced from that for σ_∞ in the intrinsic range according to the relation (4.70):

$$E_{\sigma_\infty} = \tfrac{1}{2}E_{\mathrm{DL}} + E_{\mathrm{Lm}}, \tag{5.11}$$

and this makes E_{DL} appear larger in the case of HCl than for HF. This illustrates the difficulty of quoting reliable values for these defect parameters.

5.4.3.3 Ammonia

In water ammonia is a weak base dissociating according to the equation

$$NH_3 + H_2O \rightleftharpoons NH_4^+ + OH^-.$$

It is soluble in ice at small concentrations (< 0.001 M), though subject like HF to out-diffusion. Electrical measurements on NH_3-doped ice have been reported by Arias *et al.* (1966) and by Hubmann (1978) who also studied the effect of hydrostatic pressure on the doped crystals. The main effect of the doping is to increase σ_s, which must be due to the introduction of OH^- ions. As with the addition of H_3O^+ in acid-doped crystals, this can result in a cross-over at which the OH^- ions become the dominant charge carriers, and data showing this are included in Figs 5.6 and 5.7.

As indicated in §5.4.1, it is commonly suggested that an NH_3 molecule will occupy an H_2O site in the crystal and become a source of D-defects. However, Hubmann found no change of σ_∞ with NH_3 concentration, and Arias *et al.* found only a slight increase. If L-defects are dominant in pure ice, the first effect of introducing D-defects should be to decrease σ_∞, but this was not observed in these experiments. If no D-defects are formed, it is an open question whether NH_3 enters the lattice substitutionally or interstitially, but see evidence from NH_4F and polar ice in §5.4.4 and from thermally stimulated depolarization in §5.8.

5.4.3.4 Alkali hydroxides

Doping ice with the alkali hydroxides has the remarkable effect of maintaining sufficient proton mobility at low temperatures for the transition to the ordered phase, ice XI, to occur at 72 K (see §11.2). The hydroxides from lithium to rubidium have similar effects (Abe and Kawada 1991), but potassium hydroxide is the most effective and most extensively studied. These hydroxides have very low solid solubility in ice, and to produce significant enhancement of the dielectric relaxation rate at low temperatures the ice must be frozen quickly from relatively concentrated (100–1000 mole p.p.m.) solution and then stored at a low temperature. The ice so produced is sometimes cloudy, containing inclusions of concentrated KOH solution which freeze at 210 K to a eutectic mixture of almost pure ice and $KOH \cdot 4H_2O$ (Cohen-Adad and Michaud 1956; Tajima *et al.* 1984). Annealing such a sample for a day at a typical cold-room temperature of -10 to $-20\,°C$ results in almost complete loss of the electrical activity at low temperatures, presumably due to aggregation or precipitation of the dopant.

The most detailed studies of KOH-doped ice have been reported by Kawada and co-workers (e.g. Kawada 1989*a*,*b*; Kawada *et al.* 1992; Kawada and Tutiya 1997). Above 200 K the effect of KOH is similar to that of other dopants with the dominant effect being on σ_s and some evidence of a possible cross-over from Bjerrum to ionic defects being dominant at lower temperatures. Below 200 K a freshly grown crystal exhibits a single Debye relaxation for which the susceptibility χ_s parallel to the *c*-axis has a Curie–Weiss temperature dependence as $1/(T-\Delta)$, with $\Delta \approx 10$–30 K and a strength close to that expected for polarization by ionic charge carriers according to equation (4.40) (Oguro and Whitworth 1991; Kawada and Tutiya 1997). Figure 5.10 shows a compilation of observations on the relaxation time for this polarization process in a variety of freshly prepared single and polycrystalline samples. Values of τ_D for 'pure' ice are included from Fig. 5.3 to show how enormously the relaxation rate at low temperatures is enhanced by the KOH. All these observations have the property that the relaxation time is more or less independent of temperature between 200 and 110 K but then rises at lower temperatures with an activation energy of 0.15–0.17 eV. In suitable specimens the relaxation process remains fast enough down to 72 K to facilitate the ordering transformation of ice Ih to ice XI.

Kawada and Tutiya (1997) made measurements at 130 K on a 120 p.p.m. KOH-doped specimen after successive annealing periods of a few hours at $-10\,°C$. This annealing caused the Debye relaxation to split into two components with very different relaxation times. The total permittivity remained almost constant, but an increasing fraction of it was only attained with a relaxation time several orders of magnitude longer than that observed in a fresh specimen. After a total annealing time of more than 20 h almost the whole relaxation occurred at the slower relaxation rate. At low temperatures this rate was so slow that very little polarization could be produced, and such annealed specimens would not transform to ice XI. The value of the constant total polarization achievable at 130 K corresponds to polarization by ionic defects in accordance with equation (4.40) rather than by Bjerrum defects (equation (4.42)). It is not understood why two distinct relaxation times were observed rather than there being a gradual increase in a single relaxation time as the dopant aggregated.

Figure 5.11 suggests how KOH will be incorporated in the ice lattice to release both an L-defect and an OH^- ion. The K^+ is assumed to occupy an interstitial

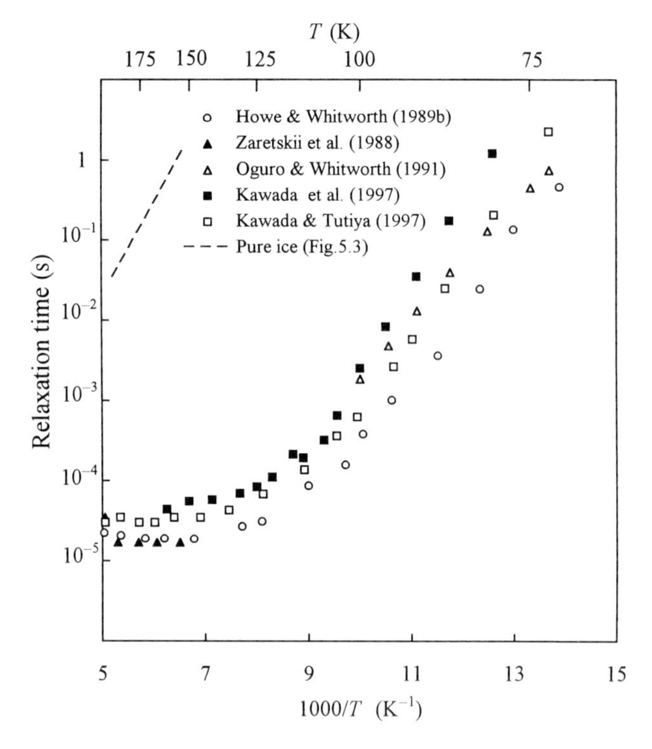

Fig. 5.10 Various measurements of the fast relaxation time at low temperatures in freshly grown KOH-doped ice.

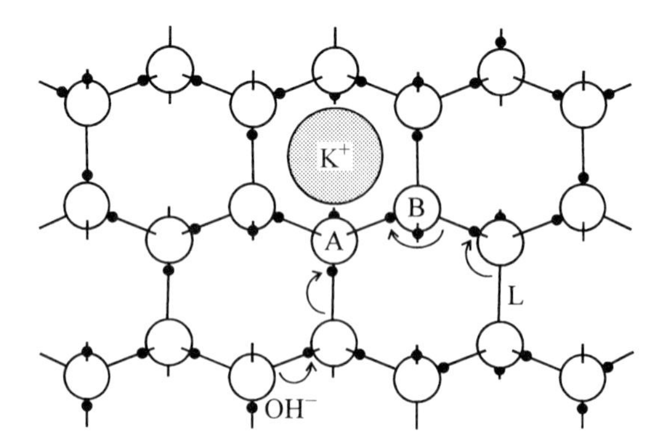

Fig. 5.11 Proposed model for the incorporation of KOH in the ice structure. The K^+ occupies an interstitial position and the OH^- is initially at A with no proton on the bond AB. Proton jumps as indicated can release either an OH^- ion or an L-defect or both.

site, though there is no direct evidence for this. As we have explained above, the dominant defects at low temperatures must be the OH^- ions. Any contribution from the L-defects, which would produce a steady-state conductivity, dies out at these temperatures. The almost constant value of the relaxation time in Fig. 5.10 between 200 and 120 K indicates that the mobility of OH^- ions is effectively independent of temperature, as is the case for the H_3O^+ ions giving the low-frequency conductivity of acid-doped ice. This temperature independence implies an activation energy for motion $\leq k_B T \approx 0.01$ eV, and the rise of τ_D below 110 K with an activation energy much larger than this has to be attributed to trapping of the ions at sites which may or may not be associated with the K^+. The process of ionic mobility, including the possibility of the ions moving by tunnelling, is discussed in §6.5.4.

5.4.4 *Salts and combinations of dopants*

When an aqueous solution is frozen the anions and cations are not necessarily incorporated in the proportions originally present in the solution (Gross *et al.* 1977, 1978). In the case of sodium chloride, for example, the Cl^- enters the ice as HCl leaving the Na^+ and OH^- in the liquid phase, and the electrical properties of ice formed in this way are indistinguishable from those of ice frozen from an HCl solution. A similar observation was reported by A. Steinemann (1957) in the case of ice grown from solutions of CsF.

A compound which is remarkable for its high solubility in ice is NH_4F. For its concentration it has very much less effect on the electrical properties than HF (Zarcomb and Brill 1956; Gross *et al.* 1978). If both the NH_4^+ and F^- ions occupy H_2O sites in the lattice there is no surplus or deficit in the number of protons and so no extrinsic ionic or Bjerrum defects will be formed. Referring back to the discussion of NH_3-doping in §5.4.3.3, this suggests that the NH_4^+ ion enters the lattice substitutionally even if the resulting D-defect is not released. Gross and Svec (1997) have shown that the presence of NH_4^+ enhances the solid solubility of many anions in ice, with a resulting increase in σ_∞ and a reduction in σ_s, as might be expected from a suppression of the H_3O^+ ion concentration.

The properties of ice containing a mixture of anions and cations have been extensively studied in connection with the characterization of polar ice cores as described in §12.5.1. For these cores both the high-frequency and steady-state conductivity data and the impurity concentrations are recorded as a function of depth. In such experiments at −10 to −30 °C it is found that σ_s is primarily dependent on the acidity of the ice (Taylor *et al.* 1992). The plot of the high-frequency conductivity associated with the Debye dispersion, however, exhibits peaks corresponding to the presence of Cl^- without any corresponding peak in acidity or σ_s (Moore *et al.* 1992). These observations clarify the interpretation of doping illustrated for the case of HF in Fig. 5.4. If HF (or HCl) is substituted for H_2O there is one proton missing, and an L-defect can be released into the lattice. The HF molecule can also dissociate to release an H_3O^+ ion. However, if F^- or

Cl^- is substituted for H_2O and its charge is compensated by a cation introduced interstitially, two protons are missing and in principle two L-defects could be released, though in practice there is probably not more than one (Moore *et al.* 1992). In this case there is no possibility of an H_3O^+ ion being formed.

A third component identified as giving an important signal in the electrical profile of the ice cores is NH_4^+. This produces a peak in σ_∞ approximately double that due to an equivalent concentration of Cl^- (Moore *et al.* 1994). The natural interpretation is that the NH_4^+ is releasing highly mobile D-defects, but this would be inconsistent with the laboratory experiments on ice doped with NH_3 or NH_4F. The empirical work on ice cores provides motivation for fresh laboratory studies using the relevant dopants.

In the case of polycrystalline ice there is evidence that large anions like SO_4^{2-} or NO_3^- are present in liquid inclusions at grain boundaries or grain boundary junctions, and that conduction is dominated by these liquid regions (see §7.7.2).

5.5 Charge exchange at ice–metal electrodes

As already stated, a metal electrode in contact with ice does not allow free conversion between the electronic charge carriers in the metal and the protonic carriers in the ice, and this leads to the observed space charge relaxation processes. The necessary chemical reactions which occur quite normally at metallic electrodes in liquid water are

$$2H_3O^+ + 2e^- \rightarrow H_2\uparrow + 2H_2O$$

$$2OH^- - 2e^- \rightarrow \tfrac{1}{2}O_2\uparrow + H_2O.$$

In the case of ice each equation must include both Bjerrum and ionic defects (see Fig. 4.11), but we must first consider the case of the liquid. The electronic energy levels are shown in Fig. 5.12, where E_F is the Fermi energy in the metal and E_{H_2} and E_{O_2} are the energies of electrons in equilibrium with the reactions given above. E_{H_2} and E_{O_2} differ by 1.23 eV, which is the energy per electron to dissociate the water molecule in the liquid. For no applied potential difference as in (a) the reactions cannot occur, but with a potential difference of 1.23 V or more as in (b) the energy levels can be shifted so that the two reactions can take place at the respective electrodes. The shifts in potential across the liquid–metal interfaces between the states (a) and (b) are produced by charged double layers consisting of an excess or deficit of electrons in the metal and H_3O^+ ions close to the cathode or OH^- ions close to the anode. The charge density is high and the layer in the liquid is at most a few molecules thick.

The electrical properties of the interface between a metal and ice have been studied by Petrenko and Chesnakov (1990*c*). In this case the application of a potential difference builds up stable space charge layers of the types described in

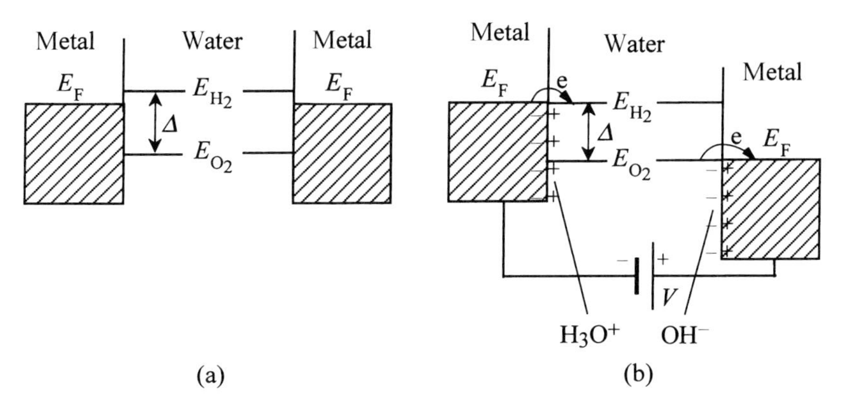

Fig. 5.12 Electronic energy level diagram for water in contact with two metal electrodes (a) without bias and (b) with d.c. bias $V = \Delta$, where $\Delta = 1.23$ V.

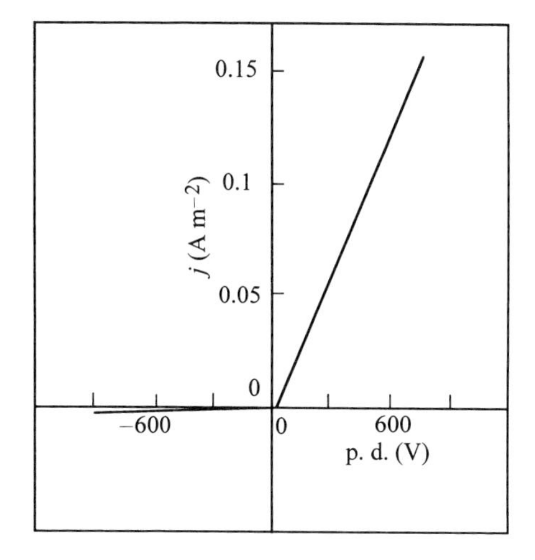

Fig. 5.13 Current–voltage characteristics of an ice specimen frozen between platinum electrodes in an electric field. Temperature −10 °C. (From Evtushenko *et al.* 1988.)

§4.6, which are typically of the order of a micrometre thick. Such layers do not provide the intimate contact with the metal necessary for the discharge of the ions and release of free atoms and the appropriate Bjerrum defect. Evtushenko *et al.* (1988) have shown that a very different situation can be attained if the water is frozen onto the electrodes while a p.d. is applied such that a current is flowing. In this case the ice–electrode assembly has rectifying properties with the current–voltage characteristics shown in Fig. 5.13. The 'forward' direction is that in which the current was flowing during freezing, and in this direction the behaviour is ohmic apart from the minimum bias of about 1.5 V required to dissociate the H_2O. In this system the double layer appropriate to the electrolysis of the liquid has been frozen into the ice. This double layer is quite inappropriate for charge flow in the opposite direction, and for such potential differences the metal behaves as a blocking electrode.

These rectifying junctions are stable for days at $-10\,^\circ C$ and longer at lower temperatures. Even more remarkable is the fact that the metal electrode can be removed from the ice by warming just to the melting point, attached to a different ice sample, and used to make ohmic measurements on that sample.

5.6 Space charge effects

As already indicated in §5.2.3, if the specimen has blocking or even partially blocking electrodes the low-frequency conductivity will give rise to a build-up of space charge at the electrodes. In an ideally simple case this will produce a single low-frequency relaxation with time constant C_e/G_s, where C_e is the effective capacitance at the blocking electrode and G_s is the static conductance of the bulk ice. In practice a space charge layer of finite thickness is established in the ice, and the theory for small applied potential differences is described in §4.6. Using equations (4.64) and the intrinsic parameters for pure ice at $-20\,^\circ C$ the shorter screening length κ_1^{-1} is of the order of 0.1 μm; the longer screening length κ_2^{-1} depends on the extrinsic concentration of ionic defects, but for nominally pure ice at this temperature it will usually be of the order of 0.1 mm. As the temperature is reduced these lengths both increase as the inverse square root of the defect concentrations.

An understanding of space charge effects is relevant to a number of kinds of experiment on the electrical properties of ice, which will be described in the following sections.

5.6.1 *Dispersion curves*

Figure 4.16 shows the form of dispersion curve expected for an ice crystal with blocking electrodes and only two kinds of active charge carrier (e.g. L-defects and H_3O^+ ions). As can be seen from equation (4.65a) the Debye dispersion is not much affected except in very thin specimens for which $L \sim \kappa_1^{-1}$. However, the low-frequency dispersion arises because of the build-up of space charge and it is described by equation (4.65b). With more types of charge carrier or with air gaps at the electrodes further low-frequency dispersions can be produced. An example from the experiments of Noll (1978) is shown in Fig. 5.14. Such curves are commonly obtained though the dispersions may not be so well resolved. By fitting such data to the full theoretical predictions of Petrenko and Ryzhkin (1984*a*) it is in principle possible to derive the concentrations and mobilities of both majority and minority charge carriers. This was done by Zaretskii *et al.* (1987*a*) who

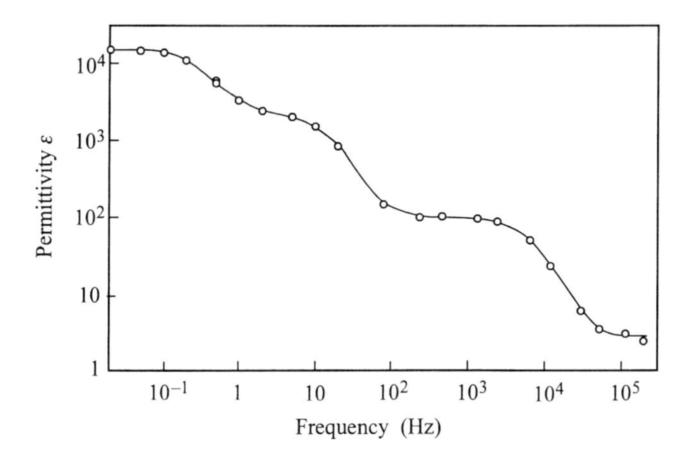

Fig. 5.14 Frequency dependence of the permittivity of a specimen of 'pure' ice with blocking electrodes at $-3\,°C$. The capacitance of the blocking layer has been removed by calculation (Noll 1978). Reproduced by courtesy of the International Glaciological Society from *Journal of Glaciology*, **21**(85), pp. 277–89, 1978, Fig. 4.

deduced that the mobility of L-defects μ_4 at $-10\,°C$ was $1.7 \times 10^{-8}\ m^2\,V^{-1}\,s^{-1}$, but obtained what were subsequently found to be less satisfactory values for μ_1. The significant point is that it becomes possible to separate the product of concentration and mobility which occurs in all direct measurements of conductivity or relaxation time.

5.6.2 *A field effect transistor made from ice*

An elegant use of the space charge properties of ice is the construction by Petrenko and Maeno (1987) of a field effect transistor (FET). This device illustrates features which ice has in common with semiconductors and permits the determination of charge carrier mobilities. The basic idea of an FET is to use a potential applied to an electrode on the surface of the conductor (the 'gate') to control the number of charge carriers, and hence the conductance, between two other electrodes (the 'source' and 'drain'). The application to ice corresponds to a MOSFET in which the gate is separated from the ice by a thin insulating layer. When a potential is applied to the gate (relative to some distant conducting contact on the ice), a charge will be formed on its surface and an equal and opposite charge will accumulate as a space charge layer within the ice. For simplicity we will suppose that there is just one type of charge carrier of charge e_1 and mobility μ_1 in the ice. The number of these carriers, N_1 per unit area of the space charge layer, is related to the applied potential V by the equation $N_1 e_1 = C_{\text{eff}}(V)V$, where $C_{\text{eff}}(V)$ is the measured effective capacitance of the layer. The conductance in the surface layer measured with a low-frequency a.c. signal between the source and drain electrodes is given by $e_1 N_1 \mu_1$ and thus depends on V. Measurement of the change of surface conductance with applied potential can thus give the mobility of the defects in the ice. If there were two types of defect of opposite sign, their effects could be separated by reversing the gate potential.

The full theory for the case of ice has been given by Petrenko and Maeno (1987) and Petrenko and Ryzhkin (1997). The effect of the applied potential on Bjerrum defects is much less than that on ionic defects because of the larger charge on the latter. This is because the concentration enhancement for a defect of type i at a point where the potential is $\phi(x)$ is given by an equation of the form

$$n_i(x) = n_{i\infty} \exp\left(-\frac{e_i \phi(x)}{k_B T} \right), \tag{5.12}$$

which is sensitive to the difference between e_1 and e_4. In pure ice the ions control the steady-state conductivity σ_s, so that it is this conductivity in the surface layer that is influenced by the applied potential.

Figure 5.15 shows how the device was realized in practice. The oxide layer and gold electrodes were fabricated on a wafer of silicon using semiconductor technology. So that the conductance between the gold electrodes is dominated by surface effects, their spacing has to be as small as possible, and because the conductance is low they were fabricated as a pair of intermeshed combs; the inter-electrode spacing was about 50 μm. The silicon plate itself served as the gate electrode. Measurements had to be made below −30 °C to avoid surface conduction in the subsurface layer described in §10.4.1. An applied potential of −1.5 V increased the surface conductance at 10 Hz by 80 times due to the drawing in of H_3O^+ ions, and the reverse potential gave an increase of 24 times which was attributed to OH^- ions. The experiment was interpreted as implying mobilities $\mu_1 = 9 \times 10^{-8}\, m^2 V^{-1} s^{-1}$ and $\mu_2 = 3 \times 10^{-8}\, m^2 V^{-1} s^{-1}$ at −33.1 °C. This technique was refined by Khusnatdinov *et al.* (1997) to study the surface charge and contact potential at ice–metal and ice–insulator interfaces (see §10.4.2).

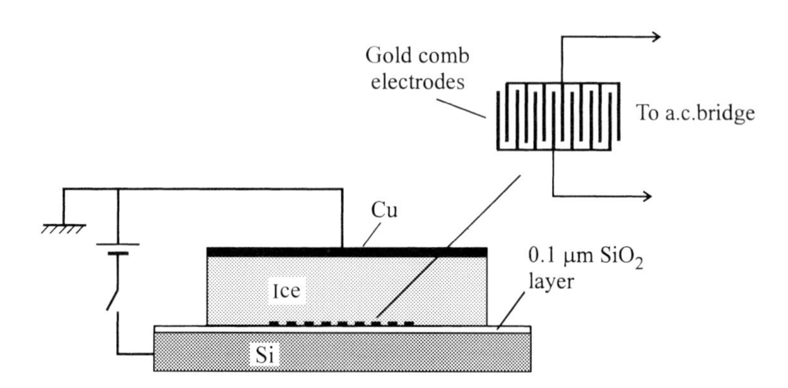

Fig. 5.15 Field effect transistor made from ice (Petrenko and Maeno 1987).

5.6.3 *Proton injection from hydrogen-loaded palladium electrodes*

The metal palladium has a high solubility for hydrogen, and when hydrogen-loaded palladium electrodes are placed on a sample of ice there is free exchange of protons between the palladium and the ice. The change between the electronic and protonic conduction mechanisms occurs within the electrode and there is no electrolysis. Such electrodes were first used on ice by Engelhardt and Riehl (1965, 1966). For small applied potentials they provide ohmic contacts to ice, but for larger potential differences they inject an excess of protonic charge carriers leading to space charge limited currents. Such phenomena are well established in the field of semiconductors and dielectrics (Lampert and Mark 1970), for which experiments of this kind yield values for charge carrier mobilities etc. Petrenko *et al.* (1983) have studied such effects in ice, where both the theory and the practical implementation are rather more difficult than for semiconductors.

The principle of space charge limited currents will first be explained by reference to Fig. 5.16 assuming just one type of charge carrier injected at the electrode where $x=0$. These charge carriers have charge e, mobility μ, and concentration $n(x)$. There is free exchange of charge carriers across the ice–electrode interface so that the effective concentration in the ice at the interface is $n(0)=\infty$, and the electric field at this point is $E(0)=0$. The field in the space charge region obeys

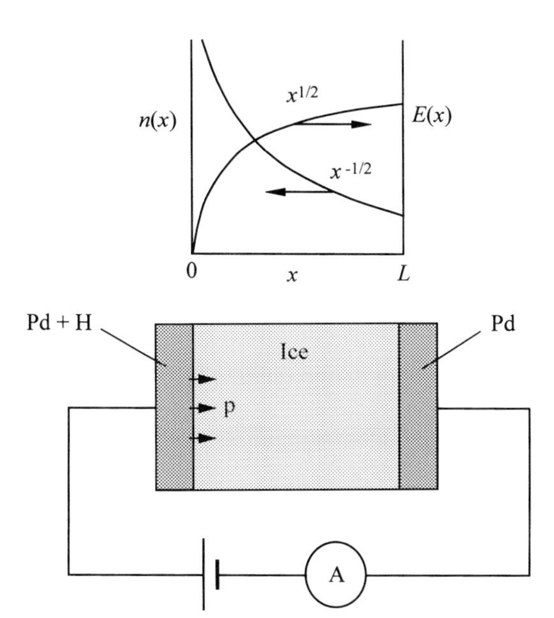

Fig. 5.16 Injection of protons from a hydrogen-loaded palladium electrode into ice. The negative palladium electrode acts as a proton acceptor. The upper diagram illustrates the spatial dependence of the charge carrier concentration $n(x)$ and the electric field $E(x)$ in the ice under space charge limited conditions.

Poisson's equation

$$\frac{\mathrm{d}E}{\mathrm{d}x} = \frac{en(x)}{\varepsilon\varepsilon_0}. \tag{5.13}$$

The current density $J = ne\mu E$ is constant through the thickness L of the specimen, and the potential difference is

$$V = \int_0^L E\,\mathrm{d}x. \tag{5.14}$$

These equations predict that $n(x)$ falls as $x^{-1/2}$ and $E(x)$ rises as $x^{1/2}$, with the current density given by

$$J = \frac{9}{8}\varepsilon\varepsilon_0\mu\frac{V^2}{L^3}. \tag{5.15}$$

This equation contains just one unknown parameter μ, which can therefore be determined from such an experiment.

If in addition to the injected charge carriers there is an equilibrium concentration n_0, then for a small applied potential the current will be given by Ohm's law

$$J = n_0 e\mu\frac{V}{L}, \tag{5.16}$$

and the transition to the space charge limited current will occur when the current given by equation (5.15) exceeds this value. This in principle enables n_0 to be determined as well as μ.

In the case of ice the theory is more complicated than that described because a steady current requires the flow of both ionic and Bjerrum defects, whose motion is correlated by the configuration vector $\mathbf{\Omega}$. The theory was first developed by Petrenko *et al.* (1983) for the case of pure ice in which the majority carriers are L-defects and the injected H_3O^+ ions remain the minority carriers. Equation (5.15) then becomes

$$J = \frac{9}{8}\varepsilon_s\varepsilon_0\mu_1\frac{e}{e_1}\frac{V^2}{L^3}, \tag{5.17}$$

in which the subscript 1 denotes as usual the H_3O^+ ions and ε_s is the normal steady-state permittivity. The general case, which includes a further stage of injection of both ionic and Bjerrum defects above their equilibrium concentrations, is described by Petrenko and Ryzhkin (1984*b*).

Experimentally Petrenko *et al.* (1983) observed both the change from ohmic to V^2 current characteristics and an associated increase in the conductance measured by a small low-frequency modulation of the p.d. Figure 5.17 shows an example of their results, and from these and related experiments they estimated

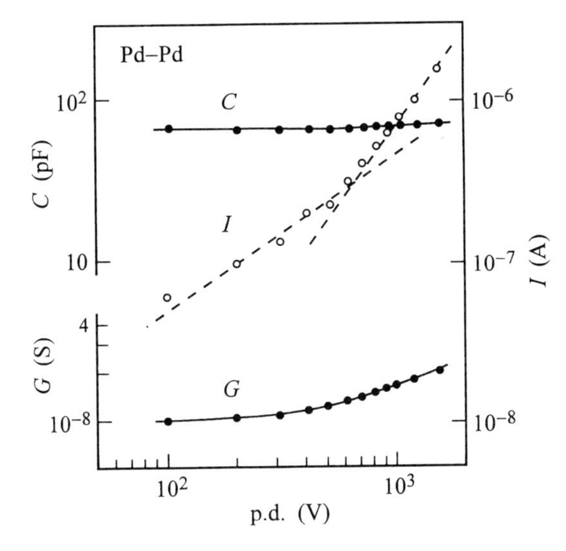

Fig. 5.17 Characteristics of ice with hydrogen-loaded palladium electrodes, showing dependence on the applied p.d. of the direct current I, capacitance C, and conductance G all plotted on logarithmic scales. Measurements were at 100 Hz at −20 °C on a specimen of area $6 \times 9\,\mathrm{mm}^2$ and thickness 0.4 mm. Broken lines of slopes 1 and 2 are fitted to the direct current characteristics. (From Petrenko *et al.* 1983, with permission from Taylor and Francis.)

positive ion mobilities μ_1 in the region of 10^{-9} to $10^{-7}\,\mathrm{m^2\,V^{-1}\,s^{-1}}$. Values obtained in this way are lower limits subject to a systematic error arising from trapping of the injected carriers in such a way that they contribute to the space charge but not the current.

A different kind of experiment was described by Eckener *et al.* (1973), who used a laser pulse to release protons from hydrogen-loaded palladium powder in contact with ice. These protons were assumed to enter the ice as H_3O^+ ions, and were swept through the ice by an electric field. The transit time yielded mobilities between $9 \times 10^{-8}\,\mathrm{m^2\,V^{-1}\,s^{-1}}$ at 95 K and $2.2 \times 10^{-7}\,\mathrm{m^2\,V^{-1}\,s^{-1}}$ at 118 K, falling with further increase in temperature. Eckener *et al.* demonstrated from the linearity of the field dependence of the transit time that the measurements were not affected by trapping. This is probably the most reliable measurement available of the ionic mobility, though for a low and limited range of temperature; the temperature dependence is not understood.

5.7 Injection and extraction of charge carriers

So far we have considered mainly experiments in which the concentrations of charge carriers have their thermal equilibrium values and the measuring current only perturbs these concentrations by changing the local electric potential within a region of space charge. We now turn to a range of experiments in which electrical properties are affected by the current flowing. These changes in the system can be monitored by observing the a.c. characteristics in the presence of an applied static potential difference. Related phenomena have been studied in more detail in semiconductors, but in ice this aspect of the electrical properties is not yet well understood.

5.7.1 *Charge carrier extraction*

When a d.c. potential difference is applied across a specimen of ice with metal electrodes, the high-frequency conductivity σ_∞ measured simultaneously between the same electrodes is decreased. This reproducible and reversible effect was studied by Petrenko and Schulson (1992) and is illustrated in Fig. 5.18 for a specimen with rectifying electrodes of the kind described in §5.5. The figure shows both σ_∞ and the direct current I. A similar reduction in σ_∞ occurs for both forward and reverse bias of the specimen, apart from a small offset which is due to the potential frozen-in when the specimen was prepared. This phenomenon has to be interpreted as a reduction in the concentration of majority carriers in the specimen, and Petrenko and Schulson suggest that it arises from field-enhanced recombination of Bjerrum defects at the electrodes.

The dependence of the effect on specimen thickness is shown in Fig. 5.19. For each thickness the open point represents σ_∞ with no applied p.d. and the solid point the value of σ_∞ with a field of 300 kV m^{-1} applied between the electrodes in

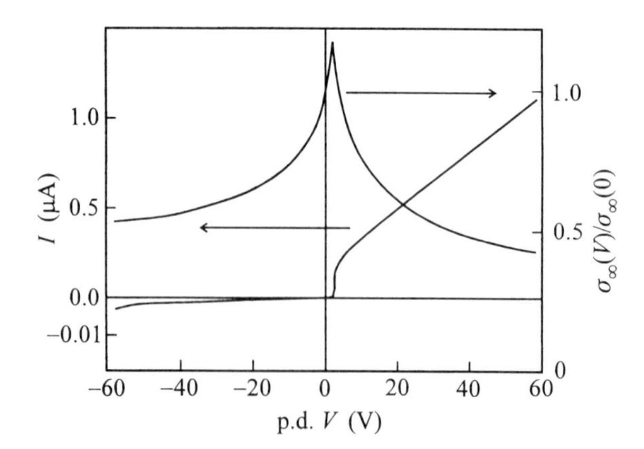

Fig. 5.18 Experiment to demonstrate charge carrier extraction. An ice specimen with rectifying characteristics passes the direct current I as a function of the applied voltage. The normalized high-frequency conductivity σ_∞ then varies with the p.d. as shown. Temperature −40 °C, specimen thickness 110 μm. (From Petrenko and Schulson 1992, with permission from Taylor and Francis.)

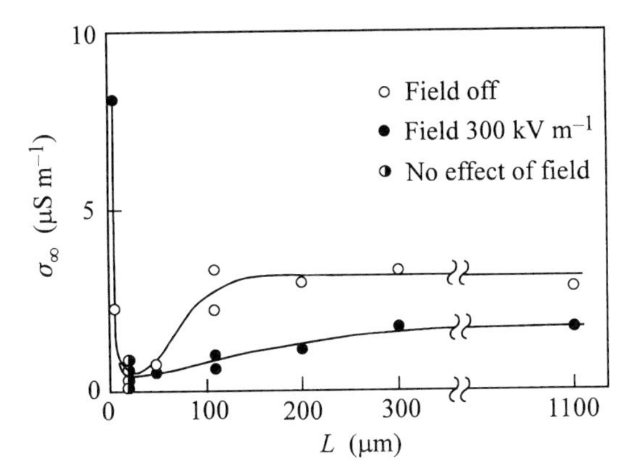

Fig. 5.19 Dependence of high-frequency conductivity σ_∞ on specimen thickness L for experiments of the kind shown in Fig. 5.18. (From Petrenko and Schulson 1992, with permission from Taylor and Francis.)

the forward bias direction. For specimens more than 100 μm thick σ_∞ falls as described above. However, for smaller thicknesses the effect of the electric field disappears, but there is a reduction in σ_∞ due to the effect of the thickness alone. Petrenko and Schulson argue that this also arises from recombination at the electrodes. The points in Fig. 5.19 showing an increase in σ_∞ and an increase with applied field for a thickness of 10 μm are related to the phenomenon described in the following section.

5.7.2 *Recombination injection*

The effect given this name was first reported for liquid water by Petrenko and Chesnakov (1990*a*). While a d.c. current is flowing in pure water, producing H_2 and O_2 at the electrodes, there is an enhancement of the a.c. conductivity in regions close to the electrodes. On switching off the d.c. current the a.c. conductivity returns to normal with a time constant of a few seconds. This effect was attributed to the creation of H_3O^+ and OH^- ion pairs from the energy released in combining H atoms into H_2 or O atoms into O_2, and hence the name given to the phenomenon. The increased concentrations are present in a layer of thickness characterized by the ambipolar diffusion length $\lambda_d = (2D\tau)^{1/2}$, where D and τ are the diffusion coefficient and the lifetime of the least mobile species, which is the OH^- ion.

The same effect has been observed in ice (Petrenko and Chesnakov 1990*d*) using forward-biased ohmic electrodes of the type described in §5.5. Figure 5.20 shows how the conductivity σ_s, measured at 30 Hz between two platinum probes, changed when a d.c. bias of 10 V was switched on or off. The low-frequency conductivity shown is due to injected ionic defects. The high-frequency

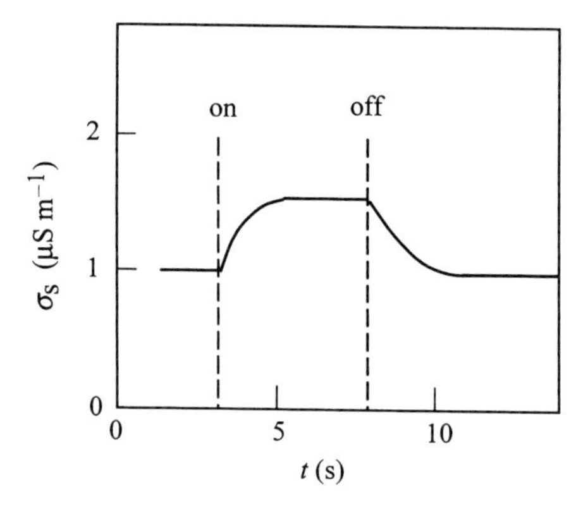

Fig. 5.20 Time dependence of low-frequency conductivity σ_s of pure ice with rectifying electrodes when a d.c. forward bias of 10 V was turned on and off. Temperature $-30\,°$C, specimen thickness 0.5 mm. (From Petrenko and Chesnakov 1990*d*.)

conductivity σ_∞ which is due to the majority charge carriers (Bjerrum defects in this case) was not detectably affected.

In the case of samples of thickness $L \ll \lambda_d$ with ohmic electrodes, the ionic concentration is modified by the current throughout the volume of the ice. This leads to current–voltage (I–V) characteristics which change from ohmic ($I \propto V$) at small currents to quadratic ($I \propto V^2$) for larger currents as illustrated in Fig. 5.21. This can easily be understood as follows. In static equilibrium the rate of thermal generation of pairs of ionic defects, G per unit volume, is in balance with the rate of recombination $\beta n_\pm^2$, where $n_\pm$ is the concentration of ion pairs. If we suppose that when a current density J is flowing excess ions are generated at the electrode at a rate $\alpha J/e$ pairs per unit area, then the balance equation becomes

$$L(G - \beta n_\pm^2) + \alpha \frac{J}{e} = 0. \tag{5.18}$$

The current density is given from eqns. (4.47) and (4.51) by

$$J = \frac{e^2}{e_\pm} n_\pm (\mu_1 + \mu_2) \frac{V}{L}. \tag{5.19}$$

For $\alpha J/e \ll GL$, $n_\pm$ is constant and this gives normal ohmic behaviour, but when $\alpha J/e \gg GL$ these equations give

$$J = \frac{\alpha}{\beta} \frac{e^3}{e_\pm^2} (\mu_1 + \mu_2)^2 \frac{V^2}{L^3}. \tag{5.20}$$

This theory thus accounts for the two regions in Fig. 5.21, and it is the explanation of the points for $L = 10\,\mu$m in Fig. 5.19.

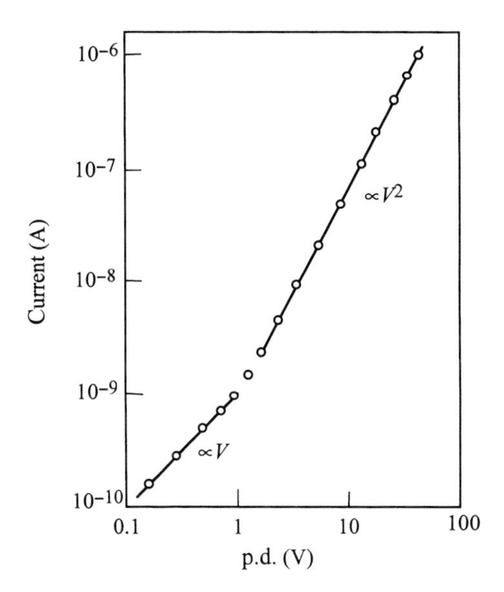

Fig. 5.21 Current–voltage characteristic on logarithmic scales of a 20 μm thick ice specimen with ohmic electrodes. Temperature −30 °C, electrode area 2 cm^2. (From Petrenko and Chesnakov 1990*d*.)

5.8 Thermally stimulated depolarization

Thermally stimulated depolarization (TSD) is a well-established technique for studying dielectric relaxation at temperatures where the process is too slow for the use of a.c. methods. The specimen is polarized in a large static electric field applied between electrodes on its surface, and it is then cooled in the field to a temperature where the polarization is frozen in. The electrodes are then shorted through a low-impedance electrometer which records the depolarization current as the specimen is warmed back to the starting temperature. For a single relaxation process this current will exhibit a peak at a temperature that depends on the heating rate, the relaxation time at this temperature, and its activation energy. The area under the current peak represents the polarization charge released.

In the case of ice polarized near 0 °C and then cooled in liquid nitrogen the thermally stimulated current shows a number of peaks (Bishop and Glen 1969; Johari and Jones 1975; Loria *et al.* 1978; Apekis and Pissis 1987; Zaretskii *et al.* 1987*b*). These indicate that there are several different polarization or depolarization processes involved, which will include the ill-defined space charge polarization at the electrodes. There is a large peak at around 110–120 K which is normally attributed to the Debye relaxation. It can usually be produced in isolation by polarizing at a temperature just above the peak, quenching in liquid nitrogen, and then warming. Figure 5.22 shows this peak as produced in 'pure'

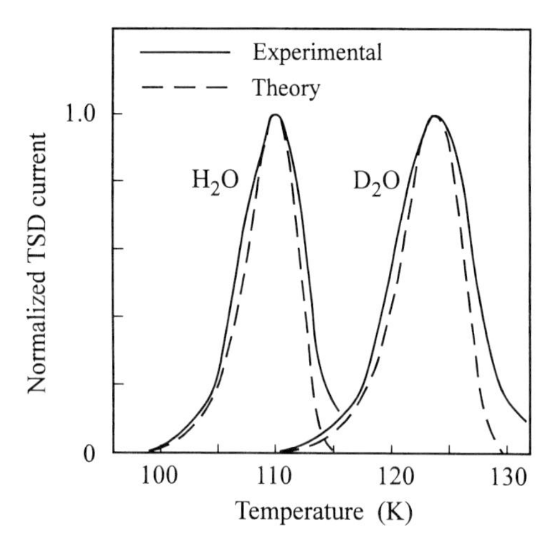

Fig. 5.22 Normalized thermally stimulated depolarization currents for H_2O and D_2O samples, with theoretical curves calculated for a single Debye dispersion. (From Johari and Jones 1975.)

H_2O and D_2O ice by Johari and Jones (1975), with a theoretical function fitted to it which corresponds in the case of H_2O to a relaxation time at 111 K of 580 s with an activation energy of 0.49 eV. This is in agreement with their own measurements using the voltage step technique described in §5.2.1. The magnitude of the relaxation time is compatible with an extrapolation of Fig. 5.3 though the temperature dependence is steeper, but the measurements are in the extrinsic range so that the properties will be specimen dependent. The use of the TSD method to extend the time scale of dielectric measurements is also well demonstrated in the work of Loria *et al.* (1978).

In interpreting TSD experiments on ice it is important to recognize that the relaxation process observed is the decay of the configuration vector $\mathbf{\Omega}$ with the time constant τ_D as discussed in §4.7, but the current observed corresponds to the change of the polarization $\mathbf{P}$. If more than one polarization process is involved there is not a simple correspondence between $\mathbf{P}$ and $\mathbf{\Omega}$. For example, if the specimen has a 'cross-over' from Bjerrum to ionic defects as the majority carriers, a state of $\mathbf{\Omega}$ introduced by the flow of L-defects at the initial high temperature and then frozen in could decay on warming by the motion of H_3O^+ ions at a lower temperature. This would give an initial TSD current in the same direction as the original polarizing field! This effect may have been observed but no detailed account has been published. It does give a warning about the interpretation of the more complicated effects that are reported in this kind of experiment. Equations for the analysis of TSD experiments in ice, taking proper account of the Jaccard theory, are given by Zaretskii *et al.* (1987*b*).

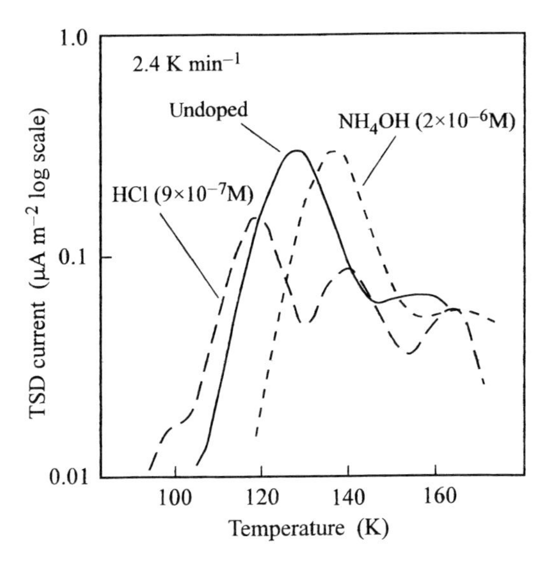

Fig. 5.23 Thermally stimulated depolarization currents for pure and doped ices. (From Zaretskii *et al.* 1987*b*.)

Figure 5.23 shows how the main low-temperature TSD peak is affected by doping the ice with HCl or NH_3. The effect of HCl is to increase the relaxation rate, as would be expected from extrapolation to lower temperatures of the experiments described in §5.4.3.2. The effect of NH_3 in *decreasing* the relaxation rate at low temperatures is a significant addition to the rather confused picture of the role of this dopant (see §5.4.3.3).

A review of the conclusions to be drawn from electrical measurements can be found at the end of Chapter 6.

6
Point defects

6.1 Introduction

A point defect in a crystal is an atomic or molecular site which differs in some way from one of the normal sites of the lattice. The defect is located at this site although it may influence surrounding sites due to any associated electric or elastic strain field. The protonic point defects introduced in Chapter 4 are types of point defect specific to ice-like materials, but the more generally familiar kinds of point defect also occur in ice. They are relevant to a range of properties, particularly where atoms or molecules move from one site to another. There are five categories of point defects in ice:

1. *Molecular defects* in which whole H_2O molecules are displaced from their normal sites. These are the *vacancy*, which is a place where there is a molecule missing from its normal site, and the *interstitial*, where an extra molecule occupies a site in one of the cavities formed by the hydrogen-bonded framework of the ice structure. The largest and thus most probable cavities are centred on the points with co-ordinates $0, 0, \frac{1}{4}$ and $0, 0, \frac{3}{4}$ in Fig. 2.6 (Goto *et al.* 1986; Itoh *et al.* 1996*a*). It is assumed that the interstitial molecules do not form hydrogen bonds to molecules of the normal lattice.
2. *Protonic defects.* These are the two Bjerrum and two ionic defects shown in Fig. 4.9.
3. *Impurity atoms.* These may be substitutional, as in the cases of the fluorine atom replacing an oxygen atom in Fig. 5.4 or the deuterium atom replacing hydrogen to form an HOD molecule with specific vibrational modes as shown in Fig. 3.10. Alternatively the impurity atom may occupy an interstitial site as proposed for K^+ in Fig. 5.11.
4. *Electronic defects* which involve trapped electrons or ionized molecules.
5. *Combined defects* occur where two or more of the above kinds of defect are linked together. A particularly neat example which may be important is the *vested* (or *dressed*) *vacancy* which consists of a vacancy with one or three OH bonds directed into it. If an H_2O molecule were placed in the vacant site an L- or D-defect would be formed so that this structure represents a combination of a vacancy with a Bjerrum defect. Other combinations involving impurities have been introduced in §5.4.2.

Point defects will be present in pure crystals in thermal equilibrium, in which case they are said to be *intrinsic*. They may also be frozen in with non-equilibrium concentrations, and finally they may be introduced with impurities in which case they are described as *extrinsic*. Their most important properties are associated with their movement by the thermally activated jumping of atoms or molecules from one site to another. In this chapter we first present the general theory of equilibrium concentrations and diffusion processes. Subsequent sections deal with the properties of molecular defects, effects involving protonic defects other than the electrical properties described in Chapters 4 and 5, the modelling of the structure of the defects, impurities, and the electronic defects arising from radiation damage. The chapter concludes with a review of our knowledge about point defects based on the whole of Chapters 4, 5, and 6.

6.2 Thermal equilibrium concentrations

The thermal equilibrium concentration of vacancies in a crystal is one of the most elementary calculations in statistical physics. For a crystal consisting of N_0 molecules the state with minimum energy is perfect with a molecule on every site. It requires an additional amount of energy E_v to form a vacancy by removing a molecule from the bulk and replacing it at a kink site (i.e. at a step on a step) on the surface. The number of surface sites is unchanged in this process, but the number of sites inside the crystal is increased by one. There are approximately N_0 ways in which this vacancy can be introduced, and the entropy therefore increases with the number of vacancies. At a finite temperature this makes states with vacancies more probable than those without. For a crystal containing n vacancies the free energy attributable to the defects is

$$F_d = nE_v - TS_c, \tag{6.1}$$

where S_c is the configurational entropy of the n vacancies on the $N_0 + n$ sites given by

$$S_c = k_B \ln \frac{(N_0 + n)!}{N_0!\, n!}. \tag{6.2}$$

The thermodynamic condition for equilibrium is

$$\left(\frac{\partial F_d}{\partial n}\right)_T = 0, \tag{6.3}$$

and using Stirling's approximation ($\ln N \approx N \ln N - N$ for large N) these equations give

$$\frac{n}{N_0 + n} = \exp\left(-\frac{E_v}{k_B T}\right). \tag{6.4}$$

In the following treatment, as in Chapter 4, we will use concentrations per unit volume, defining $N = N_0 + n$ as the number of molecular sites per unit volume. The vacancy concentration n_v is then given by the equation

$$n_v = N\exp\left(-\frac{E_v}{k_B T}\right). \tag{6.5}$$

To be more precise, the introduction of a point defect not only changes the configurational entropy but also affects the modes of vibration of the surrounding lattice. This introduces an extra entropy of formation for each vacancy S_v and leads to an additional term $-nTS_v$ in equation (6.1); equation (6.5) becomes

$$n_v = N\exp\left(\frac{S_v}{k_B}\right)\exp\left(-\frac{E_v}{k_B T}\right). \tag{6.6}$$

It is difficult to estimate S_v, but experiments indicate that (S_v/k_B) is not small and this term must be included in any absolute calculation.

Interstitials may also be introduced from the surface, but equation (6.6) has to be modified to

$$n_i = \frac{zN}{2}\exp\left(\frac{S_i}{k_B}\right)\exp\left(-\frac{E_i}{k_B T}\right). \tag{6.7}$$

The factor $\frac{1}{2}$ arises because the number of interstitial sites in ice is half the number of molecular sites, and the factor z represents the number of possible orientations of an interstitial molecule on its site. This latter factor is not known and can be empirically absorbed into S_i.

Because vacancies and interstitials can be formed independently at surfaces their equilibrium concentrations are independent of one another, but this is not the case for protonic point defects. In pure ice in thermal equilibrium the concentrations of oppositely charged ionic or Bjerrum defects in the bulk must be equal, and it is only possible to define the energies $E_{\pm}$ and E_{DL} to form pairs of these defects. (The special properties of surface layers are considered in Chapter 10.) For ionic defects an extension of the foregoing arguments then leads to the concentrations

$$n_+ = n_- = \frac{2}{3}N\exp\left(\frac{S_{\pm}}{2k_B}\right)\exp\left(-\frac{E_{\pm}}{2k_B T}\right). \tag{6.8}$$

The factor $\frac{2}{3}$ represents the ratio of the four possible orientations of an ion to the six orientations of a molecule shown in Fig. 2.9 (Ryzhkin 1985).

In the case of Bjerrum defects the equation is

$$n_{\mathrm{D}} = n_{\mathrm{L}} = N \exp\left(\frac{S_{\mathrm{DL}}}{2k_{\mathrm{B}}}\right) \exp\left(-\frac{E_{\mathrm{DL}}}{2k_{\mathrm{B}}T}\right), \tag{6.9}$$

in which N is still the concentration of molecules, but the equation conceals two factors of 2. There are in fact $2N$ bonds, but bonds without defects have two orientations whereas the bonds with defects do not.

The introduction of extrinsic concentrations of protonic defects in association with appropriate dopants has been considered in §5.4. In all such cases thermal equilibrium always requires that the products $n_{+}n_{-}$ and $n_{\mathrm{D}}n_{\mathrm{L}}$ retain the values predicted from equations (6.8) and (6.9).

6.3 Diffusion and mobility

A point defect can move to a neighbouring site by a thermally activated jump. An interstitial may just jump into a neighbouring site, but in the case of a vacancy a molecule has to jump into the vacant site leaving a new site vacant. A Bjerrum defect moves by the rotation of a water molecule as indicated in Fig. 4.12, and an ion moves from one site to another by the hopping of a proton along a hydrogen bond. To derive some basic equations describing these processes we will for simplicity consider a generalized defect moving in the one-dimensional potential of period a shown by the upper line in Fig. 6.1. The number of jumps made by the defect per second in a given direction is given by the Arrhenius equation

$$\nu_{\mathrm{h}} = \nu \exp\left(-\frac{E_{\mathrm{m}}}{k_{\mathrm{B}}T}\right). \tag{6.10}$$

Here E_{m} represents the height of the barrier and ν is of the order of the frequency of vibration of the defect in the potential well. In a three dimensional lattice ν involves frequencies of modes of the lattice with the defect both at the position of minimum energy and at the saddle point (Vineyard 1957).

If $n(x)$ is the number of defects per unit volume, the number per unit area in a single plane perpendicular to x is $an(x)$. In a concentration gradient $\mathrm{d}n/\mathrm{d}x$ the net flux of defects per unit area across the plane $x = 0$ in the $+x$ direction will be

$$j = \nu_{\mathrm{h}} a\left(n(0) - \frac{1}{2}a\frac{\mathrm{d}n}{\mathrm{d}x}\right) - \nu_{\mathrm{h}} a\left(n(0) + \frac{1}{2}a\frac{\mathrm{d}n}{\mathrm{d}x}\right) = -a^2\nu_{\mathrm{h}}\frac{\mathrm{d}n}{\mathrm{d}x}. \tag{6.11}$$

The diffusion coefficient D is defined according to Fick's law by $j = -D\,\mathrm{d}n/\mathrm{d}x$, giving

$$D = a^2\nu_{\mathrm{h}}. \tag{6.12}$$

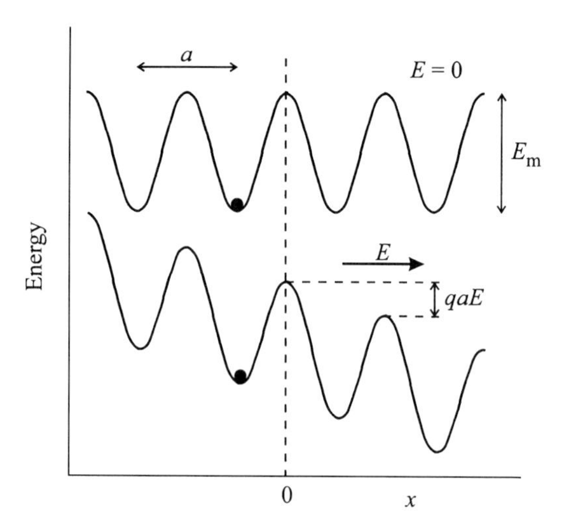

Fig. 6.1 Periodic potential for describing the thermally activated motion of a point defect of charge q, showing the effect of applying an electric field E.

If the defect has a charge q and an electric field is applied in the x direction, the potential in Fig. 6.1 will be modified to that shown by the lower line and the defect will drift in the field. The net flux of defects in the direction of the field is then

$$j = na\left[\nu \exp\left(-\frac{E_{\mathrm{m}} - \frac{1}{2}aqE}{k_{\mathrm{B}}T}\right) - \nu \exp\left(-\frac{E_{\mathrm{m}} + \frac{1}{2}aqE}{k_{\mathrm{B}}T}\right)\right]. \tag{6.13}$$

For $aqE \ll k_{\mathrm{B}}T$ and using (6.10) this becomes

$$j = \frac{na^2 q\nu_{\mathrm{h}}}{k_{\mathrm{B}}T}E, \tag{6.14}$$

and the mobility μ, which is defined as the drift velocity in unit field, is given by

$$\mu = \frac{a^2 q\nu_{\mathrm{h}}}{k_{\mathrm{B}}T}. \tag{6.15}$$

Mobilities are defined as positive irrespective of the sign of q. Comparison with (6.12) gives the well-known Einstein relation:

$$\mu = \frac{|q|D}{k_{\mathrm{B}}T}. \tag{6.16}$$

In the three-dimensional case equation (6.12) is subject to a geometrical factor that depends on the lattice, the type of defect and the precise definition of ν_{h}, but the Einstein relation (6.16) is strictly valid. For a fuller treatment the reader should consult books on diffusion theory (a relatively straightforward example is Girifalco 1973). For the interpretation of experimental observations on ice the diffusion coefficient of a particular defect of type i (equal to 1 to 4, 'v', or 'i' for protonic defects, vacancies, or interstitials) may be simply written as

$$D_i = D_{i0} \exp\left(-\frac{E_{im}}{k_B T}\right), \tag{6.17}$$

without further interpretation of the term D_{i0}.

6.4 Molecular defects

6.4.1 *Self-diffusion*

The term 'self-diffusion' refers to the diffusion of isotopically labelled molecules of the host material, and this may occur by either a vacancy or an interstitial mechanism. For the vacancy mechanism the motion of a particular molecule depends on the probability n_v/N that a site adjacent to it is vacant times the frequency of it jumping into that site. The self-diffusion coefficient is

$$D_s = f \frac{n_v}{N} D_v, \tag{6.18}$$

in which n_v/N is given by equation (6.6), D_v is given by equation (6.17), and f is a correlation factor arising from the probability that a defect may return to its original site at the next jump. For vacancies in ice or the diamond structure $f = \frac{1}{2}$ (Compaan and Haven 1956). For the interstitial mechanism the correlation factor does not arise and

$$D_s = \frac{n_i}{N} D_i. \tag{6.19}$$

In either case experimental results may be expected to fit the equation

$$D_s = D_{s0} \exp\left(-\frac{E_s}{k_B T}\right), \tag{6.20}$$

where $E_s = E_v + E_{vm}$ for vacancies and $E_s = E_i + E_{im}$ for interstitials.

Self-diffusion coefficients are determined experimentally by placing a tracer on one face of a block of material which is then held at a fixed temperature for a certain time. The block is then cut into thin sections and the quantity of the tracer in each section is determined as a function of the depth of penetration. Single crystals should be used because it is known that enhanced diffusion may occur along grain boundaries. In an early experiment on ice at approximately −2 °C Kuhn and Thürkauf (1958) showed that the diffusion coefficients of ^{2}H and ^{18}O were the same, and this indicates that self-diffusion occurs by the migration of whole molecules.

The simplest isotope to use is tritium (^{3}H) because it is radioactive, and diffusion coefficients of this isotope have been measured over the temperature range −5 to −30 °C. Typical results are those of Ramseier (1967) who gives the mean values:

$$D_s(-10\,^{\circ}\mathrm{C}) = (1.49 \pm 0.06) \times 10^{-15}\,\mathrm{m^2\,s^{-1}},$$
$$E_s = 0.62 \pm {\sim}0.04\,\mathrm{eV}.$$

Figure 6.2 shows an Arrhenius plot of this line together with a compilation of values from earlier experiments (e.g. Dengel and Riehl 1963; Itagaki 1967). There are some indications that the diffusion coefficient perpendicular to the c-axis is about 10% higher than that parallel to this axis. Delibaltas *et al.* (1966) showed that over this temperature range the diffusion coefficient of ^{18}O is within

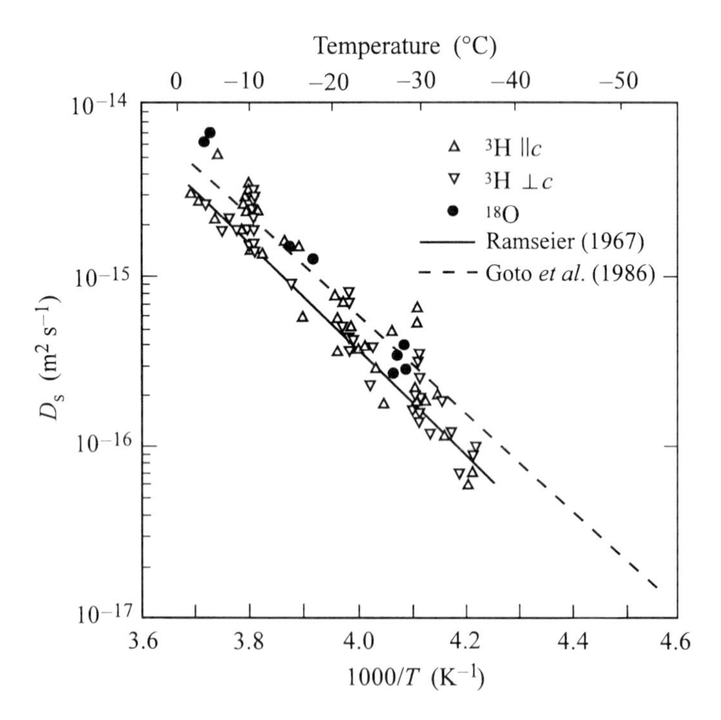

Fig. 6.2 Compilation of measurements of the self-diffusion coefficient D_s for pure ice as a function of temperature, together with the function (broken line) derived from topographic experiments on prismatic dislocation loops. (From Goto *et al.* 1986, with changes.)

experimental error the same as that determined for ^3H, and experiments on ice doped with HF or NH_4F (Blicks *et al.* 1966; Dengel *et al.* 1966) have shown no effect of doping down to $-27\,^\circ C$.

Brown and George (1996) determined the bulk diffusion coefficient of ^{18}O in ultra-thin single crystals of ice (vapour-deposited multilayers) between 155 and 165 K, obtaining

$$D_s(160\,K) = (1.5 \pm 0.5) \times 10^{-19}\,m^2\,s^{-1},$$
$$E_s = 0.72 \pm 0.07\,eV.$$

This value of D_s is about 150× larger than that obtained by extrapolation of the data in Fig. 6.2, and the reason for the difference is not yet understood. Subsequent measurements of the diffusion of HOD in H_2O multilayers by Livingston *et al.* (1998) gave values of D_s that were about 1.7× larger than those for ^{18}O.

The close similarity of the diffusion coefficients for the isotopes of hydrogen and oxygen confirm that self-diffusion is due to the migration of whole molecules. The hopping of protons from one molecule to another does not provide a faster mechanism for the diffusion of hydrogen, though the results of Livingston *et al.* (1998) have been interpreted as showing evidence of some D—H exchange associated with the molecular diffusion.

Any hopping of hydrogen isotopes between molecules must take place by the motion of protonic point defects, and the rate can be estimated from our knowledge of these defects as described in Chapters 4 and 5. Self-diffusion requires both ionic and Bjerrum defects, because separately they only reverse bonds or rotate molecules. The diffusion coefficient therefore depends on the minority carriers, which are the H_3O^+ ions ($i = 1$) in pure ice. By analogy with equation (6.19) the diffusion coefficient of hydrogen by this mechanism will be

$$D_s = \frac{n_1}{N} D_1. \tag{6.21}$$

Using the Einstein relation (6.16) and equations (4.47) and (4.51) this becomes

$$D_s = \frac{\sigma_s k_B T}{N e^2}. \tag{6.22}$$

Taking σ_s for 'pure' ice at $-10\,^\circ C$ to be of the order of 10^{-8} S m^{-1}, this gives $D_s \sim 5 \times 10^{-20}\,m^2\,s^{-1}$, which is four to five orders of magnitude smaller than that observed experimentally and attributed to interstitials. Reference to Fig. 5.5 shows that even in HF-doped crystals protonic defects will not make an appreciable contribution to the self-diffusion of hydrogen.

At this point we should note that the molecular and protonic diffusion processes operate quite independently. It is obvious that the motion of ionic and Bjerrum defects cannot exchange the positions of oxygen atoms, but the converse is also true. If an interstitial molecule replaces a molecule on a lattice site, or if a molecule jumps from one site into a vacancy, the Pauling configuration of the hydrogens on the bonds of the lattice is unaltered. Molecular diffusion cannot contribute to dielectric polarization. The only exception would be if combined protonic and molecular defects are involved; we will conclude in §6.11 that such combined defects probably exist in ice, though we do not know whether they contribute significantly to diffusion.

6.4.2 *Interstitials*

Tracer experiments provide no evidence concerning whether self-diffusion occurs by a vacancy or an interstitial mechanism. This question was resolved in a series of X-ray topographic studies of dislocation loops in ice by Goto *et al.* (1982, 1986), which are fully reviewed and beautifully illustrated by Oguro *et al.* (1988). The properties of dislocations in ice and their study by X-ray topography are described in Chapter 7. If a dislocation-free crystal of ice is rapidly cooled many small prismatic dislocation loops are formed like those in the topograph in Fig. 6.3. These loops appear because the point defects present in equilibrium at the higher temperature cannot escape to the surface, and instead condense to form dislocation loops on the basal plane. If dislocations are already present they act as sinks for point defects and their structures are visibly modified by climb. The types of prismatic loops that would be formed by vacancies and interstitials are illustrated schematically in Figs 6.4(a) and (b) respectively. From the properties of the topographic images and from the shrinkage of the loops under a compressive stress perpendicular to their plane, it was established that the loops

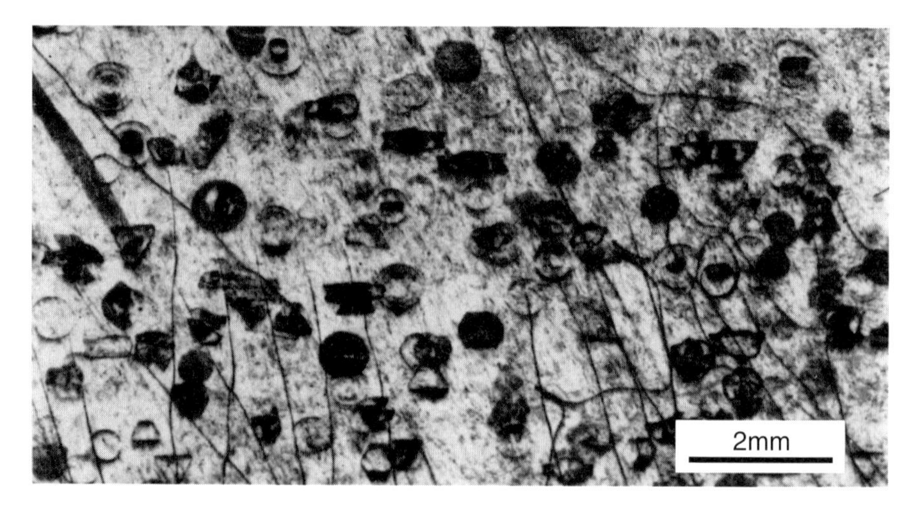

Fig. 6.3 X-ray topograph on (0001) plane showing many small faulted prismatic dislocation loops generated by cooling a crystal of ice of initially low dislocation density. (From Oguro *et al.* 1988.)

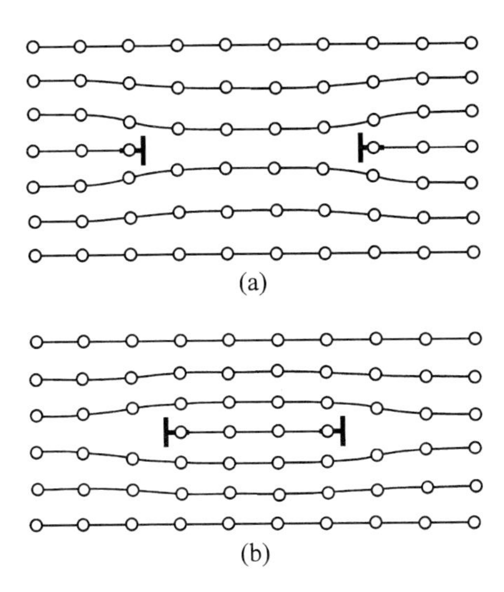

Fig. 6.4 Cross sections through prismatic dislocation loops in a simple lattice: (a) vacancy-type loop and (b) interstitial-type loop.

Table 6.1 Parameters for interstitial diffusion derived from X-ray topographic studies of prismatic dislocation loops

Energy of formation of interstitial E_i	0.40 eV
Entropy of formation S_i	$4.9k_B$
Activation energy for motion E_{im}	0.16 eV
Pre-factor for diffusion coefficient of interstitials D_{i0}	$1.8 \times 10^{-6}\,m^2\,s^{-1}$
Concentration of interstitials at 0 °C n_i/N	2.8×10^{-6}
Predicted activation energy for self-diffusion $E_s = E_i + E_{im}$	0.56 eV
Activation volume for motion γ_{im}	$\approx 0.40 V_{H_2O}$†
Activation volume for formation γ_i	$\approx -0.26 V_{H_2O}$†

Data from Goto *et al.* (1986) and Hondoh (1992).
† V_{H_2O} = molecular volume = 32.53 Å^3.

formed on cooling are of interstitial character, and that the dominant point defects at temperatures above −40 °C are therefore interstitials. The number and size of the loops formed on cooling from various temperatures yielded the equilibrium concentrations of interstitials, and the dynamics of the process yielded their diffusion coefficient D_i. The parameters finally derived are listed in Table 6.1, and from these values the coefficient of self-diffusion by the interstitial process can be calculated. The activation energy for self-diffusion is $E_s = 0.56$ eV and the values are plotted as the broken line in Fig. 6.2. The close agreement between this line and the self-diffusion coefficients determined by tracer techniques confirms that self-diffusion is an interstitial process over the temperature range where experimental observations are possible. Preliminary observations of the effect of hydrostatic pressure on interstitial diffusion are reported by Hondoh *et al.* (1991), leading to the activation volumes (see §6.5.2) included in Table 6.1.

6.4.3 *Vacancies*

Ice is expected to contain vacancies even though interstitials predominate, but it is difficult to obtain information about them. The only experiments specifically interpreted as evidence for vacancies are those by Mogensen and Eldrup (1978) and Eldrup *et al.* (1978) using the positron annihilation technique. When positrons are injected into a material they form positronium 'atoms' consisting of a positron and an electron. These are trapped and then decay by the emission of characteristic γ-rays. In ice the trapping centres are believed to be vacancies, but the interpretation of the experiments involves many assumptions. The papers estimate that $E_{vm} = 0.34 \pm 0.07$ eV and $E_v \approx 0.2$–0.35 eV, but in the light of subsequent evidence that interstitials are dominant with $E_i = 0.4$ eV this estimate of E_v is clearly too small.

6.4.4 *Modelling*

The creation of a vacancy involves the breaking of four hydrogen bonds and the formation of two others as the molecule is replaced at a kink site on the surface. This will be followed by relaxation of the molecules around the vacancy so that the energy of formation E_v should be somewhat less than the lattice energy per molecule given in §2.4 as 0.611 eV. To create an interstitial this energy of 0.611 eV is required to remove the molecule from the surface, but it is hard to predict how much energy is gained when the molecule is then bound without hydrogen bonding in its interstitial site. An early attempt by Cotterill *et al.* (1973) to calculate the formation energies E_v and E_i gave values which are too large. More recently Itoh *et al.* (1996*a*) have carried out molecular dynamics simulations of self-interstitials in ice. There are two kinds of interstitial site in the Ih structure: the larger, which they designate Tu (uncapped trigonal), at the points $0, 0, \frac{1}{4}$ or $0, 0, \frac{3}{4}$ in Fig. 2.6, and the other Tc (capped trigonal) at $\frac{2}{3}, \frac{1}{3}, \frac{1}{4}$ or $\frac{1}{3}, \frac{2}{3}, \frac{3}{4}$. They show that the larger, Tu, site is the stable one with an activation volume $\gamma_i = -0.20\,V_{H_2O}$. The negative sign means that the crystal shrinks when a molecule is transferred from the surface to an interstitial site, but the fact that $|\gamma_i| \ll V_{H_2O}$ shows that the lattice is greatly expanded around the interstitial molecule. The molecule rotates freely in its site about its axis of symmetry. The molecular interactions used in this model give elastic constants which are too large and make it unrealistic to calculate the energies of formation or motion.

6.5 Protonic point defects

The nature of the protonic point defects, which are unique to ice-like materials, has already been described in Chapters 4 and 5 in connection with their role in the electrical properties of ice. However, there are a number of other properties of ice which depend on these defects. These are described here, followed by a discussion of the nature of the defects on an atomic scale.

6.5.1 *Anelastic relaxation*

When a uniaxial stress is applied to ice the equivalence of the six possible orientations of a molecule is destroyed. A small strain additional to the elastic distortion of the lattice can then result from a favourable rearrangement of the molecular orientations. Just as in dielectric relaxation this rearrangement can only occur with the aid of protonic point defects, and there is a corresponding relaxation time τ_m which is closely related to the Debye dielectric relaxation time τ_D. The phenomenon is known as 'anelastic relaxation'. It is one of the processes contributing to internal friction (see also §8.3.5). The effect is studied by setting a specimen of ice into longitudinal, flexural or torsional mechanical resonance and observing the logarithmic decrement δ of this mode. On changing the temperature and thus τ_m the decrement passes through a maximum at the point where the resonant frequency is equal to $1/(2\pi\tau_m)$. An example of such an anelastic loss peak from the experiments of Schiller (1958) is shown in Fig. 6.5. This figure also

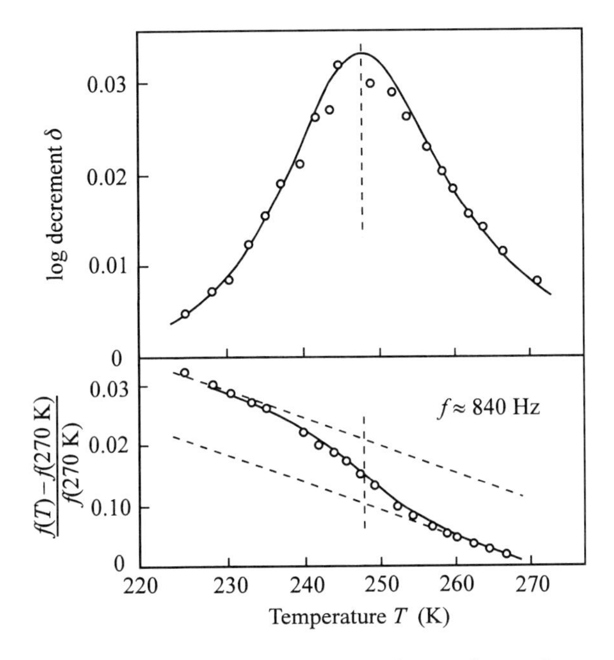

Fig. 6.5 Anelastic relaxation loss peak of a specimen of pure ice as a function of temperature, and the corresponding shift in the resonant frequency $f(T)$ of the specimen. Reproduced with permission from *Zeitschrift für Physik*, **153**, P. Schiller, Die mechanische Relation in reinen Eiseinkristallen, pp. 1–15, Fig. 5, Copyright 1958 Springer-Verlag.

Table 6.2 Temperature dependence of anelastic relaxation time τ_m in pure ice

Reference	Temperature range (K)	$10^4\tau_m$ at 253 K (s)	Activation energy (eV)
Schiller (1958)	250–268	1.1	0.58
Kuroiwa (1964)	237–253	1.5	0.57
Oguro *et al.* (1982)	240–258	1.2	0.60

shows the frequency of resonance. There is a steady fall in frequency with increasing temperature, which arises from the normal fall of the elastic modulus, together with a drop across the relaxation peak caused by the molecular rearrangements that occur when the relaxation time becomes short enough for them to do so. From the symmetry between the possible molecular orientations shown in Fig. 2.9, there should be no such effect for stresses along the [0001] axis. This is confirmed experimentally, the maximum effect being for stresses at 60° to this.

The results of three investigations of the temperature dependence of τ_m in pure ice are shown in Table 6.2, which should be compared with the electrical measurements of τ_D given in Table 5.1. All these data were obtained in the range where Bjerrum defects are intrinsic, and, although no measurements have been reported of both τ_m and τ_D on the same specimens, it is clear that the two quantities are effectively the same. The experiments of Kuroiwa (1964) include both D_2O and doped specimens, in which τ_m changes in much the same way as for dielectric relaxation. Tatibouet *et al.* (1983) have extended anelastic measurements to lower temperatures with frequencies as low as 10^{-4} Hz. The activation energy is then lower and τ_m is sensitive to the state of the specimen, but the observations correspond roughly to the dielectric data in the low temperature (extrinsic) portion of Fig. 5.3.

The theory of anelastic relaxation in ice was developed by Bass (1958) and by Onsager and Runnels (1969) who predicted that $\tau_D/\tau_m = \frac{3}{2}$, which was consistent with the experimental evidence at the time. However, their theory included several assumptions which affected the value of this ratio. A fuller treatment by Petrenko and Ryzhkin (1984*c*) of both the magnitude and dynamics of the dispersion predicted that for relaxation by Bjerrum defects $\tau_D/\tau_m \approx 2$, but the experimental results on pure ice appear to be closer to $\tau_D/\tau_m = 1$.

6.5.2 *Effects of pressure—activation volumes*

In §5.4.1 we noted that the dielectric properties of ice are sensitive to hydrostatic pressure, so that the application of a pressure could take a specimen through the cross-over condition (Fig. 5.7). This pressure dependence arises because the formation, or even the motion, of a defect affects the volume of the ice and thus makes the corresponding activation energy dependent on pressure. For the formation energy of, say, an interstitial defect one can write, to first order in the pressure p,

$$E_i(p) = E_i(0) + \gamma_i p, \tag{6.23}$$

where γ_i is called the 'activation volume'. Substituting in equation (6.7) shows that

$$\gamma_i = -k_B T \frac{\partial \ln n_i}{\partial p}, \tag{6.24}$$

and similar activation volumes can be defined for all concentrations or other measurable parameters such as the conductivities σ_∞ and σ_s.

For pure ice the intrinsic high-frequency conductivity decreases with increasing pressure, and the activation volume γ_{σ_∞} is found fairly consistently to be about $0.15V_{H_2O}$, where V_{H_2O} ($=3.25 \times 10^{-29}\,m^3$) is the volume per molecule (Chan *et al.* 1965; Taubenberger *et al.* 1973; Hubmann 1978). The conductivity in this case is dominated by intrinsic L-defects, so that

$$\gamma_{\sigma_\infty} = \tfrac{1}{2}\gamma_{DL} + \gamma_{Lm}, \tag{6.25}$$

where γ_{DL} is the activation volume for the formation of a Bjerrum defect pair, and γ_{Lm} is that for the motion of an L-defect. The activation volume for the formation of D-defects is expected to be positive because the two protons are brought close together and will push the lattice apart. An L-defect is also expected to expand the lattice because opposing oxygen atoms will repel one another in the absence of the hydrogen bond to bind them together. Taubenberger *et al.* (1973) and Hubmann (1978) also report the pressure dependence of the conductivities of HF- and NH_3-doped crystals, but the interpretation of these measurements depends on how the extrinsic concentrations are affected by pressure.

A different approach which yields the activation energies of formation alone was devised by Evtushenko and Petrenko (1991). Because the concentration of defects depends on pressure, the defects will in a pressure gradient experience an effective force $\gamma_i \nabla p$ which must be included in equation (4.45). In equilibrium this force will be balanced by an electric field set up by the displacement of some defects to the surfaces. The effect is known as a pseudo-piezoelectric effect; there is of course no true piezoelectric effect in a material with the symmetry of ice Ih. In their experiment Evtushenko and Petrenko produced a pressure gradient by bending a specimen elastically and observed the time evolution of the resulting p.d. across the specimen. From their observations on doped specimens they determined the differential activation volumes as defined by equation (6.24) for the formation of all four types of defect as given in Table 6.3.

6.5.3 D—H *isotopic exchange*

A novel technique which demonstrates the distinct roles of ionic and Bjerrum defects in ice has been developed by Devlin and co-workers (reviewed by Devlin (1990)). It is based on the optical detection of the exchange of protons and

Table 6.3 Activation volumes γ_i of formation of protonic defects as determined from the pseudo-piezoelectric effect

Dopant	Defect	γ_i/V_{H_2O}
HF	H_3O^+	1.24 ± 0.11
HCl	H_3O^+	1.04 ± 0.04
HF	L	0.31 ± 0.01
HCl	L	0.42 ± 0.09
NH_3	OH^-	0.26 ± 0.13
NH_3	D	0.22 ± 0.11

From Evtushenko and Petrenko (1991).

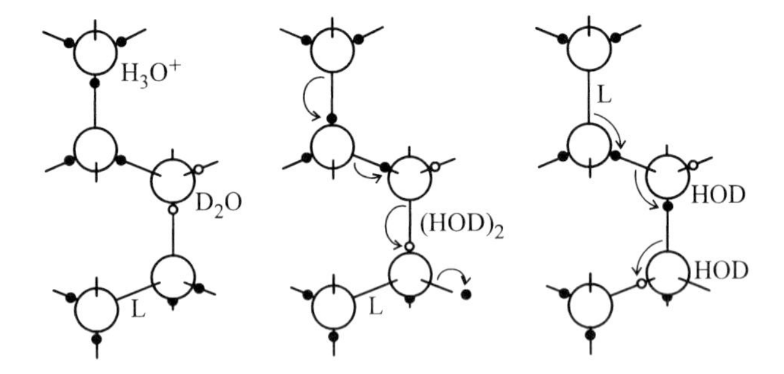

Fig. 6.6 The process of dissociation of a D_2O molecule into two HOD molecules in ice by the passage of first an ionic and then a Bjerrum defect.

deuterons between the water molecules in ice. Such exchange requires the motion of protonic defects, and Fig. 6.6 illustrates how a D_2O molecule can be broken up into two separated HOD molecules. In the first step an ionic defect (in this case H_3O^+) passes along a chain of bonds causing one of the deuterons to hop along its bond forming a pair of HOD molecules *with the deuteron on the bond joining them*; such a pair is denoted $(HOD)_2$. In the next step a Bjerrum defect (in this case L) turns one of these HOD molecules so that the deuterons are now separated by two oxygens. The deuterons can then move further apart by the alternate passage of ionic and Bjerrum defects.

As explained in §3.4.4 the replacement of a few protons by deuterons in H_2O ice introduces localized modes which produce relatively sharp lines in the infrared and Raman spectra. Ritzhaupt *et al.* (1978) have identified three distinct line profiles in the infrared spectrum of the OD-stretch mode (μ_3) corresponding to D_2O, $(HOD)_2$, and separated HOD molecules; these are shown in Fig. 6.7. In their experiments thin films of H_2O ice containing a few percent of D_2O molecules are formed by vapour deposition at about 125 K. At this temperature the ice

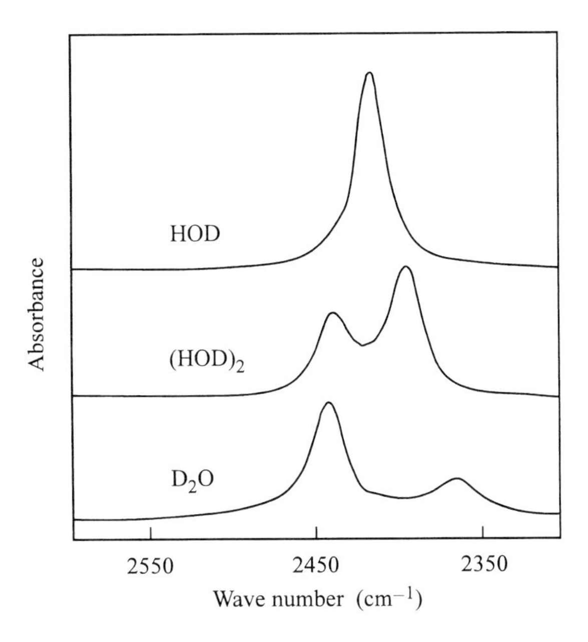

Fig. 6.7 Components of the infrared absorption spectrum of H_2O ice at 90 K arising from D_2O, $(HOD)_2$, and HOD. (From Devlin 1992.)

formed has the Ic structure (§11.8), but the difference from ice Ih is not significant for the present purpose. On warming the ice, protonic defects become mobile and the spectrum changes as the D_2O molecules break up into HOD. For 'pure' ice both $(HOD)_2$ and HOD spectra are produced at around 140 K showing that both hop and turn steps occur at similar rates. Measurements by Collier *et al.* (1984) over the temperature range 135–150 K indicated that the activation energies of the ionic and Bjerrum defect steps in Fig. 6.6 are 0.41 ± 0.03 eV and 0.52 ± 0.02 eV respectively.

The contributions from the two types of defect can be manipulated in various ways. The addition of the organic base 7-azaindole reduces the concentration of H_3O^+ ions and thus the rate of the hop step by several orders of magnitude, with the result that conversion of D_2O to HOD occurs much more slowly and the intermediate $(HOD)_2$ pairs never reach detectable concentrations (Ritzhaupt and Devlin 1980). These experiments imply that it is H_3O^+ rather than OH^- ions which are active in the reorientation process. The converse effect is that electron injection generates mobile ionic defects which form $(HOD)_2$ at 90 K with no further dissociation to isolated HOD (Devlin and Richardson 1984). In a further experiment Wooldridge and Devlin (1988) deposited the ice Ic with a small trace of 2-napthol. On irradiation with 184.9 nm ultraviolet radiation this dopant releases protons into the ice, but at 90 K these produced no conversion of D_2O to $(HOD)_2$. However, after the radiation had been stopped and the ice was warmed into the temperature range 113–130 K $(HOD)_2$ was produced in equilibrium with D_2O in the statistically expected proportion of 4 : 1. They interpret these observations as evidence for shallow traps for the ionic defects. On release from the 2-naphthol the ions become quickly and firmly held in these traps, but on warming they are released to move through the lattice and dissociate D_2O. Wooldridge and Devlin suggest that the traps are L-defects and estimate the trap depth as 0.43 eV. At 130 K the $(HOD)_2$ start to break up into separated HOD molecules. In these experiments the hop and turn processes are clearly separated, and this yields a more reliable determination of the activation energy of the turn step in ice Ic as 0.52 ± 0.03 eV. From other evidence involving NH_3-doping they deduce that the L-defects are much more mobile than the D-defects at these temperatures (Devlin 1992). The relaxation times of the order of 1200 s observed at 145 K are reasonably compatible with a very long extrapolation of the *intrinsic* Debye relaxation times determined in dielectric experiments (Fig. 5.3).

The evidence for trapping of protons or H_3O^+ ions in ice is directly relevant to the interpretation of experiments on proton injection at H-loaded Pd electrodes (§5.6.3) and is supported by the pulse radiolysis experiments to be described in §6.9. A range of possible implications of the D—H exchange experiments is considered by Devlin (1992), and we will return to some of these issues in §6.11.

6.5.4 *Atomic structure and hopping of protonic point defects*

The formal structures of the protonic defects shown in Fig. 4.9 give little indication of the actual atomic positions, and considerable relaxation is expected to minimize the energy. The extreme case is that of the D-defect, for which two protons ~0.7 Å apart on the straight line between two oxygen atoms would have a large electrostatic energy that would be a maximum with respect to rotations of the molecules. Relaxation of this structure is expected to expand the O—O distance and move the protons off the O—O line as shown in Fig. 6.8(a). The plane in which the protons are displaced will depend on the location of the protons on the six neighbouring bonds, and the possibilities are different for a defect on a bond parallel to the *c*-axis as in the figure (the 'eclipsed' or 'mirror' configuration) and one on a bond oblique to this (the 'staggered' or 'centre' configuration). In moving the defect from the bond between A and B to that between B and C the molecule B rotates through 120° passing through the mid-way position shown in Fig. 6.8(b). It is quite possible that this arrangement has a lower energy than that in (a), and Dunitz (1963) has referred to this state of a D-defect as an X-defect. Because of the potentially high energy of a D-defect and the rather uncertain evidence about it, some workers feel that D-defects may only occur in combination with other defects. Haas (1962) proposed combination with an interstitial, but a more promising possibility is the 'positive vested vacancy' DV in Fig. 6.8(c) (Onsager and Runnels 1963). In this figure there are three OH bonds pointing into the vacant site, so that if a normal H_2O molecule were inserted a D-defect would

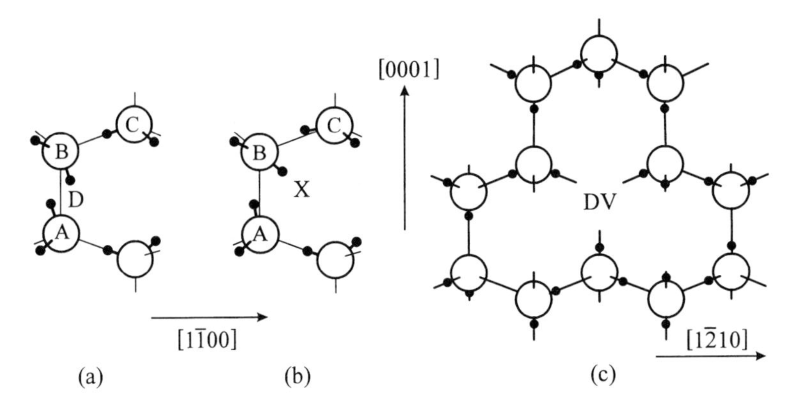

Fig. 6.8 Postulated configurations of the D-defect: (a) defect on bond AB with protons displaced off the bond to minimize the energy, (b) the X-defect configuration with molecule B rotated to a position mid-way between bonds AB and BC, and (c) the defect in combination with a molecular vacancy to form a 'positive vested vacancy'.

be created. This combined defect is therefore what will be formed by the trapping of a D-defect at a vacancy.

Modelling the protonic defects presents a difficult theoretical challenge because the energy of a relatively small cluster of molecules containing the defect has to be minimized by allowing displacements of the molecules. The extent of these displacements is governed by the constraints assumed for the molecules on the edge of the cluster. For example, a free cluster is quite different from one embedded in an ice matrix or even in an elastic continuum designed to resemble ice. The Bjerrum defects have been modelled in a 27 molecule cluster by Hassan and Campbell (1992) using empirical potentials devised in their group. They have demonstrated the importance of relaxations, which in their cluster involved displacements of up to 0.5 Å and rotations through large angles in both L- and D-defects. Plummer (1992) describes *ab initio* modelling of the Bjerrum defects, but in clusters which had to be limited to only eight molecules, which is far from realistic for bulk ice. We do not quote numerical results from either of these calculations as they have to be interpreted in the context of the specific assumptions being made.

The existence of the ionic species H_3O^+ and OH^- is well known from the liquid, but the difference between ice and the liquid is that whereas the ions in the liquid can migrate as entities, in ice the ionic state can only move from one oxygen atom to another by a proton hopping along the bond between them. The experiments described in Chapter 5 show that the mobilities of both positive and negative ions in ice are almost independent of temperature at least down to 150 K, and possibly lower depending on the interpretation of the slope of the plot of relaxation times in KOH-doped ice shown in Fig. 5.10. Although there have been attempts to model the formation of ion pairs in clusters of molecules (Plummer 1997), the greatest theoretical interest is in understanding the process of proton transfer. We illustrate this topic with reference to Fig. 6.9, which shows the results of calculations by Scheiner (1981) for the energy of an isolated $[H_2O—H—OH_2]^+$ ion as a function of the position of the proton on the central bond for various fixed separations r_{OO} of the oxygen atoms. For large separations there are clearly two degenerate states with an identifiable H_3O^+ ion at one end of the bond or at the other, but as the molecules get closer the barrier at the centre of the double potential well decreases and eventually the equilibrium state has the proton at the centre of the bond. For realistic separations the energy barriers are very small and such calculations are extremely sensitive to the model used (Luth and Scheiner 1992; Xie *et al.* 1994; Tuckerman *et al.* 1997). In the case of ice r_{OO} is determined by the surrounding lattice with appropriate relaxation around the ion. Lee and Vanderbilt (1993) attempted this calculation for an eight-molecule unit cell of the ice Ic structure with periodic boundary conditions, and with the ions introduced by adding one molecule of HF or NH_3. They obtained double potential wells with central barriers of heights of the order of 0.1 eV, but the actual values cannot be treated with any confidence.

The conventional model of distinct H_3O^+ and OH^- defects in ice presupposes that there is a barrier for proton migration along the bonds adjacent to

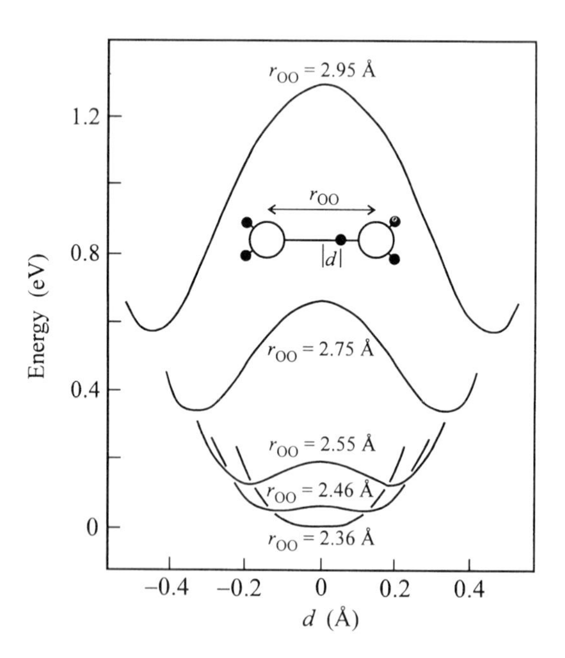

Fig. 6.9 Calculated energy of the free $[H_2O—H—OH_2]^+$ ion as a function of the distance d of the proton from the middle of the bond for various oxygen separations r_{OO}. (From Scheiner 1992.)

the defect, and the interesting question about the mobility of the defects is whether the protons cross this barrier by thermal activation or by quantum-mechanical tunnelling. The probability of a proton of mass m tunnelling through a barrier of height V_0 and effective thickness d involves a factor of the form $\exp[-2(2mV_0)^{1/2}d/\hbar]$, in which V_0 must be the value without relaxation of the surrounding molecules at the saddle point. Taking plausible values of $V_0 = 0.1$ eV and $d = 0.5$ Å, this factor is e^{-7}. The Arrhenius factor $\exp[-V_0/k_B T]$ for a thermally activated process would be equal to this at 170 K, and so both mechanisms require serious consideration. If the proton were replaced by a deuteron the tunnelling factor would change to $e^{-7\sqrt{2}}$, a decrease by a factor of 20; this would be quite compatible with the observations of Kawada (1989*a*) on KOH-doped D_2O ice referred to in §5.4.3.4. Onsager and co-workers have studied the difficult problem of formulating a quantum-mechanical theory to describe ionic migration by tunnelling in the branched and disordered network of molecules that represents ice (Chen *et al.* 1974). It must also be recognized that the concept of the proton hopping in the double potential well is a simplification. The potential well in Fig. 6.9 is sensitive to the O—O distance, and the oxygen atoms are in thermal vibration with amplitudes of the order of 0.2 Å (Table 2.7). Such displacements will cause gross fluctuations in the height of the barrier, so that the group of atoms should really be treated as a whole.

Our whole treatment so far has made the conventional assumption that ionic defects have the form H_3O^+ or OH^-. We can now see that this may not be the case, for if there were in fact no barrier at the centre of the bond in Fig. 6.9 the equilibrium state of an ionic defect would be with the proton at the centre of the bond. The positive ion would then be equivalent to a proton placed at the centre of an L-defect, and likewise the negative ion would be a proton taken out of a D-defect. The effective charges would still satisfy equation (4.28). An ion of this type would move from one bond to the next by the correlated motion of two protons, and the H_3O^+ or OH^- configurations would represent the saddle points for this process. The same questions about thermal activation or tunnelling would arise, and some equations in the theory might have to be modified by factors of the order of 2, but it is hard to see how these possibilities could be distinguished experimentally. This alternative structure for an ionic defect is analogous to the X-defect form of the D-defect (Fig. 6.8(b)) in that the minimum energy configuration is the state which has been conventionally assumed to be the saddle point for migration.

Much theoretical interest has been shown in a further development of these models in which an ionic defect (or even a Bjerrum defect) is delocalized over a linear chain of molecules as shown for a negative ion in Fig. 6.10. This 'soliton' model was introduced by Antonchenko *et al.* (1983) and is described by Davydov (1991). The model is of a dynamic defect which would be able to move very freely rather than by a series of distinct jumps, but it requires strong correlations between the protons on adjacent bonds. Atomistic modelling of linear chains of H_2O molecules by Godzik (1990) and Scheiner (1992) indicates that the coupling

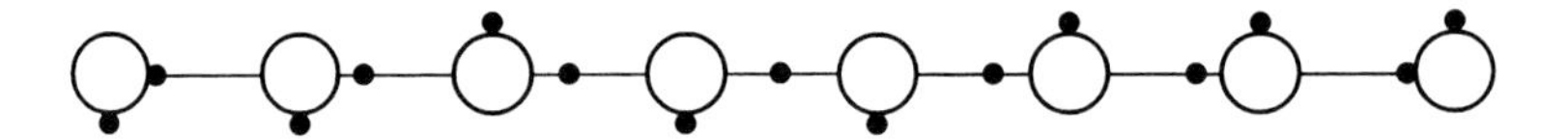

Fig. 6.10 The soliton model of an OH^- ion on a linear chain of water molecules. As the defect travels to the right the protons on each bond move slightly to the left.

is not sufficiently strong to support solitonic motion in ice, and Nagle (1992) has pointed to the difficulty of applying the concept to the branching and disordered network of bonds in ice. A detailed analysis by Zolotaryuk *et al.* (1998) has shown that free solitonic motion is not possible in two-dimensional square ice. All the issues discussed in this section, which can be examined very thoroughly for the case of ice, are highly relevant to other systems containing chains of hydrogen bonds, such as biological membranes (Nagle 1983).

6.6 Nuclear magnetic resonance

The use of nuclear magnetic resonance (n.m.r.) to provide information about the actual structure of ice was described in §2.5.1, but the technique can also be used to determine the time scale for diffusive motion of the protons. It is a non-destructive technique and does not require the use of electrodes, but the interpretation is not straightforward. In a steady magnetic field B_0 the protons have energy levels $\pm\frac{1}{2}\gamma\hbar B_0$, where γ is the gyromagnetic ratio of the proton, and in thermal equilibrium these levels are populated according to the Boltzmann factor $\exp(\mp\gamma\hbar B_0/2k_B T)$. If the field is abruptly changed the protons initially remain in the same spin states, and the populations of these states subsequently approach thermal equilibrium with the lattice with a spin–lattice relaxation time T_1. In a completely static system T_1 would be effectively infinite, because no mechanism exists to flip the proton spins from one state to the other. In practice the transition is driven by fluctuations in the magnetic field at one proton due to the magnetic moments of the others as the protons hop from one site to another in the lattice. The mean interval between hops by a given proton is the 'spin correlation time' τ_c. The fluctuating field has a broad spectrum, but only the Fourier component at the Larmor frequency $\omega_0 = \gamma B_0$ induces transitions between the spin states of the protons. The relaxation time T_1 will be a minimum for a magnetic field at which $\omega_0\tau_c \approx 1$, but for ice such fields or resonance frequencies are outside the experimentally accessible range. In typical experiments the r.f. frequency used is $\omega_0/2\pi \sim 10$ MHz, corresponding to $B_0 = 0.235$ T, and in such experiments $\omega_0^{-1} \ll \tau_c$. In this limit Fourier analysis of the fluctuating field predicts

$$T_1 = KB_0^2\tau_c, \tag{6.26}$$

with the constant K depending on the type of hopping process (ionic defects, vacancies, etc.) assumed. Techniques for determining T_1 and the theory required for its interpretation are described by Abragam (1961), and early applications to ice and water were reviewed by Glasel (1972).

Siegle and Weithase (1969) determined T_1 for 'pure' ice between -10 and $-60\,^\circ$C and have confirmed the relationship $T_1 \propto B_0^2$ for $B_0 = 0.01$ to $0.6\,$T. This implies $\tau_c \gg \omega_0^{-1}$. Subsequently Weithase *et al.* (1971) measured the related 'spin–lattice relaxation time in the rotating frame', which is denoted $T_{1\rho}$ and permits the use of smaller fields. In this case $\omega_0\tau_c$ is closer but not equal to unity and it becomes possible to fit the data to a more complete equation than (6.26). From different sets of data these workers deduced correlation times $\tau_c = 6.9$ and $9.0\,\mu$s at $-10\,^\circ$C with an activation energy of 0.603 eV. This activation energy is close to that determined by other means both for Debye relaxation and for molecular diffusion. At this temperature $\tau_D \approx 50\,\mu$s, which is so much longer than the 6.9–9.0 μs observed for τ_c that proton motions produced by the migration of Bjerrum or ionic defects cannot account for the spin–lattice relaxation. The hopping time derived for molecular diffusion from equation (6.12) (allowing for six directions of jumping) is about 8 μs, which is remarkably close to τ_c. On this basis and in line with other thinking at the time Weithase and co-workers inferred that the dominant contribution to proton motion is molecular diffusion by the vacancy mechanism. We now know that such diffusion occurs by an interstitial process, as was indeed the view at that time of Onsager and Runnels (1963, 1969). The interstitial mechanism requires some re-analysis of the theoretical calculation of K in equation (6.26) and hence of the value derived for τ_c. This has not been done, but it still seems probable that molecular diffusion is the dominant process determining T_1.

Bilgram *et al.* (1976) and Barnaal and Slotfeldt-Ellingsen (1983) report experiments which show that at low temperatures T_1 is reduced by doping ice with NH_3, HF, or other acidic impurities. As the temperature of the doped specimens is reduced, T_1 changes from its intrinsic value for pure ice to an extrinsic value with smaller activation energy, and at $-70\,^\circ$C the value of T_1 in doped ice can be two orders of magnitude smaller than that in pure ice. The self-diffusion experiments referred to in §6.4.1, however, showed no effect of doping, but they only extended down to $-27\,^\circ$C where the effect of doping on T_1 is not large. The n.m.r. experiments show clearly that at low temperatures the diffusion of protons is affected by doping which introduces protonic point defects, and Bilgram *et al.* (1976) proposed that the link between protonic and molecular diffusion mechanisms involved the presence of vested vacancies, like the one shown in Fig. 6.8(c). This topic deserves much closer study, because n.m.r. relaxation experiments are potentially a very powerful tool for the study of ice.

6.7 Muon spin rotation, relaxation, and resonance

A novel way of obtaining information about point defects in ice is to study the behaviour of positive muons μ^+ deposited in the ice. A number of accelerator laboratories now produce beams of muons which have been used for such experiments. The muons have unit charge, half-integer spin, and a mass equal to 0.1126 times that of the proton. In ice they can be thought of as ultra-light protons, but it is possible to monitor their behaviour in ways that are not possible for the protons themselves. In combination with an electron the muon forms an ultra-light hydrogen atom called 'muonium' with the chemical symbol 'Mu', and if the muon replaces one of the protons in an H_2O molecule it forms MuOH. The muon has a half-life of 2.2 μs, which may seem short but is quite sufficient for atomic processes to take place. It decays by the emission of a positron, and an important feature is that the direction of emission of this positron is closely correlated with the instantaneous direction of the muon spin vector.

An experiment which demonstrates the behaviour of the muons is called 'muon spin rotation'; it was performed by Leung *et al.* (1987) and is illustrated in Fig. 6.11. The muon beam is produced as a very short pulse with spin polarization along (actually opposite to) the direction of travel, and the muons are injected into the specimen of ice which is held in a uniform transverse magnetic field **B**. The muons are rapidly slowed down and become incorporated in the ice, where they precess in the magnetic field. The emitted positrons are detected with good time resolution by the two oppositely placed detectors D1 and D2, and the asymmetry between the signals at these detectors exhibits a periodicity arising from the precession. The signal in the figure shows two periodicities. The slower one, of which only about a quarter of a cycle is seen, corresponds to the precession rate of a free muon in the applied field. The faster one arises from Mu atoms in which the

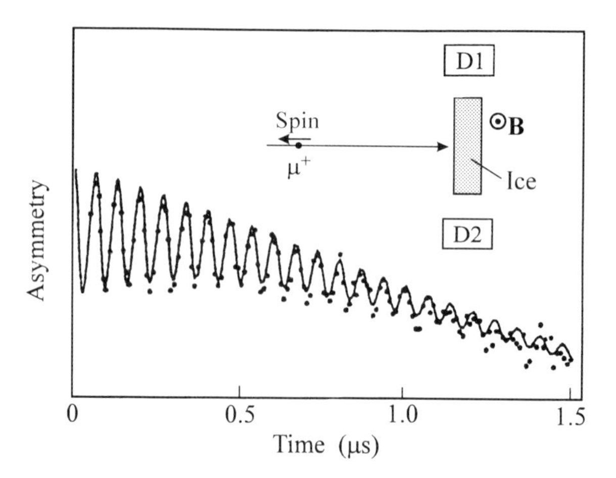

Fig. 6.11 The basic muon spin rotation experiment showing the asymmetry of positron counts in the detectors D1 and D2 as a function of time following the arrival of the muon pulse. Data reproduced from *Chemical Physics*, Vol. 114, S.-K. Leung *et al.*, Muonium diffusion in ice, pp. 399–409, Copyright 1987, with permission from Elsevier Science.

spin of the muon is tightly coupled to the spin of an electron and the whole precesses at a rate dominated by the larger magnetic moment of the electron. The muons precessing at the slower rate must be bound in situations where there are no unpaired electrons, and the obvious candidate for this state is in an MuOH molecule. When muons are slowed down within the ice we can therefore distinguish between the temperature-dependent 60–80% that become free Mu atoms in interstitial sites and the remainder that become incorporated into molecules.

The application of muon experiments to ice is described by Cox (1992). The spin relaxation time for the muonium atoms can be determined and has properties similar to T_1 in n.m.r. experiments. It yields a correlation time $\tau_c \sim 1$ ns at 50 K for the thermally activated motion of the Mu atoms between interstitial sites. In more sophisticated experiments on ice enriched with ^{17}O, enhanced spin relaxation is observed in magnetic fields for which the splitting of the energy levels of the muon is in resonance with splittings in the ^{17}O nucleus (Cox *et al.* 1990, 1994). These experiments yield features which are believed to distinguish such states as MuOH fully incorporated in the lattice and H_2O—Mu^+—OH_2, which represents a muon trapped in an L-defect. As the muon is much lighter than the proton it will have a large zero-point motion and is expected to occupy a central position in the double well shown for this situation in Fig. 6.9. An interesting question is how injected muons become incorporated in the lattice: are they added to a bond, forming a combined ion and D-defect as in Fig. 4.11, or do they move about freely until trapped in an L-defect? Both cases dissociate to yield MuOH and H_3O^+. The experiments referred to show dependencies on temperature and on doping with HF and NH_3, which strongly suggest that point defects already present in the ice are involved in the process.

This range of experiments is referred to as 'muon spin rotation, relaxation, and resonance' or 'μSR' for short. At the present time the experiments are more concerned with understanding the behaviour of the muon in ice than yielding fresh information about ice itself, but such studies may eventually aid our understanding of protons in ice.

6.8 Chemical impurities

Impurity atoms at interstitial or substitutional sites constitute the third class of point defect listed in §6.1, but, although many materials are highly soluble in water, solubilities in ice are very small. When aqueous solutions are frozen the solute is largely rejected into the liquid. These impurities may subsequently be incorporated into the ice as inclusions of concentrated solution (as in sea ice §12.2), or trapped in grain boundaries (§7.7.2), but they are not then in solid solution at sites in the ice lattice. Particular impurities such as HF or NH_3, which can be incorporated substitutionally and affect the electrical properties, were discussed in §5.4. It is known from such work that these two impurities in particular readily diffuse out of the specimens being studied, and the diffusion coefficient of HF in ice was measured by Haltenorth and Klinger (1969). For diffusion parallel to the c-axis at $-10\,°C$ they found $D_{HF} = (1.08 \pm 0.01) \times 10^{-11}\,m^2\,s^{-1}$ with an activation energy of 0.200 ± 0.002 eV. The diffusion coefficient perpendicular to the c-axis was 20% higher and that in polycrystals 25% higher due to diffusion along boundaries. Such diffusion coefficients are ~5000 times larger than the coefficient of self-diffusion, which implies that molecules of HF move comparatively easily into and between interstitial sites.

The HCl–water system has been studied and reviewed in some detail by Thibert and Dominé (1997). There is renewed interest in this topic because of its application to the behaviour of chlorine in stratospheric clouds (§12.3.3), and in this context their experiments studied the diffusion of HCl into ice from the vapour phase. The phase diagram as determined and compiled by them is shown in Fig. 6.12; it shows, for example, that the maximum solubility of HCl in ice at $-10\,°C$ is 3×10^{-6} mole fraction. The partition coefficient for crystal growth from very dilute solution was deduced to be 2.5×10^{-4}, which is smaller than some earlier determinations. The electrical measurements quoted in §5.4.3.2 do not yield reliable values for the impurity concentrations in the specimens, but certainly support partition coefficients very much less than unity. The actual determinations of the diffusion coefficient of HCl were scattered over the range 10^{-15} to $10^{-16}\,m^2\,s^{-1}$ and independent of temperature. These can only be upper limits on D_{HCl} due to diffusion via crystal imperfections, and this implies that the diffusion of HCl, is slower than that of H_2O. Similar experiments on HNO_3 (Thibert and Dominé 1998) show that its solubility is ~25 times less than that of HCl, but that the diffusion coefficient is higher, varying from $2 \times 10^{-14}\,m^2\,s^{-1}$ at $-8\,°C$ to $2 \times 10^{-15}\,m^2\,s^{-1}$ at $-35\,°C$.

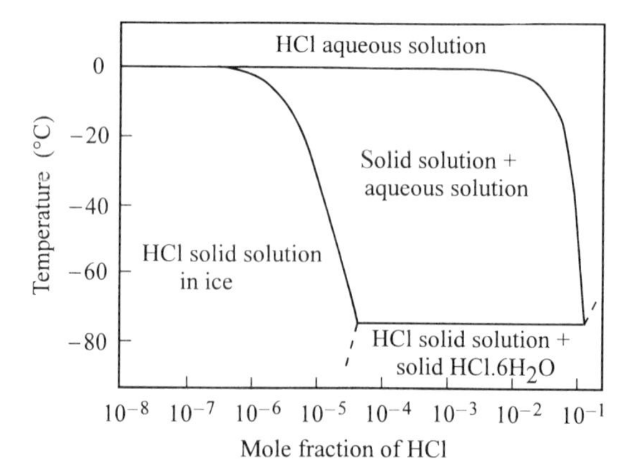

Fig. 6.12 Simplified phase diagram of the HCl–H_2O system (Thibert and Dominé 1997). Reproduced in simplified form, with permission from *Journal of Physical Chemistry*, **B101**, 3554–65. Copyright 1997 American Chemical Society.

Apart from these measurements and the effects of dopants in introducing protonic point defects, there is rather little information available about impurities in ice. This is because their concentrations are so low that they are very difficult to study. Even if the concentrations are low, the role of ice is important to these impurities in catalysing atmospheric reactions and in transporting and storing environmentally important materials.

6.9 Electronic defects

Our fourth class of point defects is made up of those generated by ionization of the water molecules in ice. In a series of experiments Warman *et al.* (1980) and de Haas *et al.* (1983) exposed ice to pulses of X-radiation or 3 MeV electrons of duration 1 ns and recorded the subsequent electrical conductivity. On this time scale it will be an excess σ_∞ that is observed. The ionization initially produces free electrons with a mobility of approximately $2 \times 10^{-3}\,m^2\,V^{-1}\,s^{-1}$ that is almost independent of temperature; this mobility is very much higher than that of the most mobile protonic point defects. These electrons are then trapped on a time scale of order 10 ns, which increases in pure ice with falling temperature with an activation energy of 0.55 ± 0.05 eV down to $-70\,°C$; it varies more slowly below this. The final state of the trapped electron produces an optical absorption band at 680 nm, which closely resembles one that is observed in the radiolysis of liquid water (Shubin *et al.* 1966; Taub and Eiben 1968; Nilsson *et al.* 1972). The state is referred to as a 'solvated' (or 'hydrated') electron. Electron spin resonance experiments (Kevan 1981) and theoretical treatments (e.g. Wallqvist *et al.* 1988) of the solvated electron in water favour a model in which the electron is trapped at a cavity with excess OH bonds pointing inwards. A comparable defect in ice might be the positive vested vacancy (Fig. 6.8(c)), but if the electron is trapped at a defect specific to ice it is remarkable that the spectra of the solvated electrons in water and ice are so similar.

Kunst *et al.* (1983) showed that the trapping rate for free electrons is greatly enhanced by doping with NH_3, and this would be explained if the NH_3 introduces D-defects that are present as vested vacancies. However, trapping is also increased by doping with HF, but in this case the detailed interpretation requires trapping at shallow traps. A small concentration of electrons remains present in equilibrium with these traps before they all finally attain the solvated state.

At temperatures above $-40\,°C$ the free electron conductivity decays to zero within 5–10 ns, but Kunst and Warman (1983) report a residual conductivity that decays in a time of the order of 100 ns. They attribute this conductivity to H_3O^+ ions (which they refer to as 'bare' protons). These ions are formed by a proton hopping along a hydrogen bond according to the second of the coupled reactions:

$$H_2O + \text{ionization} \rightarrow H_2O^+ + e^-$$
$$H_2O^+ + H_2O \rightarrow H_3O^+ + OH^\bullet$$

Here $OH^{\bullet}$ represents a free radical, but, if we remember that the effective charge of H_3O^+ in ice is $e_{\pm} = 0.62e$ (§4.4), the charge on $OH^{\bullet}$ will not be zero as in normal chemistry but $+e_{DL}$. The protonic conductivity referred to decays to a steady level as the H_3O^+ ions come into equilibrium with traps, which may be L-defects in line with the evidence from D—H exchange (§6.5.3). Eventually on a time scale of the order of 10 μs both this conductivity and the optical absorption due to the solvated electrons disappear together. If this occurs by the ion recombining with the electron at a vested vacancy they will release a neutral H atom in the vacancy. To restore the ice completely to its initial state this H atom must recombine with the $OH^{\bullet}$ radical where it will form a D-defect, with the charge that was originally at the vested vacancy where the electron was solvated.

The complex series of phenomena generated by pulse ionization are clearly related to the point defect systems in the ice and therefore provide evidence about them. However, a consistent quantitative interpretation of all the effects has not yet been produced. The evidence from the formation of solvated electrons points to the possible importance of combined defects, so that at least a proportion of the D-defects may be present in combination with vacancies.

6.10 Photoconductivity

Khusnatdinov *et al.* (1990) observed that ultraviolet radiation in the range 175–190 nm generates photoconductivity in pure ice. The excitation spectrum has a single peak at 182 nm (6.8 eV). Referring to the spectrum of ice in Figs 9.2 and 9.3, the peak lies near the bottom of the ultraviolet absorption edge, at a point where the absorption coefficient is only $\sim 10^{-6}$ of its value at the peak. Both σ_∞ and σ_s are increased by roughly equal amounts, which are small compared with σ_∞ but in a pure crystal can be many times σ_s. The time scale for the decay of the conductivity after the illumination has been turned off is of the order of 10 s, which is far too long for the effect to be due to the release of free electrons. Further features are that the photoconductive response of the ice decreases after prolonged illumination and that the ice is then found to contain H_2O_2. In further experiments by Petrenko and Khusnatdinov (1994) the photo-generated charge carriers were observed to diffuse into a non-illuminated region of the specimen, and the most mobile charge carriers were found to be negatively charged and to have a mobility of $\sim 10^{-7}\, m^2\, V^{-1}\, s^{-1}$.

The initial photoexcitation process does not create free electrons and is presumed to be to a bound state of a single molecule. The first excited state of a free molecule is known to be linear (§1.3), and it will not fit into the ice lattice, so that relaxations must occur. Petrenko and Khusnatdinov propose an 'auto-ionization' reaction to form an H_3O^+ ion, an interstitial (neutral) $OH^{\bullet}$ radical, and a complex consisting of an electron trapped at a positive vested vacancy. The negative complex has to be mobile, and the H_2O_2 is formed by the combination of a pair of $OH^{\bullet}$ radicals. However, many features of the process are not yet properly understood.

A quite different form of photoconductivity has been observed by Petrenko *et al.* (1986) in ice doped with *o*-nitrobenzaldehyde. In solution in water these molecules are known to release hydrogen ions under illumination with ultraviolet radiation at 395 nm. Photoconduction occurs at this same wavelength in ice, and is presumed to be due to the release of protons. These must be incorporated in the ice as H_3O^+ ions and D-defects as shown in Fig. 4.11. However, further complexities such as trapping will have to be invoked to account fully for the experimental observations.

6.11 Review

The past three chapters have explained the variety of point defects which occur in ice and the many important properties that depend on them. Over more than 40 years an impressive range of experiments has been performed to study these properties, and we attempt in this section to draw some conclusions.

In the case of electrical measurements the parameters χ_s, τ_D and σ_∞ describing the Debye relaxation are well established over the intrinsic range for pure ice (Table 5.1). The partitioning of the effective charges of ionic and Bjerrum defects depends only on the static susceptibility and is also well established at

$$e_{\pm} = (0.62 \pm 0.01)e \quad \text{and} \quad e_{\mathrm{DL}} = (0.38 \pm 0.01)e.$$

For *pure ice* the relaxation process has been shown to be dominated by L-defects so that

$$\sigma_\infty = n_{\mathrm{L}} e_{\mathrm{L}} \mu_{\mathrm{L}}. \tag{6.27}$$

The corresponding activation energies will then be given by

$$E_\tau \approx E_{\sigma_\infty} = \tfrac{1}{2} E_{\mathrm{DL}} + E_{\mathrm{Lm}}. \tag{6.28}$$

However, the extraction of the concentrations, mobilities, and separate activation energies requires measurements on doped ice, which introduce large uncertainties and show up some incompatibilities between experiments.

Because it is valuable to know orders of magnitude for the quantities involved, we present in Table 6.4 a set of derived parameters for pure ice at a typical temperature of $-20\,^\circ\mathrm{C}$, but *these values must be used with caution.* The mobility and activation energy of motion of the L-defects have first been taken from the experiments on HF- and HCl-doped ice (Table 5.2), but there are differences in the interpretation of these experiments as explained in §5.4.3.2. Equation (6.28) then yields a range of values for the activation energy E_{DL} of formation of the

Table 6.4 Parameters for intrinsic protonic point defects in ice at −20 °C
Caution: these values are derived from a range of electrical measurements and should be used with great care. See text

Defect	Index i	Effective charge e_i/e	Intrinsic concentration n_i/N	Activation energy of pair formation $E_{\pm}$, E_{DL} (eV)	Mobility μ_i (m^2 V^{-1} s^{-1})	Activation energy of motion E_{im} (eV)
H_3O^+	1	$+0.62$	$\leq 10^{-13}$	≥ 1.4	10^{-7}	≈ 0
OH^-	2	-0.62			3×10^{-8}	≈ 0
D	3	$+0.38$	10^{-7}	0.66–0.79	$\ll \mu_4$	?
L	4	-0.38			2×10^{-8}	0.2–0.3

Bjerrum defect pair, and equation (6.27) gives the order of magnitude of the intrinsic concentration. In intrinsic ice the L- and D-defects must be present in equal concentrations, but the D-defects are less mobile. Even in doped ice there is no reliable evidence for D-defect mobility, and it seems probable that the D-defects are trapped either as vested vacancies (Fig. 6.8(c)) or elsewhere. The presence of some such traps is supported by the evidence for the solvation of electrons in §6.9, but a much fuller interpretation will be required to reconcile the activation energy of 0.55 eV observed for trapping with data from electrical measurements.

The mobility of ionic defects has been determined from measurements with the field effect transistor (§5.6.2), from proton injection (§5.6.3), or from the conductivity of doped crystals. There does not appear to be any significant temperature dependence, at least above 150 K, and plausible mean values are listed in Table 6.4. The values of σ_s in the purest specimens then set a limit on the intrinsic concentration of free ionic defects, and hence on the energy of formation of the ion pair $E_{\pm}$. The concentration quoted is extremely low, so that residual impurities will almost always dominate the ionic concentration even in nominally pure ice. The isotopic exchange experiments described in §6.5.3 support the basic concept of ionic and Bjerrum defects, but seem to require that the H_3O^+ ions are trapped with an activation energy of 0.43 eV possibly at L-defects. The isotopic exchange experiments also suggest a much larger ratio between the mobilities of positive and negative ions than that shown in Table 6.4 (Devlin 1992).

The possibility that protonic defects may be trapped at other defects in the lattice raises a serious complication in our understanding of point defects in ice. The conductivities depend only on the concentrations of free protonic defects, and the Jaccard model is probably correct when expressed in terms of free concentrations and mobilities. However, the interaction between defects greatly complicates the interpretation of results depending on temperature, doping, and space charge effects.

Molecular diffusion occurs by the interstitial mechanism, with the parameters for interstitials as shown in Table 6.1. The detailed information here comes from one set of experiments, but they are consistent with the many measurements on bulk diffusion in Fig. 6.2. The fractional concentration of interstitials at $-20\,^{\circ}\mathrm{C}$ calculated from the data in Table 6.1 is $n_{\mathrm{i}}/N = 7 \times 10^{-7}$, and this is larger than that of Bjerrum defects in Table 6.4. These interstitials provide the mechanism for spin–lattice relaxation in n.m.r., with a spin correlation time shorter than the Debye relaxation time. The vacancy concentration must be less than that of interstitials to account for the formation of interstitial type loops on cooling (§6.4.2), but there can still be enough vacancies to trap all the D-defects as vested vacancies.

We finally conclude that the basic principles concerning protonic and molecular defects are well founded and that the theory of electrical properties developed in Chapter 4 in terms of free concentrations and mobilities is correct. However, to account adequately for the whole range of experiments used to study ice it will be necessary to include combinations of defects such as positive ions coupled with L-defects and D-defects trapped at vacancies. The need to consider such matters dates back to the 1960s and the possibilities were reviewed by Kröger (1974). Many of the standard experiments on the electrical properties of ice were performed some time ago. Such experiments should now be repeated in the light of fresh understanding and of improved techniques for the preparation of high-quality crystals and electrodes. Parallel measurements should be made using different techniques on the same samples.

7
Dislocations and planar defects

7.1 Introduction to dislocations

Dislocations are line defects in crystals, and they are most easily visualized in terms of the slip of one plane of atoms over another. This is illustrated in Fig. 7.1 in which the crystal (a) is sheared into the state (c) by slip through a single lattice spacing **b**. In a real crystal such slip cannot occur simultaneously over the whole plane. If slip starts at the corner A, Fig. 7.1(b) shows an intermediate stage in which the line BC is the boundary between the area ABC over which slip has occurred and the rest of the slip plane. This boundary line is a *dislocation line* or *dislocation* for short. The presence of the dislocation results in an elastic distortion of the whole crystal, but the disruption of the lattice is greatest along the core BC. The lattice vector **b** by which slip has occurred on the plane ABC is known as the *Burgers vector* of the dislocation. Where the dislocation lies parallel to **b** as at B it is said to have *screw* orientation, and where it is perpendicular to **b** as at C it has *edge* orientation. In general a dislocation will have both screw and edge components. An edge dislocation has dangling bonds at its core, as seen in the enlarged region around C in Fig. 7.1. It can alternatively be thought of as the line at which an added half-plane of atoms terminates within the crystal, and edge dislocations are commonly represented in diagrams by the symbol $\perp$ in which the horizontal and vertical lines indicate the glide plane and the extra half-plane respectively. There are no dangling bonds on a screw dislocation, but the lattice around it is distorted. Imagine a path proceeding step by step from one atom to the next that would close in a perfect crystal. If this path encircles a screw dislocation it takes the form of a helix with pitch equal to the Burgers vector.

From the way in which we have introduced them it is clear that the motion of dislocations on the slip plane is associated with plastic deformation of the crystal. Such motion of the dislocation is known as *glide*, but it represents only one aspect of the properties of dislocations. Dislocations may be incorporated into a crystal as it is grown, affecting the topology of the whole lattice but with little effect on plastic deformation. The core of an edge dislocation may be displaced perpendicularly to the glide plane by the addition of atoms to or the removal of atoms from the end of the extra half-plane, and this process is known as *climb*. It requires molecular diffusion within the material, which glide does not.

The properties of dislocations are described in many books and articles. Read (1953) provides a very clear description of the basic geometry, and a comprehensive treatment including many aspects relevant to ice is that of Hirth and

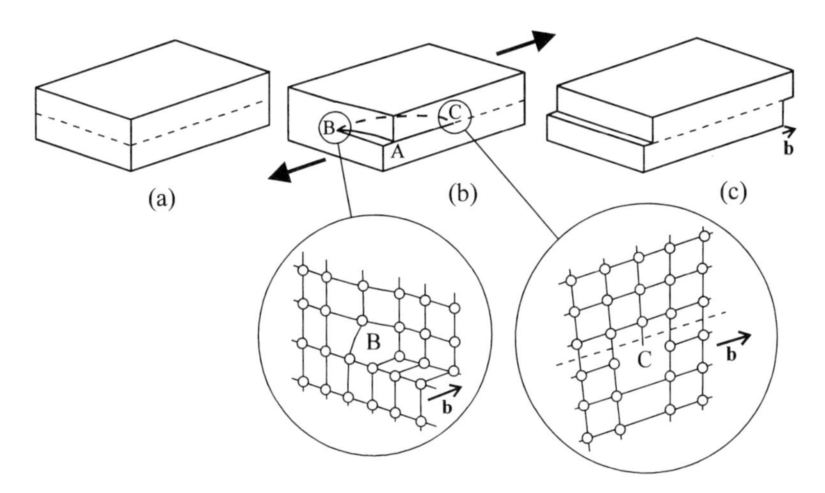

Fig. 7.1 The production of slip in a crystalline material by the glide of a dislocation BC across the slip plane, starting at A. The amount of slip is given by the Burgers vector **b**. The magnified portions show the arrangement of atoms around the screw component at B and the edge component at C.

Lothe (1982). Higashi (1988) has produced a beautifully illustrated book describing dislocations in ice.

7.2 Dislocations in the ice structure

7.2.1 *Basal dislocations*

It is now well established that crystals of ice deform by slip on the basal plane (0001) (Glen and Perutz 1954), and that macroscopic slip on any other plane is difficult (§8.2.2). The Burgers vectors for slip on the basal planes are the three lattice vectors, which can be written in the form $(a/3)\langle 2\bar{1}\bar{1}0\rangle$. In macroscopic experiments slip can, however, occur in any direction on the basal plane by a combination of dislocations with these three vectors (Kamb 1961). We will consider initially only basal dislocations of this type. Because of the hexagonal symmetry the simplest dislocations are those that lie parallel or at 60° to the Burgers vector; these are called screw and 60° dislocations respectively.

Figure 7.2 shows the structure of ice (without the hydrogens) projected on a $(1\bar{2}10)$ plane with the basal plane horizontal. Slip can in principle take place between two kinds of planes, called the *glide set* and the *shuffle set* and so marked on the figure. The planes of the shuffle set are more widely spaced and were originally thought to be the natural slip planes. However, the planes of the glide set fit over one another in a way that resembles the packing in close-packed

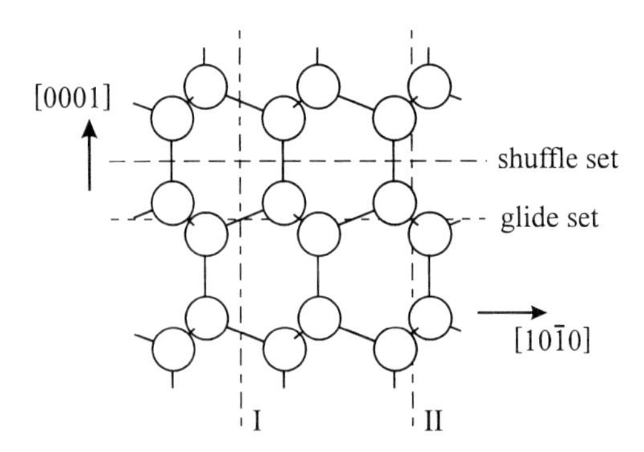

Fig. 7.2 Projection of the structure of ice (without hydrogens) on the (1$\bar{2}$10) plane, showing basal planes of the shuffle set and the glide set. I and II are different kinds of (10$\bar{1}$0) planes.

metals, and on such planes a dislocation may lower its energy by dissociating into two *Shockley partial dislocations* separated by a stacking fault (see §7.6). This dissociation is illustrated in Fig. 7.3, in which the open and shaded circles in (a) represent oxygen atoms on adjacent basal planes of the glide set. There are no atoms in the normal structure at the sites marked C. In Fig. 7.3(b) a screw dislocation of Burgers vector $\mathbf{b} = (a/3)[\bar{1}2\bar{1}0]$ at the top is shown dissociated at the bottom into two partial dislocations of Burgers vectors $\mathbf{b_1}$ and $\mathbf{b_2}$ of the type $(a/3)\langle\bar{1}100\rangle$ separated by the shaded area of stacking fault. Returning to Fig. 7.3(a) slip by $\mathbf{b_1}$ corresponds to motion of the shaded atoms from the positions labelled B to those marked C. As this is not a normal position for this plane in the structure a stacking fault is formed. Further slip by $\mathbf{b_2}$ brings the layer back to a B location, removing the fault and resulting in a net slip by the full Burgers vector $\mathbf{b} = \mathbf{b_1} + \mathbf{b_2}$. Both of the partial dislocations associated with a screw dislocation have Burgers vectors at 30° to the line, and they are called 30° dislocations. Figure 7.3(c) shows the dissociation of a 60° dislocation into an edge and a 30° partial dislocation. These dissociations lead to a reduction in the elastic strain energy of the dislocations, and this reduction is balanced by the energy required to create the stacking fault. From estimates of the stacking fault energy described in §7.6.2 it has been calculated that in ice dislocations on planes of the glide set will be dissociated into Shockley partial dislocations separated by 50–100 lattice spacings (Fukuda *et al.* 1987).

In many semiconductors with structures resembling that of ice the dissociation of dislocations on the basal plane has been observed directly in the electron microscope, and slip is therefore assumed to occur on planes of the glide set (see George and Rabier 1987). There is no such direct evidence for ice, and as the bonding is quite different we cannot presume that the same conclusion applies. However, as we will see in §7.4.2, indirect evidence suggests that dislocations do in fact glide on planes of the glide set.

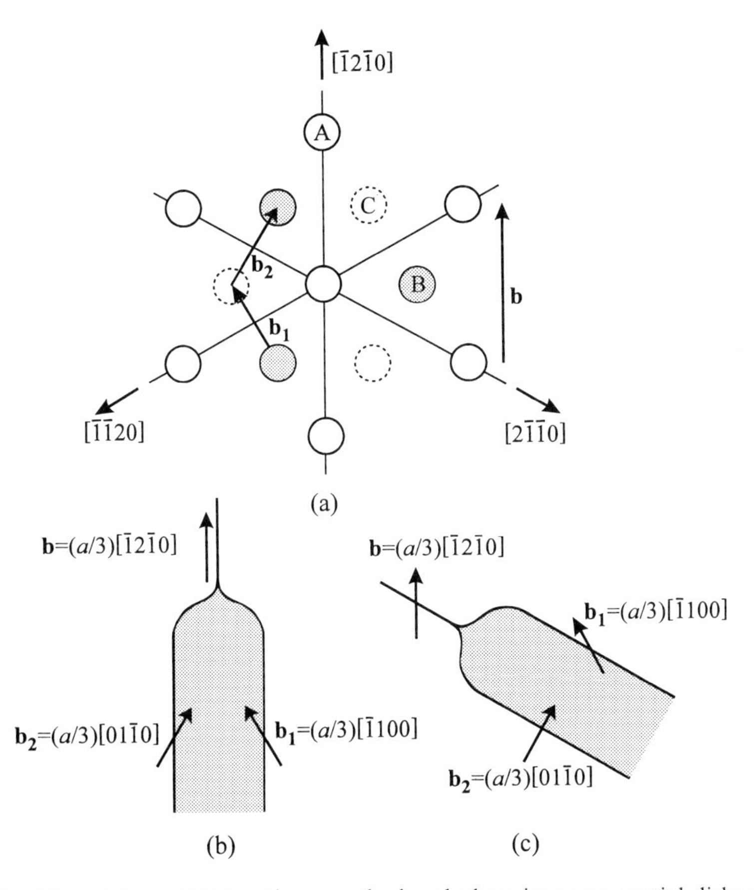

Fig. 7.3 Dissociation of dislocations on the basal plane into two partial dislocations separated by a stacking fault. (a) Positions A, B, and C of possible (0001) layers of molecules in the ice structure, indicating the full Burgers vector **b** and the partial Burgers vectors $\mathbf{b}_1$ and $\mathbf{b}_2$. (b) Dissociation of a screw dislocation. (c) Dissociation of a 60° dislocation.

If a straight dislocation lying on its glide plane moves forward by one Burgers vector over part of its length, the step so produced is called a *kink*. An example of a kink in an edge dislocation is shown in Fig. 7.4. The most elementary step in the process of glide is for a kink to move along the dislocation by one lattice spacing, as shown by the broken line. Figure 7.4 also shows a *jog* at which the dislocation makes a step from one glide plane to another. For a dislocation with an edge component the formation of a jog or its motion along the dislocation requires that molecules be added to or removed from the end of the extra half-plane, and this is what is involved in climb. It occurs by molecular diffusion, and in this process the dislocation acts as a source or sink of vacancies or interstitials as described in § 6.4.2. In the ice structure the simplest jog would take the dislocation from a

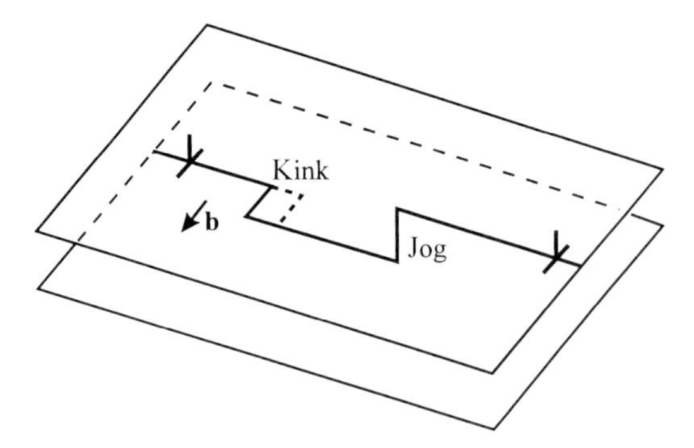

Fig. 7.4 Edge dislocation containing a kink in its glide plane and a jog at which it makes a step from one glide plane to another. The broken line shows an elementary step in the motion of the kink.

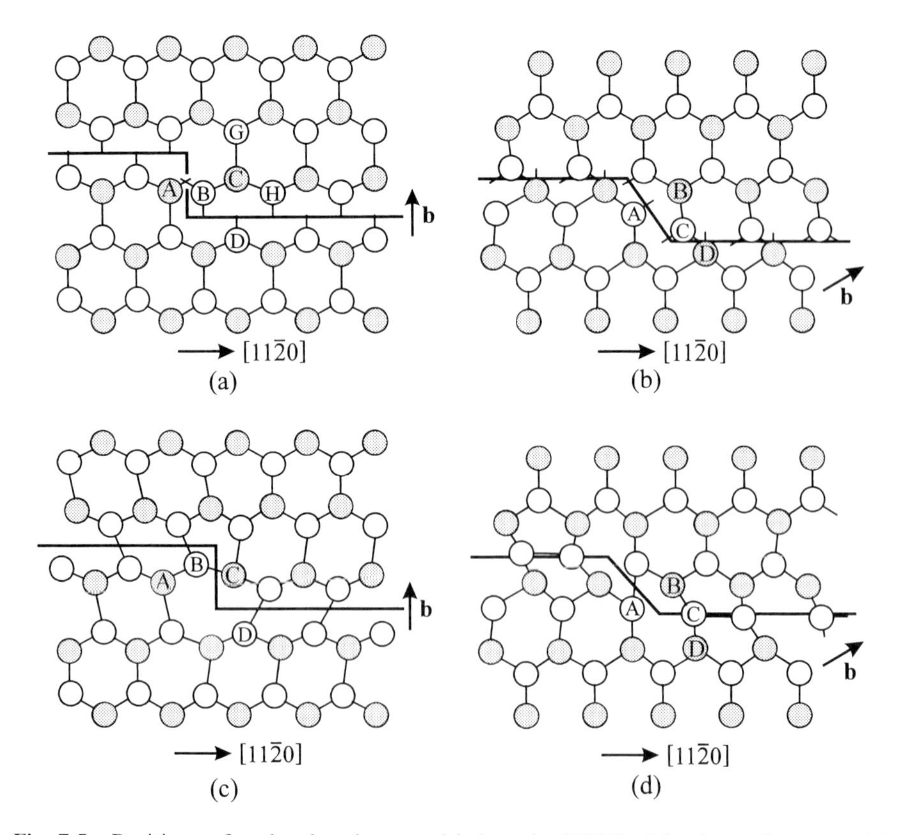

Fig. 7.5 Positions of molecules above and below the (0001) glide plane of two partial dislocations, each containing a kink. (a) Edge and (b) 30° partial dislocations with dangling bonds. (c) and (d) Possible ways in which the cores of these dislocations could be reconstructed to remove the dangling bonds.

glide to a shuffle plane; as can be seen in Fig. 7.2 a full jog must span two layers of molecules.

Kinks and jogs are in general quite different in character. However, a pure screw dislocation has the property that it can in principle glide on any plane containing its Burgers vector, and in this case whether a particular step behaves as a kink or a jog depends upon which plane is being considered as the glide plane.

Figures 7.5(a) and (b) show the arrangement of molecules on the two sides of the basal glide plane at an edge partial and a 30° partial dislocation in the ice structure (Whitworth 1980). Open circles represent molecules above the glide plane and shaded circles represent those below. The extra half-plane is above the glide plane, and the dislocations in both diagrams contain kinks. There can be seen to be dangling bonds on the molecules above the glide plane at the edge dislocation and on both sides of the glide plane for the 30° dislocation. With some changes to bond lengths and bond angles, and with some local elastic distortion it is possible to link up these dangling bonds as shown in Figs 7.5(c) and (d). This is known as *reconstruction*, which is generally believed to occur in semiconductors. Calculations by Heggie *et al.* (1992) suggest that it probably happens in ice as well, although there is more than one possibility about how the reconstruction might occur for the 30° dislocation.

7.2.2 *Non-basal dislocations*

Dislocations with the basal Burgers vector $(a/3)\langle 2\bar{1}\bar{1}0\rangle$ can in principle glide on non-basal planes that include this vector, such as $\{01\bar{1}0\}$ or $\{01\bar{1}1\}$, and in §7.3.4 we will report observations of edge dislocations that glide on such planes. There is again more than one possible type of glide plane, such as the sets of $\{10\bar{1}0\}$ planes marked I and II in Fig. 7.2, but there is no evidence to show which of these is operative.

In addition we may expect dislocations in which the Burgers vector has a [0001] component. Such dislocations have been observed in ice, usually in the form of loops lying in the (0001) plane. These are called *prismatic* loops and are formed by the condensation of point defects. There is no evidence that these dislocations can glide or that ice can be deformed plastically by slip in the [0001] direction.

7.3 Direct observation of dislocations

7.3.1 *General*

Since the earliest experiments of Webb and Hayes (1967), X-ray topography has been extensively used for the study of dislocations in ice. It is undoubtedly the most suitable technique for use with this material, and of all materials ice has been the one most fruitfully studied by this method. We will first try to explain why this is so.

For other materials the most powerful technique for observing dislocations is usually transmission electron microscopy. However, in the case of ice the preparation and handling of suitable thin specimens presents enormous difficulties, and even when prepared the specimens have very limited life in the electron beam. Although there have been reports of such experiments on ice (Unwin and Muguruma 1972; Falls *et al.* 1983), no significant information has been obtained in this way. It is possible to produce etch pits on ice that can be useful for orienting crystals (Higuchi 1958), but these do not normally correspond to the points of emergence of dislocations. Particular kinds of etch pits and other etch features do appear to be related to dislocations (e.g. Muguruma and Higashi 1963; Sinha 1977; Wei and Dempsey 1994), but the interpretation of these observations is difficult and the information obtainable is very limited compared with that which can be obtained from etching materials like lithium fluoride.

In contrast, X-ray topography has been used to reveal dislocations in the interior of ice crystals that are a few millimetres thick, and to observe the motion of these dislocations under stress and during annealing. This is possible because ice, having a low molecular weight, is sufficiently transparent to X-rays of wavelength less than about 0.9 Å, and because ice crystals can be grown with dislocation densities that are sufficiently low for individual dislocations to be clearly distinguishable even in thick specimens. A further reason why ice is suitable for dynamic experiments is that dislocations can be moved slowly under stress, whereas in most materials dislocations move suddenly over large distances once some critical stress has been reached.

7.3.2 *X-ray topography*

The form of X-ray topography that is the simplest and most easily explained became possible only with the availability in the 1980s of intense highly collimated beams of 'white' X-radiation from a synchrotron source. If such a beam falls on a single crystal as shown in Fig. 7.6, it produces Laue diffraction spots in which there is a one-to-one correspondence between a position on the spot and the position at which the X-rays passed through the crystal. Local misorientations of the lattice within the crystal, such as occur at dislocations, change the diffraction conditions slightly and result in contrast in the Laue spot. Each spot is therefore an image of the crystal in which the dislocations are visible, and these images are called topographs. The topographs can be recorded on high-resolution photographic films or plates and observed in real time at lower resolution with an X-ray sensitive video camera.

The imaging conditions depend on the diffraction vector $\mathbf{g}$ of the particular Laue reflection in such a way that dislocations of Burgers vector $\mathbf{b}$ will not appear in a topograph for which both $\mathbf{g} \cdot \mathbf{b} = 0$ and $\mathbf{g} \cdot (\mathbf{b} \times \mathbf{l}) = 0$, where $\mathbf{l}$ is a unit vector parallel to the dislocation line. By comparing topographs obtained with different $\mathbf{g}$ it is possible to identify the character of each dislocation.

The resolution depends on the degree of collimation of the incident beam, which can only be satisfactorily achieved with a synchrotron. Without this collimation it is necessary to use monochromatic radiation, and with a conventional

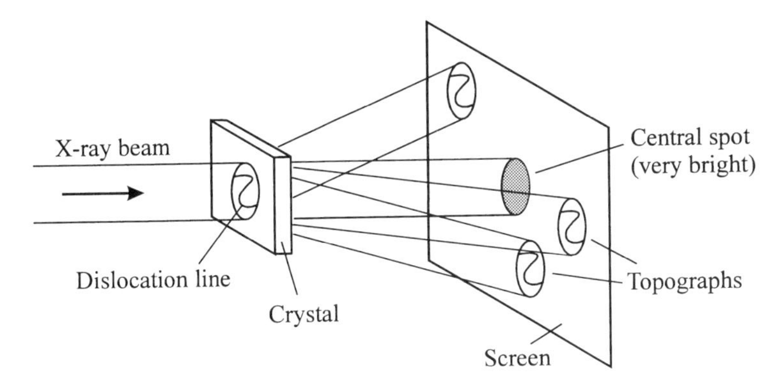

Fig. 7.6 Principle of synchrotron white-beam X-radiation topography.

X-ray source the beam divergence for a reasonable intensity is then too large for Bragg's law to be satisfied over the whole crystal at once. Provision has to be made for scanning the crystal across the beam and the most commonly used arrangement is the Lang (1959) camera. This technique has been used in all but the most recent work on ice, as reviewed by Higashi (1988). Because of the scanning the exposure time is typically a few minutes or hours depending on the power of the X-ray generator, whereas with a synchrotron it is only a few seconds. This difference is very important for dynamic experiments.

Dislocations observed by topographic methods fall into three groups: dislocations grown into the crystal, dislocations produced during plastic deformation, and dislocations introduced or modified by diffusion processes. The conclusions obtained for the last of these groups have already been described in §6.4.2.

7.3.3 *Grown-in dislocations*

All crystals produced without special procedures, and grains within polycrystals, contain stable networks of grown-in dislocations, which often form arrays constituting low-angle grain boundaries. In ice almost all such dislocations have the basal Burgers vector and their concentrations are appreciably lower than those found in metals and other commonly occurring solids. Examples of high-quality ice from the Mendenhall Glacier with $\sim 10^4$–10^5 dislocations per square centimetre are described by Fukuda and Higashi (1969), but other naturally occurring ice is usually much less perfect (Fukuda and Shoji 1988). For the study of individual dislocations by topographic methods much lower dislocation densities are needed, and much attention has been given in Japan to the growth and examination of such crystals (Higashi *et al.* 1968; Higashi 1974; Oguro 1988). Suitable crystals prepared by the methods shown in Fig. 2.2 contain less than 100 dislocations with $(a/3)\langle 2\bar{1}\bar{1}0\rangle$ Burgers vectors per square centimetre. There are usually also a few large circular or spiral loops with [0001] Burgers vector lying on

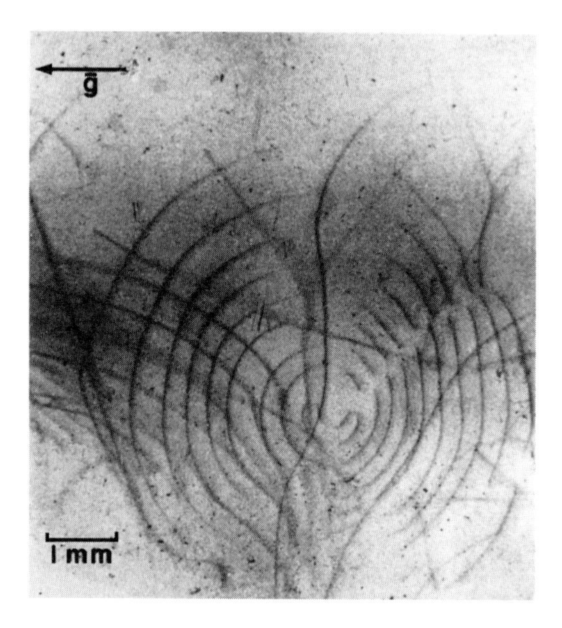

Fig. 7.7 Topograph projected on the (0001) plane showing concentric dislocation loops in an as-grown crystal of ice. The diffraction vector **g** is marked and the variation of contrast round the loop is evidence of its prismatic character. The other dislocations seen are typical of the low density random network in a good as-grown crystal. (From Oguro 1988.)

or close to the basal plane; an example is shown in Fig. 7.7. These loops have the prismatic character illustrated in Fig. 6.4, and have been studied by Oguro and Higashi (1981). It is now believed that they were formed by the condensation of interstitials (Oguro *et al.* 1988). Crystals with significant concentrations of impurities are usually much less perfect (Oguro 1988), but the grains within polycrystalline ice grown slowly under carefully controlled conditions can be remarkably good (Liu *et al.* 1992).

7.3.4 *Dislocations formed by plastic deformation*

Many experiments have been described in which dislocations were observed to glide and to multiply under an applied stress. Such motion represents an extremely early stage of deformation, because once any significant macroscopic strain has been produced the dislocation density is far too high for topographic observations to be possible. In early experiments (e.g. Fukuda and Higashi 1973; Jones and Gilra 1973; Maï 1976; Fukuda *et al.* 1987; Fukuda and Higashi 1988) much information was lost because, as was found subsequently, the dislocation structure changed after unloading in the time required to obtain the topographs. In more recent work, especially that using synchrotron radiation, it is much easier to identify different kinds of dislocations in the topographs.

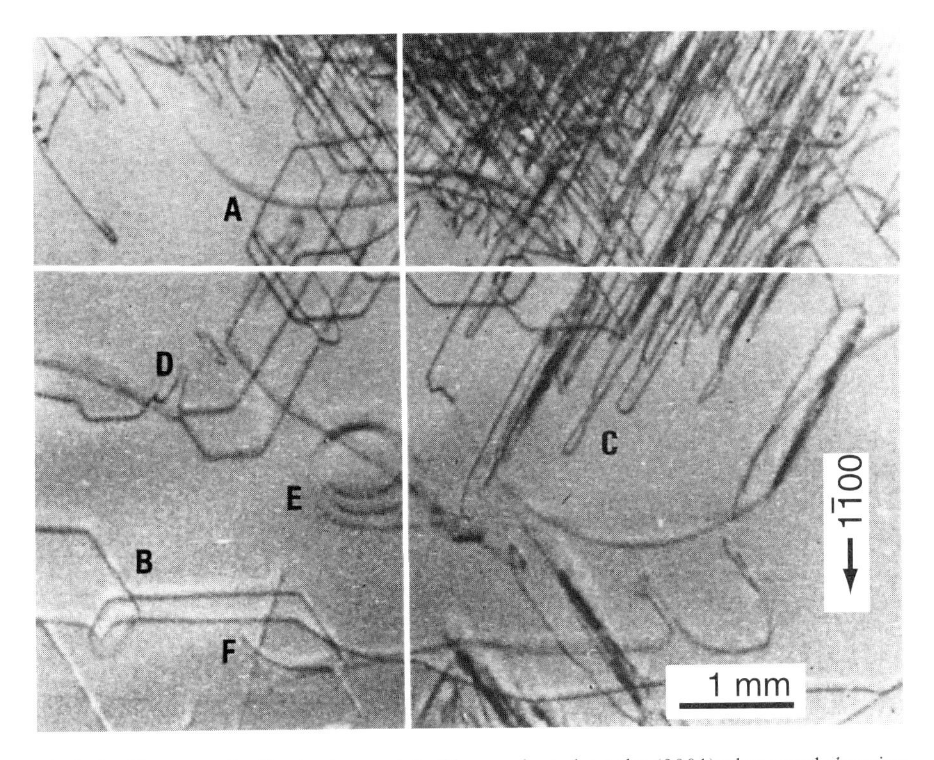

Fig. 7.8 Topograph of a single crystal of ice projected on the (0001) plane and showing dislocations introduced by a compressive loading that produces a shear stress on this plane in the $[1\bar{1}00]$ (i.e. vertical) direction. The diffraction vector is $1\bar{1}00$. A and B are loops on the basal plane. C and D show edge dislocation segments which have moved rapidly on planes almost perpendicular to the plane of the topograph. Prismatic loops like those in Fig. 7.7 are seen at E. (From Ahmad and Whitworth 1988, with permission from Taylor and Francis.)

Figure 7.8 is a synchrotron radiation topograph obtained by Ahmad and Whitworth (1988) using the technique shown in Fig. 7.6. The dislocations are projected on the basal plane and the crystal had been subjected to a compressive stress in the vertical direction. Features with 120° angles as at A and B are dislocations gliding on the basal plane. These dislocations glide as almost straight segments in the screw and 60° orientations, but the corners rapidly become curved after the stress is removed. Very occasionally, for example in a collapsing loop, dislocations in edge orientation have also been seen. The long narrow loops like those at C and D have basal Burgers vectors parallel to their lengths and lie on non-basal planes oblique to the plane of the topograph. The tip of the loop is an edge dislocation and the long segments seen in the figure are screw dislocations. The loop E is a prismatic loop of the type shown in Fig. 7.7.

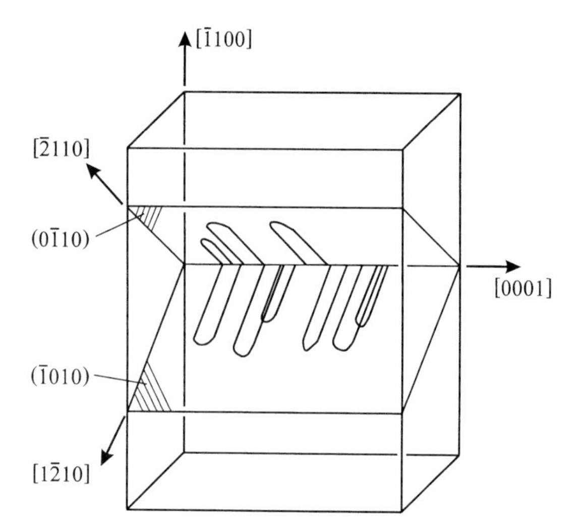

Fig. 7.9 Geometry of the topographs in Fig. 7.10. The specimen is compressed in the $[\bar{1}100]$ direction so that there is no stress on the (0001) plane. Dislocations were nucleated from a scratch on the back face and propagate on the $\{0\bar{1}10\}$ slip planes with $\langle\bar{2}110\rangle$ Burgers vectors.

Similar features to the non-basal segments at C and D were observed in the work of Fukuda *et al.* (1987) (see also Fukuda and Higashi 1988), but the nature of this non-basal glide is best revealed in crystals stressed in the orientation shown in Fig. 7.9, for which there is no resolved stress on the normal (0001) slip plane. The sequence of topographs in Fig. 7.10 shows dislocations with the basal Burgers vectors $(a/3)[1\bar{2}10]$ and $(a/3)[\bar{2}110]$ propagating on the $(\bar{1}010)$ and $(0\bar{1}10)$ planes from a scratch on the back of the crystal as illustrated in Fig. 7.9 (Shearwood and Whitworth 1989). The edge dislocations move forward easily on the $\{\bar{1}010\}$ planes, but the screw dislocations that they leave behind do not glide to right or left.

We conclude from these observations that all glide dislocations have Burgers vectors in the basal plane, and that loops expanding on this plane take up a hexagonal form made up from screw and 60° segments. The screw components cannot cross-glide on non-basal planes, as they would do in many materials, but edge dislocations can glide easily on non-basal planes containing their Burgers vector. These characteristics are probably unique to ice and are relevant to its high degree of plastic anisotropy, which we will consider further in Chapter 8.

7.3.5 *Dislocation multiplication*

The simple glide of a dislocation across a crystal as envisaged in Fig. 7.1 cannot lead to macroscopic plastic deformation. A few pre-existing dislocations would

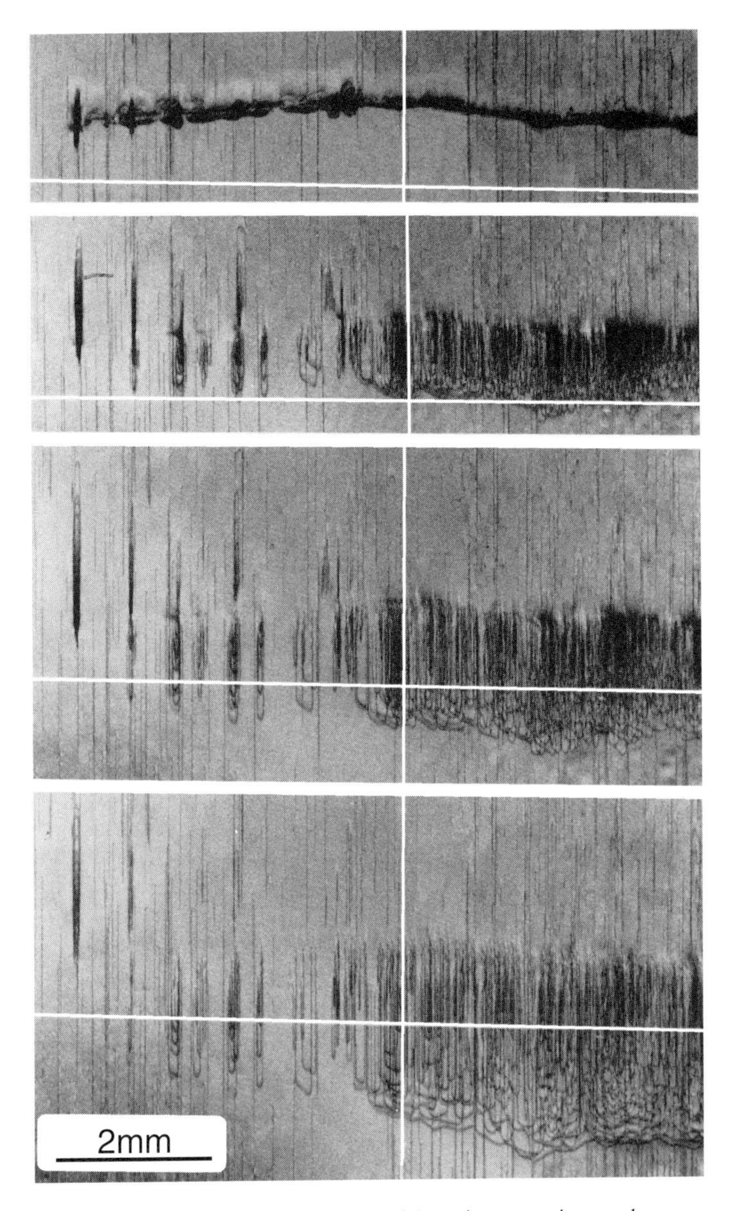

Fig. 7.10 Sequence of topographs showing dislocations moving under stress on non-basal planes as shown in Fig. 7.9. Diffraction vector $1\bar{1}00$ and plane of projection parallel to front face of specimen (Shearwood and Whitworth 1989). Figures 7.9 and 7.10 are reproduced by courtesy of the International Glaciological Society from *Journal of Glaciology*, **35**(120), pp. 281–3, 1989, Figs 1 and 2.

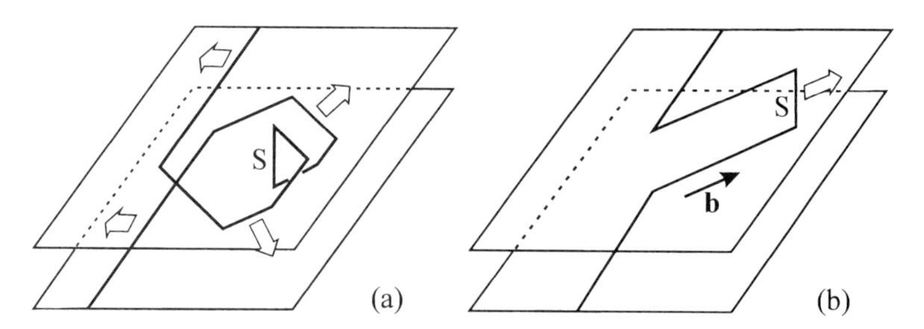

Fig. 7.11 (a) Operation of a Frank–Read source with two segments spiralling on basal planes around a fixed step S, and (b) what commonly happens in ice if the segment S glides rapidly in edge orientation on a non-basal plane.

just glide out of the crystal and leave it dislocation-free. To produce macroscopic deformation there have to be processes by which fresh dislocations are generated and by which slip can be transferred from one slip plane to another. The standard mechanism is the Frank–Read source (see Read 1953). An example of such a source is illustrated in Fig. 7.11(a), in which dislocation segments spiral around the fixed points at which the dislocation line makes a step from one glide plane to another. Ahmad *et al.* (1986) observed such a source in topographic experiments on ice, and Fig. 7.12 shows a more recent example. In most materials many such sources are generated by the cross-slip of screw dislocations off their primary glide planes, but the immobility of screw dislocations on the non-basal planes in ice means that this cross slip does not occur (at least under the conditions of topographic experiments). However, a feature of ice is the high mobility of edge segments on non-basal planes. This means that the segment S which acts as the centre of the Frank–Read source in Fig. 7.11(a) may not remain fixed but more often glides away from the original dislocation as illustrated in Fig. 7.11(b). This leads to long narrow loops like those at C in Fig. 7.8, and subsequently to complex processes of dislocation multiplication (Ahmad and Whitworth 1988; Ahmad *et al.* 1992; Shearwood and Whitworth 1993). Dislocations are commonly generated at points of stress concentration, and topographic experiments on polycrystalline specimens have shown that generation at grain boundaries is particularly significant (Hondoh and Higashi 1983; Liu *et al.* 1993, 1995).

7.4 Dislocation mobility

7.4.1 *Experimental observations*

Ice is one of very few materials on which, for the reasons outlined in §7.3.1, systematic observations have been made of the velocities at which dislocations glide as functions of both stress and temperature. The most detailed measurements are those of Shearwood and Whitworth (1991), who obtained sequences of

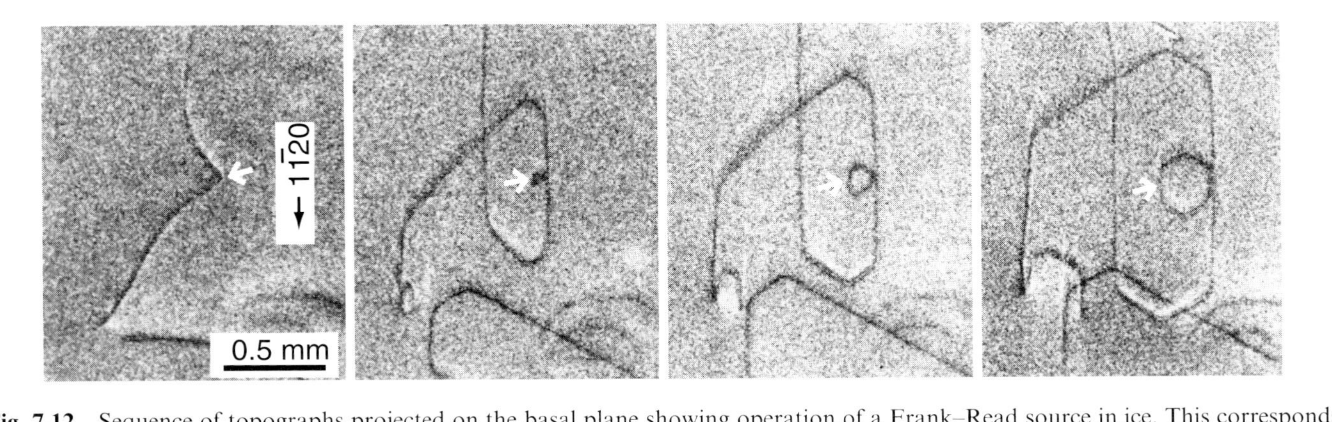

Fig. 7.12 Sequence of topographs projected on the basal plane showing operation of a Frank–Read source in ice. This corresponds to Fig. 7.11(a). (From Ahmad *et al.* 1992.)

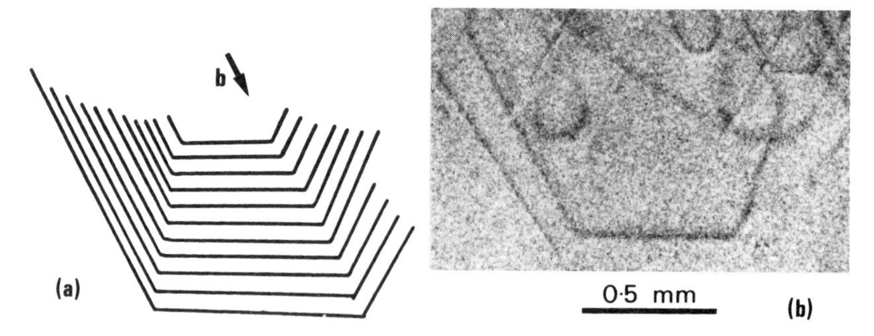

Fig. 7.13 Hexagonal dislocation loop expanding on the basal plane in ice under successive loadings at −4.3 °C. A sequence of tracings is shown in (a), and (b) is the actual topograph for the sixth of these tracings. (From Shearwood and Whitworth 1991, with permission from Taylor and Francis.)

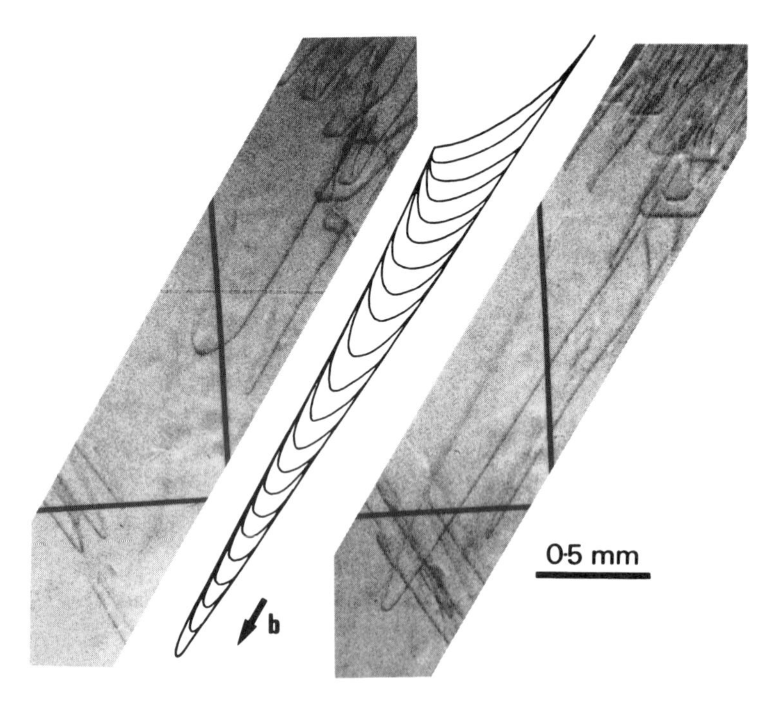

Fig. 7.14 Two topographs and a sequence of tracings showing the motion of a pointed dislocation loop gliding at −23.8 °C towards the lower left on a non-basal plane oblique to the plane of the topograph. (From Shearwood and Whitworth 1991, with permission from Taylor and Francis.)

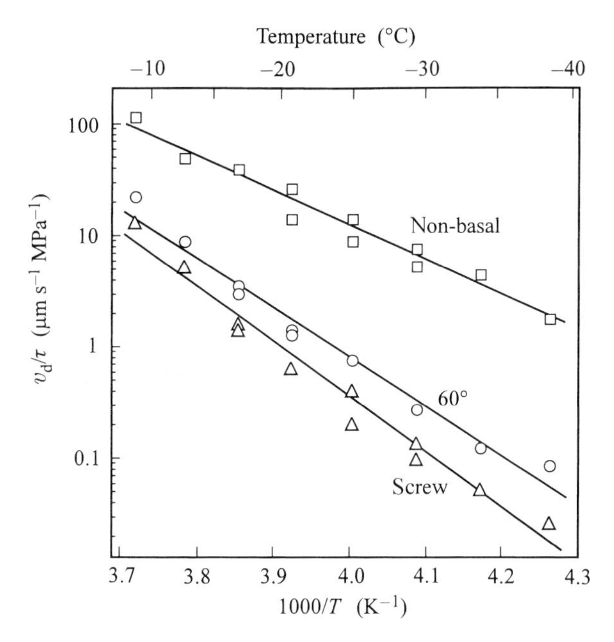

Fig. 7.15 Dislocation velocities per unit stress v_d/τ in pure ice as functions of inverse temperature $1/T$ for straight screw and 60° segments on the basal plane and for edge segments on non-basal planes. (From Shearwood and Whitworth 1991, with permission from Taylor and Francis.)

topographs showing the positions of dislocations between successive applications of stress. Figure 7.13 shows tracings of such positions for a loop expanding on the basal plane, and Fig. 7.14 shows the projection on the basal plane of an edge dislocation moving to the lower left on a non-basal plane. For the stresses up to 1 MPa used in these experiments the dislocation velocity v_d was found to be directly proportional to the resolved shear stress τ. The velocities per unit stress for basal screw, basal 60°, and non-basal edge dislocations in pure ice are plotted as functions of temperature in Fig. 7.15. The results for non-basal dislocations are consistent with the observations of Hondoh *et al.* (1990). Earlier results, such as those of Maï (1976) and Yamamoto and Fukuda (quoted by Fukuda *et al.* 1987), do not distinguish between different types of dislocations and were probably subject to recovery in the time taken to obtain the topographs (Ahmad and Whitworth 1988). There is no evidence to support the non-linear dependence of velocity on stress reported by Maï. In so far as the Arrhenius plots in Fig. 7.15 are straight lines, the activation energies for dislocation glide (actually from plots of $\ln(v_d T/\tau)$ vs. $1/T$) are as in Table 7.1.

Another technique which in principle provides information about dislocation mobility is internal friction. Many experiments (e.g. Vassoille *et al.* 1978; Tatibouet *et al.* 1986) have observed a contribution that increases with

Table 7.1 Activation energies for dislocation glide in pure ice

	Activation energy (eV)
Basal screw (straight segments)	0.95 ± 0.05
Basal 60° (straight segments)	0.87 ± 0.04
Non-basal edge	0.63 ± 0.04

From Shearwood and Whitworth (1991).

temperature, is enhanced by plastic deformation, and often depends on amplitude. This effect can reliably be attributed to dislocations and might be usable for distinguishing between the motion of kinks, nucleation of kink pairs, and breakaway of dislocations from pinning points, but there are too many disposable parameters for us to draw conclusions from these experiments.

7.4.2 *Peierls model for basal dislocations*

In theory a straight dislocation will have minimum energy when it lies along a particular line within the crystal lattice. The Peierls barrier is the energy barrier for it to glide through one lattice spacing to the next stable position. The fact that under stress dislocations on the basal plane take up a hexagonal form and then glide as straight segments is strong evidence for motion which is limited by a Peierls barrier. This behaviour has also been observed in materials such as silicon (George and Rabier 1987; Nadgornyi 1988). A step in a dislocation between two positions of minimum energy constitutes a kink, as in Figs 7.4 and 7.5. The kinks can glide along the dislocation, and under stress their motion will carry the dislocation forward until it consists of almost straight segments along the directions of minimum energy. In ice these are the screw and 60° segments seen in Figs 7.8 and 7.13. To advance further, pairs of kinks must be thrown forward across the Peierls barrier and then move sideways until they reach the end of the segment or are annihilated by other kinks moving in the opposite direction. The dislocation velocity is then given by

$$v_{\mathrm{d}} = h n_{\mathrm{k}} v_{\mathrm{k}}, \tag{7.1}$$

where n_{k} is the number of kinks per unit length, v_{k} is their velocity, and h is the separation of the Peierls troughs (a lattice spacing in the appropriate direction).

According to the standard theory of this kind of dislocation motion (Hirth and Lothe 1982), in the limit of low stress ($\tau bah \ll k_{\mathrm{B}}T$, where a is the lattice parameter and b is the magnitude of the Burgers vector), v_{d} is directly proportional to stress. It depends on both the activation free energy to form an isolated kink F_{k} and that to move it F_{m} according to the equation

$$v_{\mathrm{d}} = \nu \frac{2\tau bah^2}{k_{\mathrm{B}}T} \exp\left[-\frac{F_{\mathrm{k}} + F_{\mathrm{m}}}{k_{\mathrm{B}}T}\right]. \tag{7.2}$$

The quantity ν is a characteristic frequency that cannot be more than the Debye cut-off frequency. The free energies can be written in terms of the energies and entropies of formation or migration (i.e. $F_k = E_k - TS_k$ etc. cf. §6.2), and the measured activation energy for glide is then the temperature independent part of $(E_k + E_m)$. However, with the experimental values in Table 7.1 and other known parameters, the entropies of activation turn out to be remarkably high (for details see Shearwood and Whitworth 1991). The model indicates that dislocations seem able to move more easily than expected from theory. The contributions arising from kink nucleation and kink migration could in principle be separated by studying the motion of curved segments (Hondoh 1992), but this has not yet been successfully achieved.

There is no evidence for a Peierls barrier to the motion of non-basal edge dislocations, and this is consistent with their lower activation energy for glide (Table 7.1). The most significant fact about non-basal slip is the complete absence of any glide of screw dislocations on the non-basal planes under the conditions used in topographic experiments. This is strong evidence that screw dislocations in ice are dissociated on the basal plane, because the partial dislocations have edge components that cannot glide off this plane. A consequence is that, as in most materials with similar structure, basal slip must occur on planes of the glide set.

7.4.3 *Proton disorder*

A unique feature of the structure of ice, first recognized by Bjerrum (1952) and then developed by Glen (1968), is that the disorder of the protons presents in principle an obstacle to the glide of dislocations. This arises quite simply because, if two planes of molecules are linked by randomly oriented hydrogen bonds, they cannot be sheared over one another and still link up correctly. The idea is easily understood by considering a 60° dislocation on planes of the shuffle set as shown in Fig. 7.16, where the positions of the protons are disordered according to the ice rules. For the dislocation to glide to the left the bond DD′ must be broken and D′

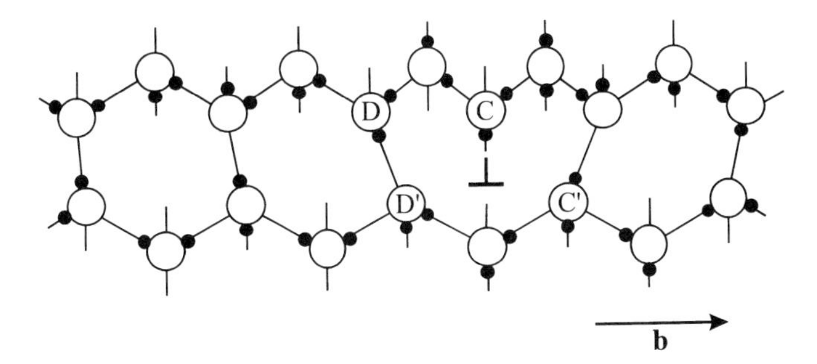

Fig. 7.16 Section in the $(1\bar{1}00)$ plane of a 60° dislocation on a plane of the shuffle set in the structure of ice, illustrating the consequence of the disorder of the protons according to the model of Glen (1968).

joined to C. This presents no problem, but for motion to the right C must link to C′ which would create a D-defect. There is not sufficient energy available from the stress to form this defect, and there is in general no local rearrangement of bonds that will avoid the formation of a defect somewhere. In practice we expect glide to occur by motion of kinks along the dislocation, with each step of the kink involving a single exchange of bonds like the one just considered. There will be a 50% chance of bonds being mis-matched at each step.

Glen (1968) therefore proposed that the limiting factor for dislocation motion may be the rate at which bonds are randomly reoriented by ions or Bjerrum defects, just as in the process of dielectric relaxation. It is important to realize that the stress cannot force the reorientation of the required bonds as the dislocation approaches. This idea was quantified for kinks on the shuffle plane by Whitworth *et al.* (1976) and Frost *et al.* (1976). The stresses involved are always such that they impose only a small bias on the random motion of kinks; the kinks are not pushed up against the mismatched bonds. With this assumption the kink velocity v_k for bond reorientations by Bjerrum defects is given by

$$v_k = \frac{3\sqrt{3}a^4}{\tau_b k_B T}\tau, \tag{7.3}$$

where τ_b is the mean time between successive reorientations of a bond.

For dislocations on planes of the glide set the obstacle presented by proton disorder may be less severe. Provided the core is not reconstructed, molecules B and C in Fig. 7.5(a) are free to rotate about their single bonds perpendicular to the glide plane as they switch linkages from G and H to A and D, and similarly in Fig. 7.5(b). This freedom increases the probability of the bonds being able to match correctly. For a kink on an unreconstructed partial dislocation Whitworth (1980) has shown that

$$v_k = \frac{310\sqrt{3}a^4}{\tau_b k_B T}\tau. \tag{7.4}$$

If the partial dislocation were reconstructed the barrier presented by proton disorder would be much greater, but this situation has not been analysed theoretically. In all cases where the Peierls barrier is present proton disorder will affect the rate of double kink nucleation as well as kink mobility, but it can only make the theoretical dislocation velocity less than that predicted by equation (7.2). A theory has also been developed for a flexible dislocation line (Whitworth 1983). This does not seem to be applicable to dislocations that glide as straight segments, but it may apply to the more mobile non-basal edge dislocations.

The quantity τ_b in these models is the mean time for the reorientation of bonds close to the dislocation core, and if we assume that the ice rules are applicable this reorientation will take place only by the motion of protonic point defects. In bulk crystal this time must be approximately the same as the Debye relaxation time τ_D, but Shearwood and Whitworth (1991) have shown that for dislocations of any kind to glide at the observed rate the appropriate value of τ_b must be much shorter than this.

Jones (1967) made the remarkable observation that ice crystals doped with HF deform much more easily than pure crystals, and, as it was known that HF-doping reduces τ_D (§5.4.3.1), this led Glen (1968) to propose that the rate-limiting process for dislocation motion was the rate of bond reorientation by protonic defects (see Jones and Glen 1969*b*). A critical experiment to test this idea is therefore to measure dislocation velocities in doped ice. This was first attempted by Maï *et al.* (1978), who found a small effect but much less than predicted. Using HCl doping Shearwood and Whitworth (1992) found no significant change in the velocities of basal or non-basal dislocations under conditions where τ_D was known to have been reduced by more than a factor of 10.

From all that we know about the disorder of the protons in ice it is essential that there be a process for reorienting bonds at the dislocation core to retain compatibility with the ice rules in the surrounding material. From what has been said above it is clear that this is not simply reorientation at the rate determined from dielectric relaxation in the bulk, but it could arise from an enhanced concentration or mobility of ions or Bjerrum defects near the core. Interstitials or vacancies should not be directly involved as they do not change the state of proton disorder. Perez *et al.* (1978, 1980) postulated that the dislocation core was in some sense non-crystalline, thereby avoiding the obstacle presented by the ice rules. However, the fact that dislocations glide as straight segments which are indicative of a Peierls barrier shows that the dislocation core must retain much of its crystalline character. The model of Perez *et al.* has several adjustable parameters, and these take unphysical values that depend on subsequently unsubstantiated non-linearities in their observations of $v_d(\tau)$. The precise way in which dislocations overcome the barrier presented by proton disorder is at present unknown, and as we will see in §8.2.1 the evidence about the effects of doping on plastic deformation of ice is not entirely clear.

Weertman (1963) proposed a further mechanism by which the disorder of the protons will inhibit the glide of dislocations. This is anelastic relaxation (§6.5.1), which dissipates energy in the stress field around the moving dislocation and predicts a limiting velocity proportional to τ_m^{-1} (which is effectively the same as τ_D^{-1}). Any such effect constitutes a barrier additional to those already considered, but it seems likely to be comparatively small.

7.5 Electrical effects

Dislocations with an edge component have dangling bonds at the core, and if the numbers of these bonds with and without protons are unequal the dislocation will carry an electric charge. In equilibrium any such charge will be screened by a surrounding cloud of excess protonic point defects of opposite sign (Whitworth 1975), but during deformation a dislocation may become separated from this charge cloud. Petrenko and Whitworth (1983) observed small transverse electric currents associated with the tensile deformation of previously bent crystals, and interpreted these currents as due to dislocations carrying a net positive charge of at least 0.002 protonic charges per atomic length. Itagaki (1970) has reported X-ray topographic experiments in which dislocations appear to move in an alternating electric field. In cases where the sign of the charge could be identified it was positive, but it is difficult to deduce magnitudes from such experiments. A problem with any experiment of this kind is that the field should be maintained for sufficient time to move the dislocations by an observable amount, but in this time the field is largely eliminated by polarization and conduction in the ice. If the dislocation cores are indeed reconstructed as suggested in §7.2.1, then any core charge will be confined to kink sites or perhaps to Bjerrum-type defects trapped in the reconstruction. Another possibility is that point defects are trapped in the elastic strain field around the dislocation line.

On applying electric fields to thin specimens of ice undergoing deformation in shear on the basal plane Petrenko and Schulson (1993) observed reductions in both the strain rate and the high-frequency conductivity σ_∞. It was shown in §5.7.1 that such fields reduce the concentration of charge carriers, and if this experiment is interpreted as changing the plastic creep rate of the ice it provides some support for the idea that dislocation velocities are limited by the availability of defects capable of reorienting bonds. However, the observations may alternatively be interpreted in terms of sliding at the ice–metal interface (§13.1).

7.6 Stacking faults

7.6.1 *Structure of stacking faults in ice*

The crystal structure of ice consists of (0001) planes of molecules stacked above one another as indicated in Fig. 7.2. If the positions of these layers in their planes are denoted by the letters A, B, and C as defined in Fig. 7.3(a), the hexagonal structure of ice Ih follows the sequence

AABBAABBAABB....

The C positions are unoccupied, resulting in empty channels running through the structure in the [0001] direction. Stacking faults arise if the stacking across the planes marked 'glide set' in Fig. 7.2 departs from the perfect sequence. General explanations of such faults can be found in Hirth and Lothe (1982) or similar texts, and the geometry appropriate to the ice structure was analysed by Frank and Nicholas (1953). To simplify the discussion we will denote each pair of layers such as AA by the single letter A, and the stacking of ice Ih is then that familiar in hexagonal close-packed metals:

ABABABAB....

Cubic ice (ice Ic §11.8) has the stacking sequence

$$\text{ABCABCABC}\ldots,$$

illustrated in Fig. 11.14, but it is unstable at all temperatures relative to ice Ih.

A stacking fault is a planar defect normally lying on a (0001) plane. It must extend to the surface or terminate at a partial dislocation with a Burgers vector equal to the displacement required to create the fault. To distinguish among the various kinds of faults it is useful to introduce Frank's notation, which concentrates not on the absolute location of the layers A, B, and C but only on the stacking relative to the layer below. Thus the equivalent stackings of B on A, C on B, and A on C are all denoted by the symbol Δ, while the inverse stackings of

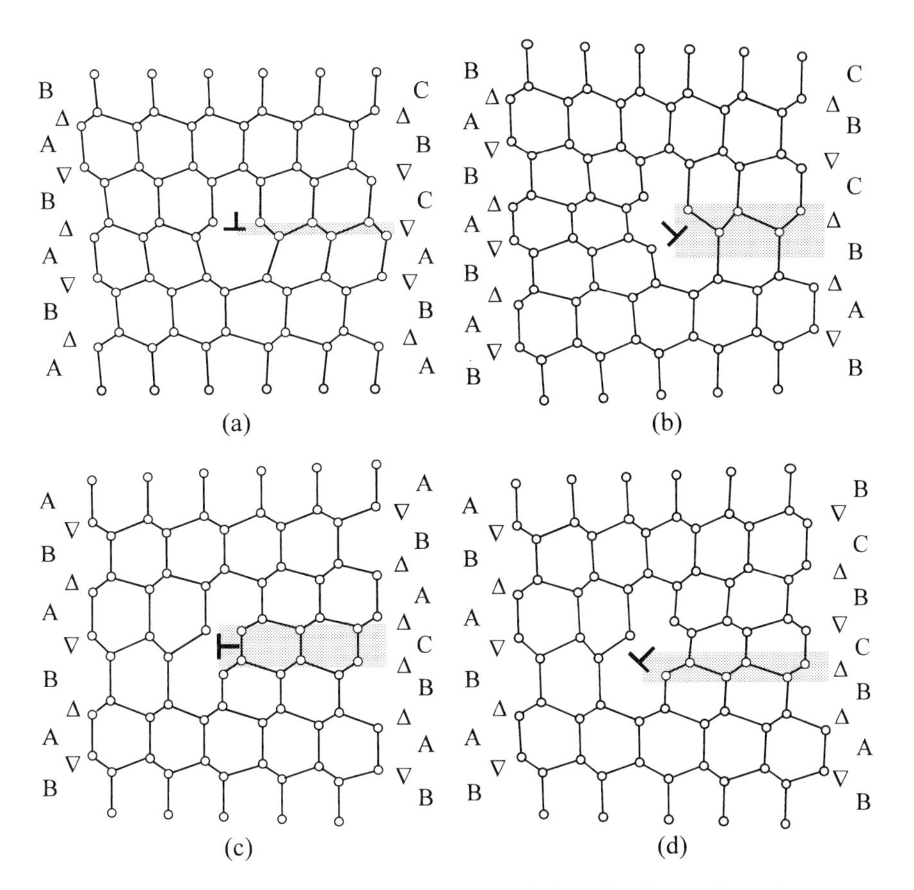

Fig. 7.17 Stacking faults and partial dislocations in ice. The figures show the structure projected on a $\{11\bar{2}0\}$ plane, with correct stacking on the left and a fault on the right terminating at the appropriate partial dislocation. (a) Shockley partial dislocation produced by glide. (b) An A-layer removed followed by shear. (c) A C-layer added. (d) As (c) followed by shear on the upper face of the added layer.

A on B, B on C, and C on A are given the symbol ∇. The hexagonal stackings ABABABAB..., BCBCBCBC..., or CACACACA... are then all denoted by

Δ∇Δ∇Δ∇Δ...,

whereas cubic stacking would be

ΔΔΔΔΔΔΔ... or ∇∇∇∇∇∇∇....

There are four different ways of introducing stacking faults into ice and these are illustrated in Fig. 7.17. The first is to shear a B layer over an A layer into a C position. This changes a Δ into a ∇, giving the fault

↓
BABACBCB....
∇Δ∇∇∇Δ∇...

This fault is illustrated in Fig. 7.17(a), in which it terminates at a partial dislocation of Burgers vector $(a/3)\langle 10\bar{1}0\rangle$. When a perfect $(a/3)\langle 11\bar{2}0\rangle$ dislocation dissociates as in Fig. 7.3 into two Shockley partial dislocations, this is the type of stacking fault that will be formed.

The other types of faults require the addition or removal of a layer A, B, or C, each of which actually contains two planes of molecules joined by bonds parallel to the [0001] axis. These are the kinds of faults associated with the precipitation of molecular point defects referred to in §6.4.2. If an A layer is removed the B layers on opposite sides of the gap can only link together if there is a displacement of one face over the other by an amount $(a/3)\langle 10\bar{1}0\rangle$, generating the fault

↓
ABABCBCB....
Δ∇ΔΔ∇Δ∇...

This fault is shown in Fig.7.17(b). The dislocation terminating it has both prismatic and glide components, and its Burgers vector is $(1/6)\langle 20\bar{2}3\rangle$.

Unlike the case just described, the addition of a C layer into the perfect ABAB... structure does not necessitate any shear and the fault generated is

↓
ABABCABAB....
Δ∇ΔΔΔΔ∇Δ

This fault is illustrated in Fig. 7.17(c). It is terminated by a prismatic dislocation loop of Burgers vector $(c/2)[0001]$. The fault contains four consecutive Δ-type stackings, which constitute a thin layer of cubic ice. It can lower its energy by a shear between the C and the A layer on the upper face of the added C layer in Fig. 7.17(c), yielding the fault

which is shown in Fig. 7.17(d). Examination of the sequence shows that this is the same stacking fault as in Fig.7.17(b), but the nature of the bounding dislocation corresponds to an interstitial rather than a vacancy-type prismatic loop. Stacking faults are often classified as *intrinsic* if perfect stacking is maintained up to the plane of the fault and *extrinsic* if it is not. On this basis the faults (a), (b), and (d) in Fig. 7.17 are intrinsic and only the higher energy fault (c) is extrinsic.

7.6.2 *Observations of stacking faults*

Stacking faults of macroscopic dimensions can be observed by X-ray topography and have been studied extensively in ice by the group at Sapporo. Their work is reviewed by Fukuda *et al.* (1987), Oguro and Hondoh (1988), and Oguro *et al.* (1988). Stacking faults in ice are energetically unstable and never remain in crystals that have been well annealed and observed at the temperature of annealing. However, a few large area faults may occur in freshly grown crystals (Hondoh *et al.* 1983), and their formation is enhanced by doping, particularly with ammonia (Oguro and Higashi 1973).

Figure 7.18 is an example of a topograph of pure ice cooled to $-45\,^{\circ}\mathrm{C}$. The dark areas are stacking faults parallel to the basal plane. The large irregular faults originate from the growth of the crystal, while the smaller patches are prismatic loops formed during cooling. The presence of contrast within the area of the loop shows that the fault vector $\mathbf{f}$ satisfies the condition $\mathbf{g}\cdot\mathbf{f} \neq 0$ or an integer, where $\mathbf{g}$ is the diffraction vector of the topograph. In this case $\mathbf{g} = \langle 10\bar{1}0\rangle$, and this implies that these faults have a shear component. The bounding dislocations have a [0001] component, and careful analysis shows that the faulted loops formed on cooling are interstitial in character (Hondoh *et al.* 1983). It appears that an interstitial prismatic loop with Burgers vector $(c/2)[0001]$ does in fact lower its energy by taking the sheared form of Fig. 7.17(d) rather than that of Fig. 7.17(c). The stable prismatic loops shown in Fig. 7.7 do not contain stacking faults and must be bounded by perfect dislocations of Burgers vector $c[0001]$ including two double layers of molecules within the loop. Such dislocations would have a very high energy and will almost certainly be dissociated, but the stacking fault ribbon between them is unresolvable in topography.

As described in §6.4.2 detailed studies of the growth and shrinkage of faulted and unfaulted prismatic dislocation loops have yielded the best available parameters for the formation and diffusion of interstitials in ice. The energy of a $(1/6)\langle 20\bar{2}3\rangle$ fault has been determined by Hondoh *et al.* (1983) to be $0.31\ \mathrm{mJ\,m^{-2}}$, which is $3\times10^{-4}\,\mathrm{eV}$ per molecule in the plane. This is assumed to be the lowest energy type of fault because it has only two adjacent Δs in the stacking sequence. The shear fault with three adjacent Δs is postulated to have twice this energy, and on this basis the separation of the Shockley partial dislocations formed by the dissociation of a perfect $(a/3)\langle 11\bar{2}0\rangle$ dislocation on a basal plane has been estimated to be 20 nm for a screw dislocation and 46 nm for an edge dislocation (Fukuda *et al.* 1987). Observation of this dissociation would require electron microscopy, and this has not yet been achieved in ice.

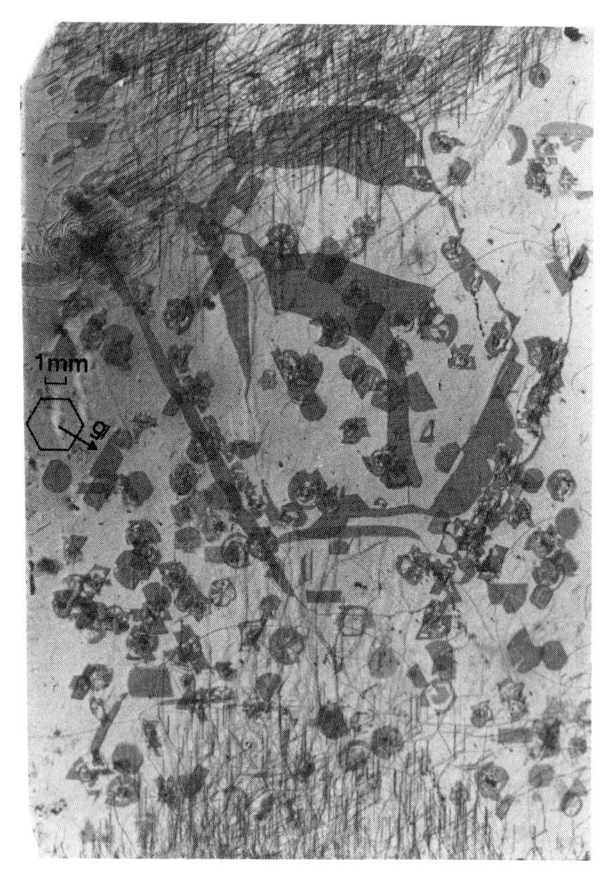

Fig. 7.18 X-ray topograph projected on the (0001) plane of a crystal of pure ice cooled rapidly to −45 °C. Large stacking faults were formed during growth and small faulted prismatic loops were produced by the condensation of interstitials. (From Oguro *et al.* 1988.)

7.7 Grain boundaries

7.7.1 *Structure and properties*

Natural ice is polycrystalline and the boundaries between the individual crystals (or 'grains') can be considered as planar defects within otherwise perfect material. We will briefly describe the structure and intrinsic properties of this class of defects.

Examination of polycrystalline ice samples shows grains of many shapes, sizes, and orientations, depending on the history of the material (e.g. Matsuda and Wakahama 1978). In well-annealed ice the boundaries are fairly flat, which is a necessary condition for minimizing the surface energy. Within a single grain, or within a piece of ice that is nominally a single crystal, there are often sub-boundaries across which there is a lattice misorientation of at most a few degrees and often very much less. These sub-boundaries are made up of arrays of dislocations in accordance with geometrical principles that apply to any crystalline material (e.g. Read 1953; Hirth and Lothe 1982). Except for elastic interactions with one another to form a stable structure, these dislocations behave as individual dislocations within a single crystal. A tilt sub-boundary composed of edge dislocations will migrate by the glide of these dislocations under an appropriately oriented stress, and this has been observed in ice by Higashi and Sakai (1961).

For large angle boundaries the concept of an array of dislocations is not applicable, and there must be an interface across which bonding departs greatly from that in a perfect crystal. For some particular orientations the bonding may be less irregular than for others, and the grain boundary energy will depend on the relative orientations of the grains and the boundary between them, as observed by Suzuki and Kuroiwa (1972).

A commonly assumed condition for the formation of any special type of low energy boundary is that the lattice points for the two half-crystals should match up with one another in a periodic way along the plane of the boundary. The 'coincidence-site lattice' (CSL) model has this property and has been described in relation to ice by Higashi (1978) and Hondoh (1988), but it imposes further constraints which are not generally thought to be significant (Sutton 1984). A favourable configuration for ice is one in which the grains are rotated relative to one another by 34.1° about the $[10\bar{1}0]$ direction, and such a boundary at which the grains are joined across their $(1\bar{2}11)$ planes is illustrated in Fig. 7.19(a). Hondoh and Higashi (1978) have grown bicrystals containing this type of boundary, and have shown by X-ray topography that the boundary is not flat but is made up of facets that make small angles to one another. They propose that a facet that makes a small angle to the exact $(1\bar{2}11)$ plane will contain an array of intrinsic grain boundary dislocations of the small Burgers vector shown in Fig. 7.19(b). Between these dislocations the lattices will match exactly as in Fig. 7.19(a). Experiments on the diffusional motion of boundaries during strain-free annealing have shown variations with the type of boundary and some tendency to form facets with particular orientations (Hondoh and Higashi 1979; Nasello *et al.* 1992).

Other examples of misorientations that satisfy the CSL conditions are 47° about $[10\bar{1}0]$ and 21.8° about [0001]. The latter is thought to lead to certain types of 12-branched snowflakes (Kobayashi and Furukawa 1975). These special cases can also be thought of as growth twins, though in ice they are not usually formed in the ways commonly associated with twinning. It must be remembered that the coincidence of a number of lattice points across an interface does not ensure that the hydrogen bonds link up in a manner even approximating to that in the perfect crystal, and there must be considerable disorder of the molecules in the boundary. The thickness of this non-crystalline, or even 'liquid-like', region is not known.

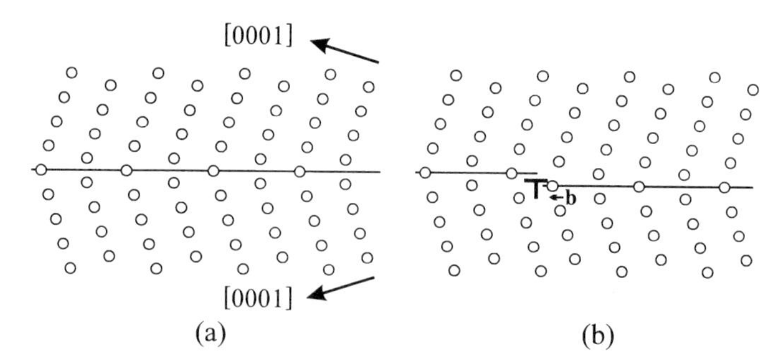

Fig. 7.19 (a) The boundary on a $(1\bar{2}11)$ plane between two grains rotated relative to one another by 34.1° about the $[10\bar{1}0]$ axis perpendicular to the diagram. The circles are lattice points not molecules. (b) The same boundary containing an intrinsic boundary dislocation

In this connection it is relevant to note that close to the melting point liquid water is present in polycrystalline ice in 'veins' which lie along the lines of intersection of grain boundaries (Mader 1992; Nye 1992), and this liquid will be in equilibrium with the internal structure of the boundaries themselves.

When a grain boundary meets a free surface its energy acts as a surface tension. This produces a groove on the surface, and Ketcham and Hobbs (1969) found that in equilibrium at $-3\,°C$ the included angle $\theta = 145 \pm 2°$. This angle is related to the free energies γ_{sv} and γ_{gb} of the ice–vapour interface and the grain boundary by the equation

$$\gamma_{gb} = 2\gamma_{sv}\cos(\theta/2). \tag{7.5}$$

Using $\gamma_{sv} = 69.2\,\text{mJ m}^{-2}$ from §10.7, this equation gives $\gamma_{gb} = 42\,\text{mJ m}^{-2}$. This should be compared with the much smaller stacking fault energy of $0.31\,\text{mJ m}^{-2}$ quoted in §7.6.2.

Grain boundaries can move in a number of ways according to principles that are generally applicable to all materials. Under a shear stress a pair of suitably oriented grains may slide over one another on the boundary (§8.3.1). In other cases shear parallel to the boundary can be produced by the migration of the boundary perpendicular to its plane, as has been reported for a 34° $[10\bar{1}0]$ boundary by Hondoh (1988) and explained in terms of the glide of intrinsic boundary dislocations of the type shown in Fig. 7.19(b). As this dislocation glides to the left in the plane of the boundary one row of lattice points of the lower crystal is displaced to sites belonging to the upper one, and the boundary therefore moves downwards by one layer.

The motion of grain boundaries or of dislocations within them leads to a characteristic peak in the low-frequency internal friction of ice (Perez *et al.* 1979; Tatibouet *et al.* 1987). Boundaries also move by diffusive processes in which molecules in one grain are rearranged in the lattice structure of the other, and this is what happens when ice recrystallizes during large-scale plastic flow (§8.3.4).

Being places where there are irregularities in the lattice, grain boundaries can act as sources and sinks of point defects. We have referred in §6.4.2 and §7.6.2 to the formation of prismatic dislocation loops by the condensation of interstitials on cooling. Near a boundary there is a zone where these loops are not formed because the excess interstitials are lost to the boundary (Hondoh *et al.* 1981).

7.7.2 *Electrical properties of grain boundaries in doped ice*

Electrical measurements on monocrystalline and polycrystalline samples of ice lead to the conclusion that these types of ice differ significantly even when the concentrations of impurities averaged over the volume are the same, and this leads to the conclusion that grain boundaries make an appreciable contribution to the conductivity.

The most obvious reason why this may be so is that the impurities segregate to the boundaries, and particularly to the lines along which three grain boundaries meet. Mulvaney *et al.* (1988) investigated the impurity distribution in polycrystalline Antarctic ice using a scanning electron microscope with a cold stage and an X-ray microanalysis facility with a spatial resolution of 10 nm and a detection limit of 5 mM (equivalent to 490 p.p.m. for H_2SO_4). They found that, although the volume average impurity concentrations were less than 1 p.p.m., the concentration of SO_4^{2-} at triple junctions was 2.5 M over an area of 1 μm^2. This concentration is of the order of the eutectic composition for H_2SO_4 which freezes at $-73\,°C$, so that the triple junctions will form a network of veins of concentrated liquid down to very low temperatures. Wolff and Paren (1984) have suggested that the low-frequency conductivity σ_s of polar ice could be accounted for by the presence of acidic liquid layers at the grain boundaries, and most particularly at grain boundary intersections. Using reliable data for the concentrations of H_2SO_4, HNO_3, and HCl at the South Pole they derived the correct magnitude and temperature dependence for the conductivity of such ice.

It is possible that at temperatures near the melting point grain boundaries have an enhanced conductivity due to a quasi-liquid layer similar to that on a free surface (§10.4.1), but this is just a speculation and measurements on genuinely pure boundaries would be difficult to achieve.

8
Mechanical properties

8.1 Introduction

It has been known for hundreds of years that glaciers flow down mountains. The way in which this could be explained in terms of the properties of ice was a matter of controversy during the nineteenth century (e.g. Tyndall 1860; Hopkins 1862), and at that time Reusch (1864) described observations of both the ductility and the brittleness of ice on a laboratory scale. Subsequently McConnel and Kidd (1888) showed that ice can be deformed plastically in tension or compression provided this is done sufficiently slowly for the ice not to fracture. They found also that single crystals would not deform under stresses parallel or perpendicular to the c-axis. McConnel (1891) showed that a single crystal bar could be bent through 90° in 6 hours at about −1 °C. From their early but quite detailed experiments they concluded that ice deforms by slip on the basal planes, as has already been described in Chapter 7.

Common experience in handling pieces of ice tells us that ice is a brittle material. However, for large blocks of ice this brittleness does not necessarily result in complete rupture of the block; it is possible for a polycrystalline mass of ice to accommodate many deformation-induced cracks without falling apart. In situations where ice is subjected to large rapidly applied stresses its brittle properties are more relevant than its ability to creep, though as we will see the two processes are not independent on the microscale. The mechanical properties of ice are very important for the engineering of structures to be built on or with ice, for the design of ice-breaking ships, and for the impact of floating ice on drilling platforms or other structures. On a larger scale of size and time there is also great interest in modelling the flow of glaciers and ice sheets. Under the confining pressure of their own weight such ice masses deform entirely plastically without cracking. 'Ice mechanics' has become one of the largest areas of research on ice, but in this chapter we will deal only with the basic physical principles involved.

The review of the creep of ice by Weertman (1973) includes a catalogue of work and proposed interpretations up to that time. The books on 'Ice Mechanics' by Michel (1978) and Sanderson (1988) deal with the subject from the point of view of the engineering applications, while the review by Budd and Jacka (1989) describes the rheological properties required for modelling the flow of ice sheets. Goodman *et al.* (1981) collated and extrapolated the data and interpretations available at the time to form 'deformation mechanism maps' covering a wide range of temperature and strain rate.

Mechanical tests of materials are performed in certain standard ways of which two are of particular concern to us. In the first the test sample is subjected to a constant stress, in the case of tension or compression often simply by hanging a weight on it. It will then elongate or contract, yielding a graph of the change in length as a function of time; this process is known as 'creep'. The deformation may accelerate, decelerate, or settle down to a constant rate, though at large strains it may be necessary to take account of changes in the cross sectional area. In the second kind of test the specimen is deformed at a constant strain rate, and the load required is recorded, giving a load–time or stress–strain graph. At any point during such a test at low strain rates the specimen itself is deforming just as it would in a creep test at that stress and accumulated strain. In both types of test the specimen will eventually fracture, but at high strain rates there may be virtually no ductility so that the specimen fractures almost immediately. This is true in compression as well as under tension. For a given type of specimen, geometry of deformation, and temperature, the transition from ductile to brittle behaviour occurs at a certain strain rate. This is illustrated schematically in Fig. 8.1, which also makes the distinction between ductile deformation with and without the formation of stable microcracks within the ice. As in this figure, the failure stress has its maximum value at the strain rate corresponding to the transition from ductile to brittle behaviour.

In the case of ice tensile tests were traditionally performed on long relatively thin specimens with the ends frozen into metal cups, and sometimes with the test section machined to be narrower than the ends. In more recent experiments end caps of fibre-based phenolic (carpet-like) material are often incorporated in the

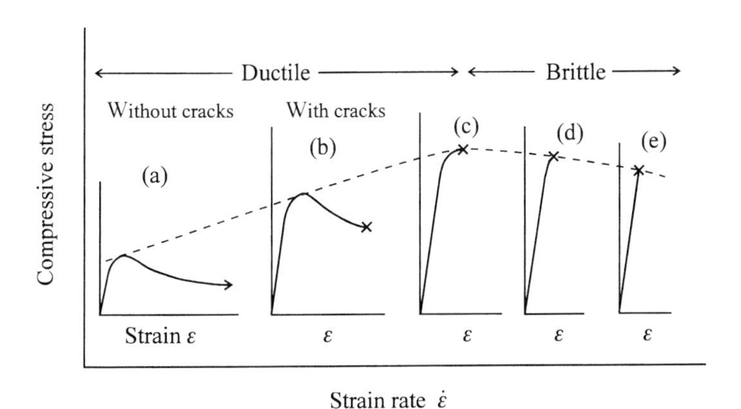

Fig. 8.1 Series of stress–strain curves illustrating the ductile-to-brittle transition with increasing strain rate. The crosses indicate the point of fracture, but at the lowest strain rates creep can continue indefinitely. Reproduced with modifications from *Acta Metallurgica et Materialia*, **30**, E.M. Schulson, The brittle compressive fracture of ice, pp. 1963–76, Copyright 1990, with permission from Elsevier Science.

ice as it is frozen, and this greatly reduces the risk of fracture at the ends (Cole 1979). Compression samples with a smaller length/width ratio appear more stable against fracture, but they are always subject to concerns over the alignment of the end plates and any constraints imposed on the deformation by transverse displacements, or lack of them, at the ends.

Because the plastic deformation of ice is confined almost entirely to slip on the basal plane, the mechanical properties of a single crystal are highly anisotropic. In a polycrystalline sample the individual crystals cannot easily deform in a compatible manner, and the properties of polycrystalline ice are very different from those of single crystals. The grain boundaries serve to nucleate dislocations and to block the slip bands so produced. In addition they can both nucleate and stabilize cracks. In this chapter we first describe the ductile properties of single crystals, then those of polycrystalline ice, and finally the brittle fracture of the polycrystalline material.

8.2 Plastic deformation of single crystals

8.2.1 *Basal slip*

Provided the stress is kept small (~0.1–0.5 MPa) and creep is allowed to take place slowly, single crystals of ice are very ductile. Glen and Perutz (1954) deformed rods with the c-axis at ~45° to their length in tension, producing local extensions of as much as 600%. Such strains converted the rod into a ribbon with the c-axis almost perpendicular to its length, but Glen and Perutz could find no preferred crystallographic direction for slip on the basal plane. Our understanding of the slip process from Chapter 7 is that each individual dislocation produces slip by one of the $(a/3)\langle 11\bar{2}0\rangle$ lattice vectors, but macroscopic slip can result from the motion of a number of dislocations which may have two or even three of these vectors. A theoretical analysis by Kamb (1961) of the net effect of slip in the three possible directions on the basal plane showed that the direction of the strain resulting from a given stress depends on the power law relating the strain rate to the stress. In the case of ice the slip direction is always too close to the direction of the maximum resolved stress for any crystallographically preferred direction of slip to be detectable.

The ductility of ice in slip purely on the basal planes is elegantly demonstrated by bending crystals of different orientations. The early experiments of this kind by McConnel (1891) were repeated with better specimens and under more controlled conditions by Nakaya (1958), and his observations are summarized in Fig. 8.2. With the c-axis at ~45° to the length of the bar as in Fig. 8.2(a) the bar bends smoothly as shown. The bent specimen will contain a very high density of edge dislocations introduced by slip on the basal planes, and traces of the slip planes become visible in shadow microscopy as shown in Fig. 8.3. With the c-axis parallel to the direction of loading, the ice still deforms by slip on the basal planes,

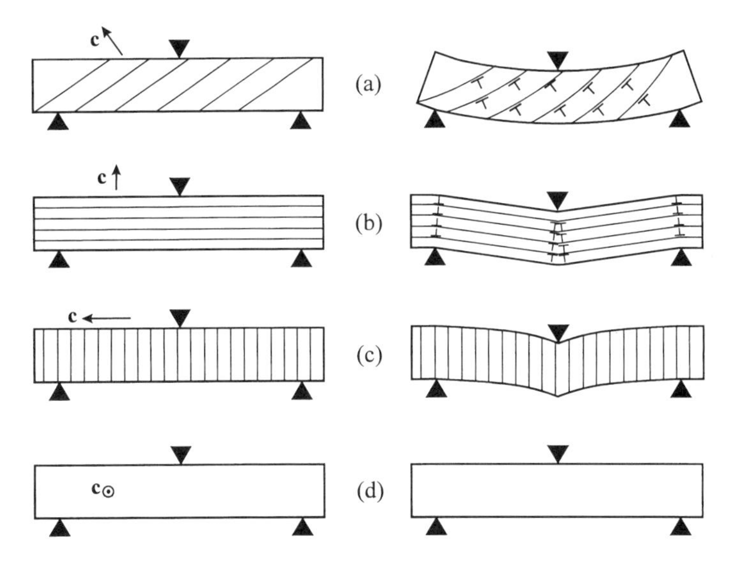

Fig. 8.2 Modes of plastic deformation in three-point bending of single crystals of ice with differently oriented c-axes (Nakaya 1958).

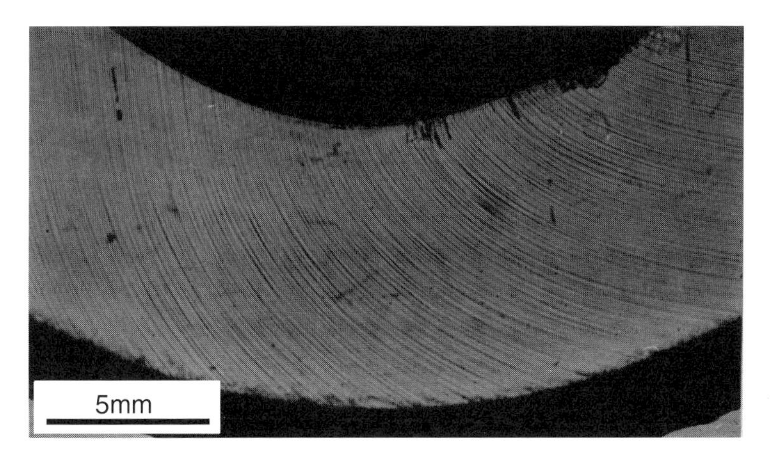

Fig. 8.3 Shadow micrograph of the surface of a single crystal of ice bent as in Fig. 8.2(a). (From Nakaya 1958.)

but edge dislocations cannot enter or leave at the surfaces. Instead they accumulate in regions close to the loading wedges to form one or more tilt boundaries, so that the specimen is sharply bent at these three points as shown in Fig. 8.2(b).

With the c-axis along the length of the bar, slip on the basal plane produces the shape shown in Fig. 8.2(c), but for the c-axis perpendicular to the plane of bending as in Fig. 8.2(d) there is no resolved shear stress on the basal plane and no deformation occurs. In all these experiments the deformation of ice is closely analogous to that of a stack of cards, except that in some cases such as Fig. 8.2(a) the cards themselves become bent. In such cases the amount of slip varies across the face of the card, which is equivalent from Fig. 7.1 to dislocations being introduced on the planes between the cards.

Figure 8.4 shows a series of stress–strain curves obtained in constant strain rate tensile tests on previously undeformed crystals oriented for basal slip. According to dislocation theory for deformation on a single slip system the strain rate is given by the equation

$$\dot{\varepsilon} = \phi N_{\mathrm{d}} b v_{\mathrm{d}}, \tag{8.1}$$

where N_{d} is the density of mobile dislocations (length per unit volume), b is the magnitude of the Burgers vector, v_{d} is the dislocation velocity, and ϕ is an orientation factor. The observed peak followed by a yield drop in the stress–strain curve arises because N_{d} is initially small. The constant deformation rate then requires a high initial value of v_{d} and hence of the stress. As the deformation proceeds the dislocations multiply, and the stress necessary to maintain the constant strain rate decreases. Eventually the stress–strain curve tends to a constant stress with little work hardening. In this steady-state region there will be a balance between the rate of dislocation multiplication and the combined effects of emergence from the surface, immobilization, and recovery processes. Muguruma (1969) has shown that the size of the initial peak in the stress depends

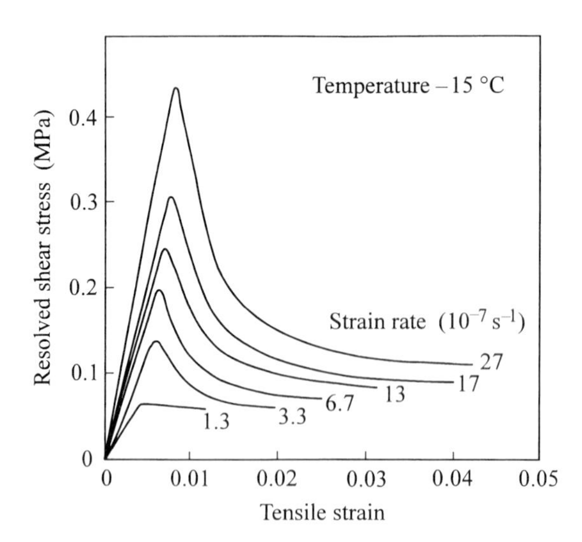

Fig. 8.4 Stress–strain curves in constant-strain-rate tensile tests on single crystals of ice from the Mendenhall Glacier oriented with [0001] at 45° to the tensile stress. (From Higashi *et al.* 1964.)

on the state of the surface, being larger for polished and well-annealed crystals in which there will be fewer sources from which dislocation multiplication can develop.

In a creep test the behaviour corresponding to Fig. 8.4 would be an initially accelerating creep leading to a constant strain rate. Higashi *et al.* (1965) observed such behaviour in bending experiments in the geometry of Fig. 8.2(b), but simple tensile tests give continuously accelerating creep as shown for a range of stresses at $-50\,^{\circ}\mathrm{C}$ in Fig. 8.5. Such measurements are not reproducible from specimen to specimen, and this is seen from the fact that the curves do not lie in order of the increasing stress; this would appear as scatter in a plot of strain rate against stress. The creep curves generally have the approximate form $\varepsilon(t) \propto t^{1.5}$ (Jones and Glen 1969*a*).

The creep rate given by equation (8.1) depends both on the dislocation velocity, which is a function of internal stress, and on the dislocation density, which will depend on stress, accumulated strain, and the elapsed time during which recovery may occur. There have been a number of studies of the deformation of single crystals of ice. These have been performed under different conditions and analysed in different ways, but there has been no systematic study such as has been done on single crystals of many other simple materials. All workers use some criterion, such as comparing strain rates at a constant strain (Homer and Glen 1978), to fit their measurements at different stresses σ and temperatures T to the single equation

$$\dot{\varepsilon} \propto \sigma^{n} \exp(-E_{\mathrm{s}}/k_{\mathrm{B}}T). \tag{8.2}$$

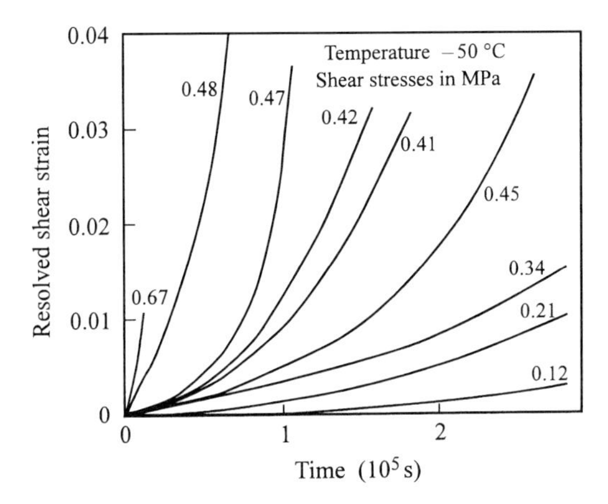

Fig. 8.5 Tensile creep curves under different loads for single crystals of ice oriented with [0001] at 45° to their length (Jones and Glen 1969*a*). Reproduced by courtesy of the International Glaciological Society from *Journal of Glaciology*, Vol. 8, No. 54, 1969, pp. 463–73, Fig. 1.

For single crystals n is then generally found to be 1.5–2.5, and Table 8.1 lists values quoted for the activation energy E_s, but such data are not sufficient to describe adequately the creep of ice. Figure 8.6, which is based on a compilation by Duval *et al.* (1983), compares the creep rates of single crystals and polycrystalline specimens of ice at $-10\,^{\circ}$C. Because single crystals show accelerating creep it is not possible to quote precise rates for a given stress, and for

Table 8.1 Activation energies for the plastic deformation of single crystals of ice in basal slip

Type of experiment	Reference	Temperature range (°C)	Activation energy E_s (eV)
Tensile, constant $\dot{\varepsilon}$	Higashi *et al.* (1964)	-15 to -40	0.69
Bending creep	Higashi *et al.* (1965)	-3 to -44	0.68
Tensile, constant $\dot{\varepsilon}$	Readey and Kingery (1964)	0 to -42	0.62
Tensile creep	Jones and Glen (1969*a*)	-10 to -50	0.68 ± 0.04
		-50 to -70	0.41 ± 0.03
Tensile creep	Homer and Glen (1978)	-5 to -20	0.81 ± 0.04

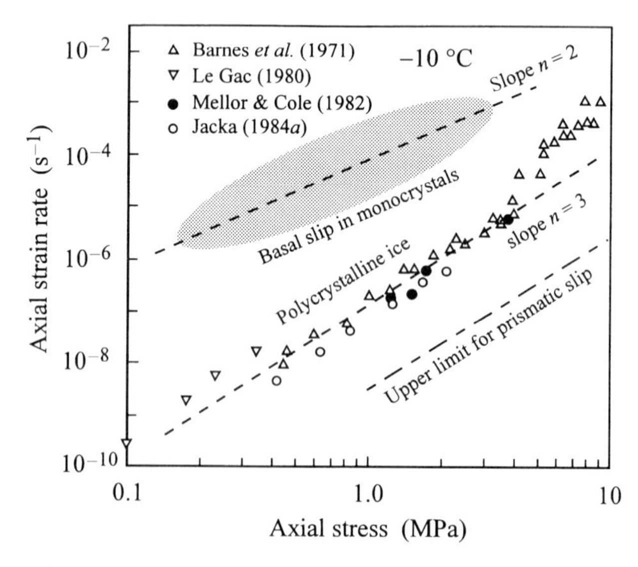

Fig. 8.6 Compilation of data on the plastic strain rate of granular polycrystalline ice at $-10\,^{\circ}$C, compared with typical strain rates for basal and non-basal slip in single crystals. Based on Duval *et al.* (1983); the data from Le Gac are quoted in that paper.

them the figure shows only a zone indicating an order of magnitude, together with a line of slope $n=2$.

In §7.4.1 it was shown that the velocity v_d of isolated dislocations is directly proportional to stress, so that values of $n=1.5$ to 2.5 in equation (8.2) imply that the density of mobile dislocations increases with stress. Weiss and Grasso (1997) reach the same conclusion from studies of the acoustic emission during the creep of single crystals. The values of E_s in Table 8.1 are all less than those for the motion of straight dislocation segments on basal planes in Table 7.1. Extensive experiments on silicon reveal no such difference in activation energies, but a difference has been observed in some other semiconductors (Rabier and George 1987). There are two possible reasons for the difference. Either the dislocation density for steady-state creep increases with falling temperature (in addition to any increase directly attributable to the stress) or the dislocations do not glide as straight segments under the conditions of the deformation experiments. If it is easier to create kinks within the deforming material than under the near perfect conditions of topography experiments, the activation energy for dislocation motion should be reduced. Macroscopic deformation is a very complicated process and for ice we do not have data which can separate the effects of stress, accumulated strain, and temperature on N_d.

The effect of hydrostatic pressure on the creep of ice is a question of interest because the ice in glaciers and ice sheets is subject to large confining pressures. The topic was first investigated by Rigsby (1958) who observed a small effect attributable to the effect of pressure on the temperature measured relative to the melting point. More recent experiments by Cole (1996) on large single crystals oriented for basal slip and deformed in compression at $\sim -10\,°C$ showed no consistent effect of pressures up to 20 MPa.

At high strain rates single crystals are brittle, and the brittle fracture of specimens of various orientations has been studied by Rist (1997). Suitably oriented crystals fracture by shear on a basal plane, but cleavage on $\{1\bar{1}00\}$ planes is also observed.

Jones (1967) found that doping ice with HF produced a remarkable softening both in creep tests (Fig. 8.7) and in constant strain rate tests at -60 to $-70\,°C$. The softening was observed both in crystals doped before deformation and when HF was diffused into a specimen part way through the test (Jones and Glen 1969*b*). NH_4OH produced a small hardening. Nakamura and Jones (1970) deformed HCl-doped ice at higher temperatures and observed a softening similar to that with HF. These observations led to Glen's hypothesis that proton disorder is the rate limiting process for dislocation glide as discussed in §7.4.3, but topographic experiments at low doping levels and higher temperatures have not confirmed this interpretation of the effect. A complication is that the strain rate depends on the dislocation density as well as the velocity, and Jones and Gilra (1972) found that diffusing HF into ice, as was done in some of the experiments referred to, produced a large increase in dislocation density. To date the large effect of HF on the plastic properties of ice has not been satisfactorily explained, and there have

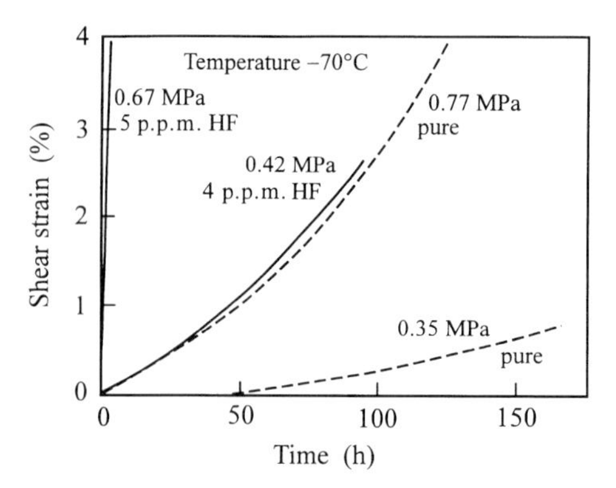

Fig. 8.7 Tensile creep curves for HF-doped single crystals of ice (solid lines) and corresponding curves for pure ice (broken lines). Reproduced from *Physics Letters*, **25A**, S.J. Jones, Softening of ice crystals by dissolved flouride ions, pp. 366–7, Copyright 1967, with permission from Elsevier Science.

been no recent experiments to confirm the phenomenon in the light of the knowledge we now have about the properties of dislocations.

8.2.2 *Non-basal slip*

In a tensile test a crystal oriented with its *c*-axis perpendicular to the length has no resolved shear stress on the basal plane, but Muguruma *et al.* (1966) found that such crystals could be deformed plastically by up to 6%. At −19 °C and a constant strain rate of $3 \times 10^{-6}\,s^{-1}$ the tensile stress required was ~10–20 MPa, which is two orders of magnitude larger than that required to deform a crystal oriented favourably for basal slip. These specimens deformed by slip in the usual $\langle 11\bar{2}0\rangle$ directions but on the non-basal $\{1\bar{1}00\}$ planes. Tensile deformation was accompanied by the formation of voids visible in an optical microscope, but voids were not formed in the deformation of similarly oriented crystals in compression. Mae (1968) has proposed that the voids are formed by the condensation of vacancies generated in the climb of edge dislocations. The topographic observations of dislocation glide on $\{1\bar{1}00\}$ planes (Figs 7.10, 7.14 and 7.15) showed that edge dislocations move easily, but at the comparatively low stresses used screw dislocations did not glide on these planes. The difficulty of producing macroscopic slip on $\{1\bar{1}00\}$ planes must be related to the difficulty of moving the screw dislocations, but the possible relevance of this more recent information to the problem of void formation has not been addressed.

There have been no observations of any plastic deformation in ice crystals loaded parallel to the *c*-axis. In this orientation there is no resolved stress on either the basal or the prismatic planes. Deformation would therefore require the glide of dislocations on pyramidal planes such as $\{10\bar{1}1\}$ or $\{11\bar{2}2\}$ with a component of the Burgers vector in the [0001] direction. Examination of a model of the structure reveals no simple slip process of this kind, and it seems that such slip is not possible in practice. Duval *et al.* (1983) have examined the evidence for slip in ice on non-basal slip systems, and Fig. 8.6 shows their upper limit to the possible tensile strain rate at −10 °C. It should be noted that slip on non-basal systems is in practice very difficult to produce, because the basal systems are so much softer that small misalignments or stress inhomogeneities at the ends will cause slip to occur on the basal planes instead.

8.3 Plastic deformation of polycrystalline ice

The plastic deformation of polycrystalline ice has been studied in a very large number of experiments. The first such experiments to adopt a modern approach were those of Glen (1955) and S. Steinemann (1958). Subsequent work has investigated the relationships between stress, strain, and time in specimens of different grain structure and, more directly, what is actually happening within the grains of the ice. The issues relevant to the flow of glaciers and ice sheets were reviewed by an international group in 1979 (Hooke *et al.* 1980), and current understanding of the rheological properties of such ice is reviewed by Budd and Jacka (1989). We will start, however, with the physical processes that occur within the ice.

8.3.1 *Physical processes*

The plastic properties of a single crystal of ice are very anisotropic, and this implies that when several differently oriented grains form a polycrystalline aggregate they cannot easily be deformed in a mutually compatible way. It is a general principle (Taylor 1938) that a crystal must possess five independent slip systems if it is to be deformed by homogeneous shear in a completely general way, and this condition is satisfied for the face-centred cubic metals which are very ductile even in polycrystalline form. The application of the principle to hexagonal crystals such as ice is explained by Groves and Kelly (1963). Slip in any direction on the basal plane constitutes only two independent slip systems. It can produce shear on the basal plane, but it cannot produce extensions parallel or perpendicular to the *c*-axis. Slip in $\langle 11\bar{2}0\rangle$ directions on the $\{1\bar{1}00\}$ prismatic planes could provide two further slip systems, but the fifth system which would be necessary to produce an extension parallel to the *c*-axis is never observed. Hutchinson (1977) showed that in practice a uniform strain on *four* slip systems would be sufficient to maintain cohesion between the grains during the deformation of a polycrystalline specimen, because the inability of one grain to deform in a particular way can be accommodated by its neighbours deforming around it. In ice, however, slip on the prismatic planes is so hard that it plays little role in the deformation of polycrystalline samples, and the properties are dominated by slip on the basal planes alone. The important feature of the deformation of polycrystalline ice is that, contrary to the assumption of Taylor or Hutchinson, the deformation within the grains need not be homogeneous.

When polycrystalline ice is subjected to stress the grains that are favourably oriented for basal slip will start to deform, but their deformation is blocked by other grains and this builds up internal stresses both in the deforming grains and in those around them. These stresses can be relieved in a variety of ways, all of which play some part in the deformation of ice.

1. Non-uniform deformation within a single grain can produce bending like that in Fig. 8.2(a), the formation of a single tilt boundary, or the formation of pairs of opposite tilt boundaries. Such opposite pairs result in 'kinking' similar to that in Fig. 8.2(b). The formation of several tilt boundaries is referred to as 'polygonization'.
2. Some slip may occur on the prismatic planes in regions of exceptionally high stress concentration, such as may be produced close to the grain boundaries.
3. Grains may slide over one another on the plane of the boundary. If two adjacent grains deform differently there may have to be some partial sliding on the boundary to maintain continuity between them. In other cases whole grains are displaced relative to one another, and this constitutes what is known in fracture mechanics as a mode II crack.
4. Grain boundary migration will occur, causing some grains to grow at the expense of others. This actually happens by the rearrangement of molecules on one side of the boundary into the structure of those on the other side. The process requires molecular diffusion, but not necessarily at a rate characteristic of bulk material.
5. There may be diffusion-controlled recovery within the deforming grains by the climb of dislocations off their glide planes.
6. In regions which are highly deformed recrystallization occurs, with the nucleation and subsequent growth of new grains, often more favourably oriented for basal slip. This can occur repeatedly during deformation, when it is called 'dynamic recrystallization' in contrast to the slower recovery that often occurs on annealing after deformation has ceased.
7. If the stress becomes too large at any point a crack may be nucleated. This relaxes the stress, and may be quite stable for a time, but eventually the growth of such cracks leads to brittle fracture of the whole specimen.

Both dislocation glide and processes 4, 5, and 6 are thermally rate-limited, and they can only occur if the ice is deformed sufficiently slowly. At higher strain rates the localized stresses become larger and will ultimately lead to the formation of cracks. This is why the ductile-to-brittle transition shown in Fig. 8.1 depends on the strain rate.

The evolution of the grain structure of thin specimens of polycrystalline ice during compression in the plane of the specimen was carefully studied in the polarizing microscope by Wakahama (1964). He observed basal slip, lattice bending, the formation of tilt boundaries, grain boundary sliding, extensive grain boundary migration, and the formation of new grains by recrystallization. More recently Wilson and Zhang (1994, 1996) made similar observations, and Fig. 8.8 is an example from their work showing how the grain structure is changed by compressions of 3% and 8%. Some grains grow at the expense of others, the grain boundaries become convoluted, new grains form, and at 8% strain few of the original grains are recognizable. Zhang *et al.* (1994*b*) used a computer to model the deformation of a two-dimensional polycrystalline aggregate with only one slip plane, and Fig. 8.9 shows examples of what happens under the requirement that cohesion is maintained across the grain boundaries. The grains become highly distorted and on average they are rotated into orientations less favourable for further shear on these planes. Very large stresses are built up within the grains and cannot be relaxed within the assumptions of the model. Wilson and Zhang (1994, 1996) have applied this model to the actual grain structures observed in some of their experiments. For small strains the basal slip model corresponded well to what was observed, but as the deformation proceeded the stresses in the real ice were relaxed by grain boundary migration and recrystallization. When grain boundary sliding was introduced into the computer model (Zhang *et al.* 1994*a*) intergranular cracks opened up. This corresponds well to what will be described for the brittle regime in ice, where

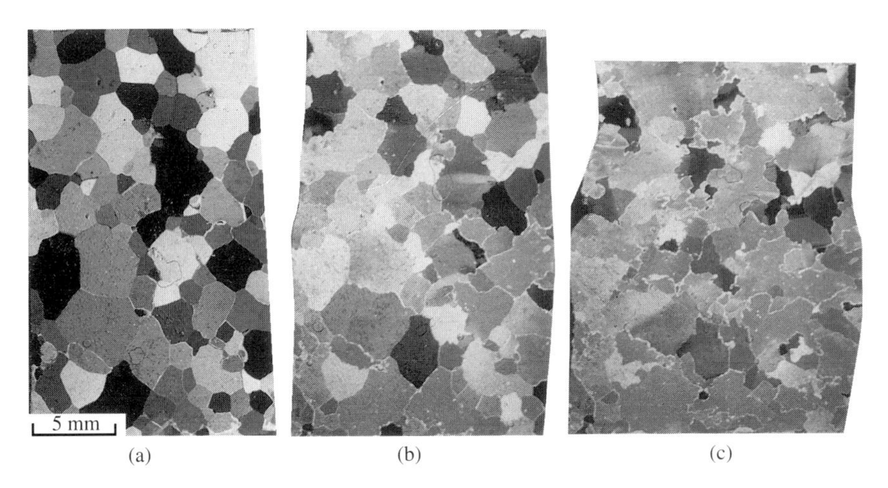

Fig. 8.8 Grain structure, observed between crossed polaroid filters, of a 0.7 mm thick polycrystalline specimen of ice held between a pair of glass plates at −1 °C and deformed in compression parallel to the plane of the plates by (a) 0%, (b) 3%, and (c) 8%. The compression plates were at the top and bottom edges of the figures and the side edges were unconstrained. Photographs provided by C.J.L. Wilson and Y. Zhang. (a) and (c) are reproduced by courtesy of the International Glaciological Society from *Annals of Glaciology*, Vol. 23, 1996, pp. 293–302, Fig. 1.

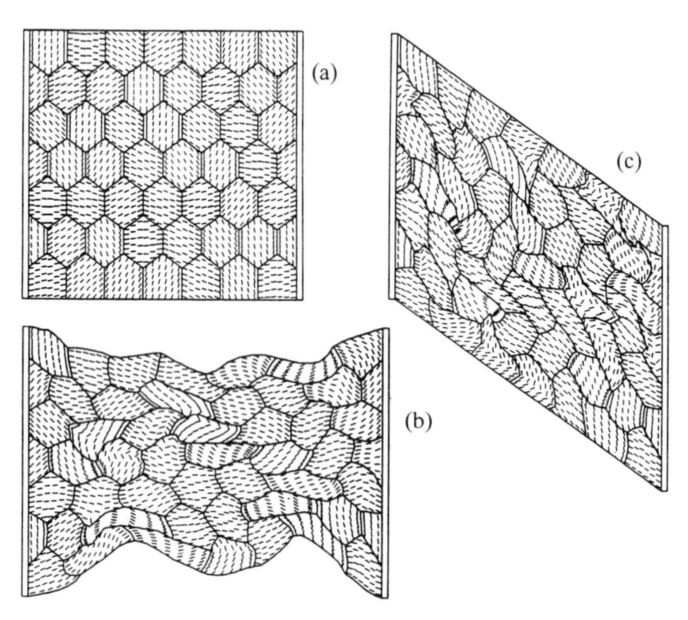

Fig. 8.9 Computer modelling of the plastic deformation of a two-dimensional randomly oriented polycrystalline aggregate of grains with only one slip system. The slip planes are marked by broken lines drawn in each grain. (a) The initial aggregate, (b) after extension by 33%, and (c) after simple shear to a strain of 0.72. The model assumes isotropic elasticity and flow above a critical shear stress corresponding to a strain of 2×10^{-6}. It also requires that contact be maintained across the grain boundaries. Reproduced from *Journal of Structural Geology*, **16**, Y. Zhang *et al.*, A numerical simulation of fabric development in polycrystalline aggregates with one slip system, pp. 1297–313, Copyright 1994, with permission from Elsevier Science.

insufficient time is available for the relaxation of the stresses by diffusion-controlled processes.

Gold (1963, 1966) made observations of the surface of specimens of columnar-grained (S2) ice (see §12.1) deformed perpendicular to the columns, also demonstrating the changes within the grains. An example is shown in Fig. 8.10, in which the fine lines are traces of the slip planes and the central grain shows the formation of two tilt boundaries. The ice is highly distorted in the regions close to the grain boundaries. These are regions where slip may be occurring on non-basal planes, but it is not possible to identify such slip positively. It is also more likely to happen internally than in grains like these which are on a free surface.

It is hard to make direct observations of grain boundary sliding, though one case is described by Ignat and Frost (1987). Nickolayev and Schulson (1995) deformed columnar-grained ice oriented to favour sliding on the boundaries between the columns, and they observed cracking consistent with sliding on these boundaries. At low deformation rates grain boundary sliding is limited by steps in

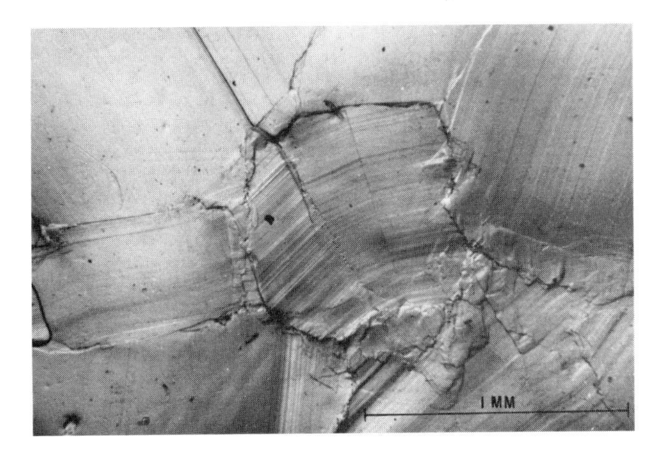

Fig. 8.10 Image of the surface of a specimen of columnar ice deformed in compression by 5% at −9 °C. The fine lines are traces of the slip plane. Several tilt boundaries can be seen in the central grain, and there is heavy deformation possibly involving non-basal slip near the grain boundaries. (From Gold 1966.)

the boundary. These have been observed as sites for the nucleation of slip bands (Liu *et al.* 1995), and the amount of sliding will be closely related to the deformation occurring in these slip bands. At high strain rates or small grain sizes the grain boundaries are the weak places at which yielding occurs.

In the slow steady deformation of ice to large strains as in the flow of an ice sheet, and particularly under confining pressure which prevents the formation of cracks, the dominant processes are basal slip followed by recrystallization to relieve the internal stresses. The recrystallization occurs when the stored energy in the grains becomes so large that the free energy can be reduced by nucleating a new grain. Once nucleated this grain will grow, but the processes require thermally activated molecular diffusion.

8.3.2 *Creep curves*

Figure 8.11 illustrates (not to scale) the shape of a typical creep curve $\varepsilon(t)$ for polycrystalline ice under a constant stress. It is very different from that for a single crystal of ice (Fig. 8.5), but it exhibits the regions known as primary, secondary, and tertiary creep which are observed in many polycrystalline materials (e.g. Boresi *et al.* 1993). On application of the stress the immediate strain AB is due to the elasticity of the specimen together with any effects arising from the bedding down of the specimen in the testing apparatus. This is followed by a region of decelerating 'primary creep' extending to the point of inflection C, after which the creep accelerates and eventually reaches a constant rate DE. Early work on ice

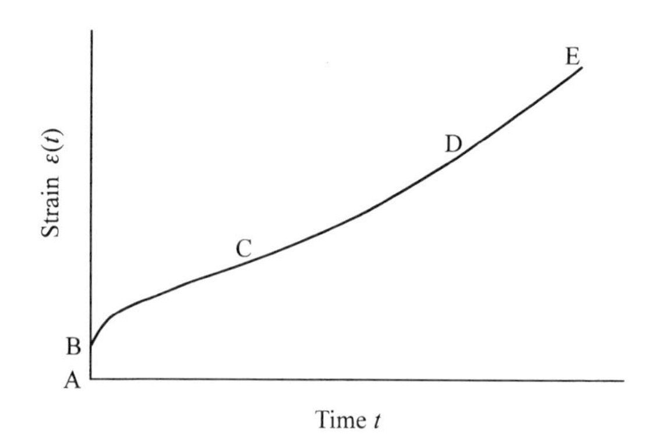

Fig. 8.11 Schematic creep curve for polycrystalline ice under constant load.

by Glen (1955) and P. Barnes *et al.* (1971) identified a region of steady-state or 'secondary' creep around the point C. Later experiments, like those to be described in the following section, show only a broad minimum in $\dot{\varepsilon}(t)$, but the minimum strain rate $\dot{\varepsilon}_{\min}$ represents a very important quantity in the analysis of the creep data. Deformation beyond the minimum is called 'tertiary creep'. The final steady state $\dot{\varepsilon}_{\max}$ marked DE in Fig. 8.11 is hard to attain in laboratory experiments, but this is the region of most significance in glaciology.

Constant strain rate deformation experiments are equivalent to creep experiments carried out under a slowly varying stress, and the correspondence has been closely examined by Mellor and Cole (1982). Examples of compression tests by Cole (1987) at various strain rates at $-5\,°\mathrm{C}$ are shown in Fig. 8.12. The broad peaks marked C correspond roughly to the stresses that would be required to produce this minimum value of $\dot{\varepsilon}$ in a creep test. The small peaks seen in the early part of the curves in Fig. 8.12 arise from dislocation multiplication as in single crystals of initially low dislocation density (cf. Fig. 8.4). Constant strain rate tests are mostly performed at strain rates higher than those obtained in creep experiments, and for the higher strain rates in Fig. 8.12 ($\dot{\varepsilon} \geq 10^{-5}\,\mathrm{s}^{-1}$) Cole reports some micro-cracking in the specimens. This has been confirmed by the detection of acoustic emission especially in the early stages of the deformation (St Lawrence and Cole 1982). Experiments in which cracking plays a dominant role will be considered in §8.4.

The experiments being considered here are all for type T1 (see §12.1) or 'granular' ice prepared with randomly oriented grains (e.g. Cole 1979; Jacka and Lile 1984). Initially such ice is macroscopically isotropic, but it deforms differently under different types of stress. For example, a given compressive stress produces a minimum strain rate that is about one fifth of the shear strain rate that would be produced under a shear stress of the same magnitude. For the purposes of modelling the flow of ice masses under complex stress systems it would be useful to have a single relation between strain rate and stress, and for this purpose

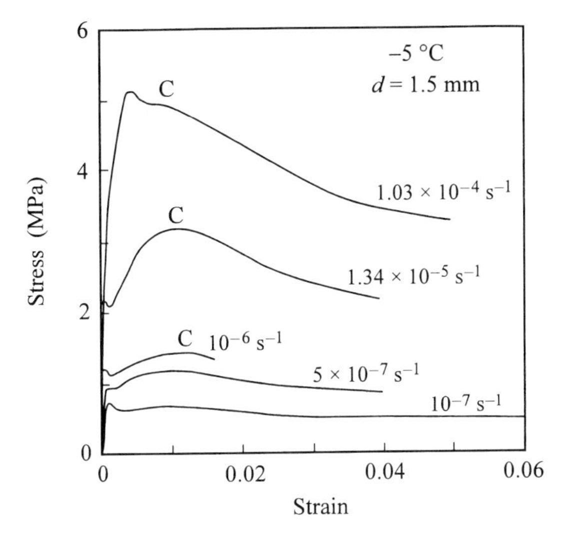

Fig. 8.12 Stress–strain curves for granular polycrystalline ice at various strain rates (Cole 1987). Reproduced by courtesy of the International Glaciological Society from *Journal of Glaciology*, Vol. 33, No. 115, 1987, pp. 274–80, Fig. 2.

Nye (1953) proposed describing the deformation by the second invariant of the stress tensor. The first invariant is just the hydrostatic pressure, and this second invariant is a measure of the shear component of the stress system. It is commonly expressed in terms of the 'octahedral shear stress' defined by the equation

$$\tau_{\text{oct}} = \tfrac{1}{3}\left[(\sigma_1 - \sigma_2)^2 + (\sigma_2 - \sigma_3)^2 + (\sigma_3 - \sigma_1)^2\right]^{1/2}, \tag{8.3}$$

where σ_1, σ_2, and σ_3 are the principal stresses (Jaeger 1956; Boresi *et al.* 1993). (Note that τ_{oct} differs by a factor of $(2/3)^{1/2}$ from the 'effective shear stress' used by Nye.) Although there is no question of resolving stresses on particular slip systems as in a single crystal, for a uniaxial stress σ_1 the stress τ_{oct} is the shear stress on the octahedral planes (i.e. planes making equal intercepts on the three principal axes). Budd and Jacka (1989) and Li *et al.* (1996) have shown that this approach works reasonably well for the early stages of the creep of ice. The general principles required for such modelling were formulated by Glen (1958). These procedures are applicable to macroscopically isotropic material, but not to specimens which are anisotropic because of the way they were grown, or which have become anisotropic as a result of previous deformation and are then deformed under a different stress system. However, in line with the procedures of Budd and Jacka (1989), many of the results in this section will be given in this form. The following conversions apply:

for a uniaxial stress σ_1

$$\tau_{oct} = \frac{\sqrt{2}}{3}\sigma_1 \qquad \varepsilon_{oct} = \frac{1}{\sqrt{2}}\varepsilon_1, \tag{8.4}$$

and for a shear stress τ_{12}

$$\tau_{oct} = \sqrt{\tfrac{2}{3}}\tau_{12} \qquad \varepsilon_{oct} = \sqrt{\tfrac{2}{3}}\varepsilon_{12}. \tag{8.5}$$

8.3.3 *The minimum strain rate (secondary creep)*

True steady-state secondary creep is not observed in ice. Figure 8.13 shows examples at −17.8 °C from the many creep curves obtained by Jacka (1984*a*),

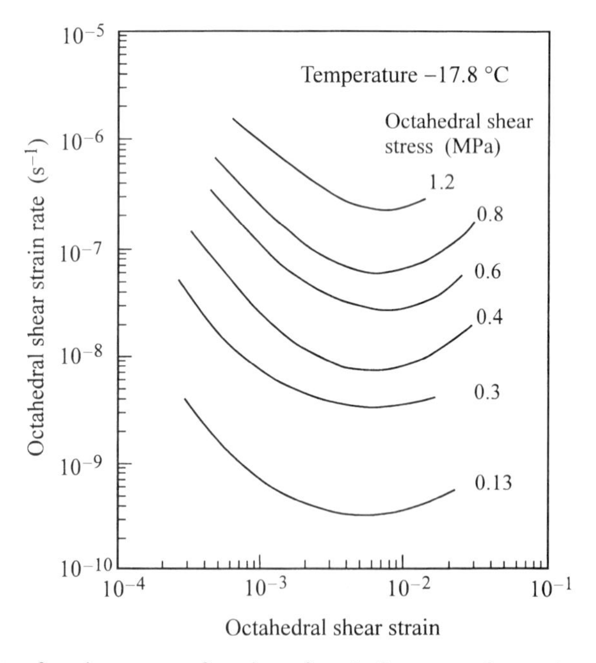

Fig. 8.13 Plots of strain rate as a function of strain for creep of granular polycrystalline ice in uniaxial compression under various stresses. In this and the following three figures stresses and strains have been converted to octahedral values according to equation (8.4). Reproduced from *Cold Regions Science and Technology*, **8**, T.H. Jacka, The time and strain required for development of minimum strain rates in ice, pp. 261–8, Copyright 1984, with permission from Elsevier Science.

plotted here in the form of strain rate as a function of strain using logarithmic scales. These curves show well-developed minima, but note that to achieve the minimum at the lowest stress the deformation had to be followed for more than a year! Many early experiments did not reach the strain rate minimum, and secondary creep rates reported in the literature may be misleading. Nevertheless secondary creep rates, whether obtained at the true minimum or not, have been extensively used in making comparisons between experiments performed at different stresses and temperatures. The important thing about the point of inflection C on the creep curve (Fig. 8.11) is not the balance between decelerating primary creep and accelerating tertiary creep. It is that in this approximately steady-state situation plastic flow of the grains occurs at a rate that is in balance with the processes which relieve the internal stresses so produced. These processes, which may include dislocation climb and grain boundary sliding or migration, are the rate limiting factors for secondary creep in polycrystalline ice.

The points plotted in Fig. 8.6 are the secondary creep rates of polycrystalline ice deduced in various ways from various tensile and compressive tests at $-10\,°C$, and the figure shows that this creep rate is intermediate between those for basal and non-basal slip in single crystals. Over an intermediate range of the stresses shown on the figure the secondary creep of polycrystalline ice obeys the power law proposed by Glen (1955)

$$\dot{\varepsilon} = B\sigma^n, \tag{8.6}$$

where B is a constant and $n \approx 3$. The logarithmic plot of Fig. 8.6 includes a line with this slope. In the experiments of P. Barnes *et al.* (1971) there was evidence that $n > 3$ at high stresses, and more recently Rist and Murrell (1994) obtained $n \approx 4$ from constant strain rate compression tests at high stress. Values of n greater than 3 have often been associated with the formation of microcracks in the ice. However, Manley and Schulson (1997) have produced evidence suggesting that n is correlated with the ratio of the climb to the glide forces acting on the dislocations.

For the creep curves in Fig. 8.13 all the minima occur at the same strain $\varepsilon_{\text{oct}} = 0.67 \pm 0.10\%$ (equivalent to $\varepsilon = 0.95\%$ in compression). This strain for minimum strain rate is the same within experimental error for temperatures from -5 to $-32.5\,°C$ in the experiments of Jacka (1984*a*) and in the independent measurements of Mellor and Cole (1982). The dependence of the minimum strain rate $\dot{\varepsilon}_{\min}$ on stress for various temperatures is shown in Fig. 8.14 (Budd and Jacka 1989). These stress dependencies all fit Glen's equation (8.6) with $n = 3$, and reports of smaller values of n at low stresses are probably consequences of failure to attain the true minimum rate. According to Jacka (1984*b*) the value of $\dot{\varepsilon}_{\min}$ is independent of the grain size in the ice, but the available range of grain sizes is quite small and there is conflicting evidence on this point (Baker 1978; Cole 1987).

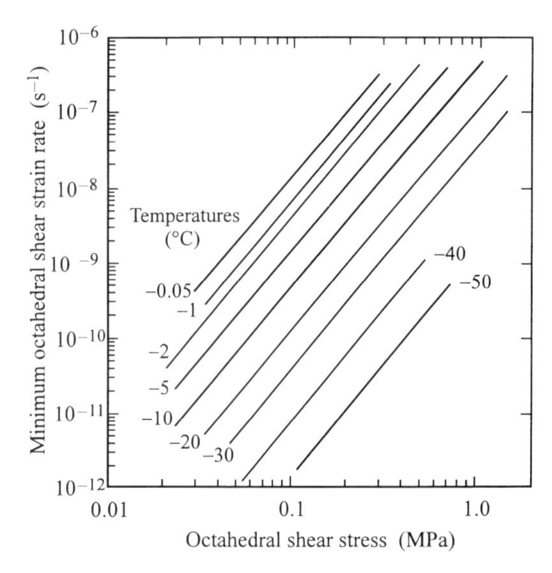

Fig. 8.14 The minimum strain rate in uniaxial creep tests on granular polycrystalline ice as functions of stress at various temperatures. Reproduced from *Cold Regions Science and Technology*, **16**, W.F. Budd and T.H. Jacka, A review of ice rheology for ice sheet modelling, pp. 107–44, Copyright 1989, with permission from Elsevier Science.

The temperature dependencies of the data in Fig. 8.14 are shown as Arrhenius plots in Fig. 8.15. At temperatures below −10 °C the lines are straight and parallel in accordance with the relation proposed by Glen (1955)

$$\dot{\varepsilon}_{\min} = A\sigma^n \exp(-E_{\rm p}/k_{\rm B}T), \tag{8.7}$$

with $E_{\rm p} = 0.72\,\text{eV}$ and $n = 3$. Above −10 °C $\dot{\varepsilon}_{\min}$ rises more rapidly with increasing temperature and cannot be described by this equation. Measurements to within 0.01 °C of the melting point are described by Morgan (1991), and show that $\dot{\varepsilon}_{\min}$ increases ever more rapidly as the melting point is approached. There have been suggestions that this is due to the presence of water in the ice, and it may be related to pre-melting phenomena at the grain boundaries such as are described for the free surface in Chapter 10. Pressure melting as a direct effect of the applied stress would not be relevant more than a few hundredths of a degree from the melting point, but in polycrystals pressure melting at the most critical points of extreme stress concentration might occur over a wider range of temperatures. Jacka and Li (1994) have shown that the value of $E_{\rm p}$ and the shape of the curves in Fig. 8.15 are very similar to those for the rate of grain growth in ice in the absence of stress.

As already stated the secondary creep rate in Jacka's experiments appears to be independent of grain size, and there are only small changes to the grain structure

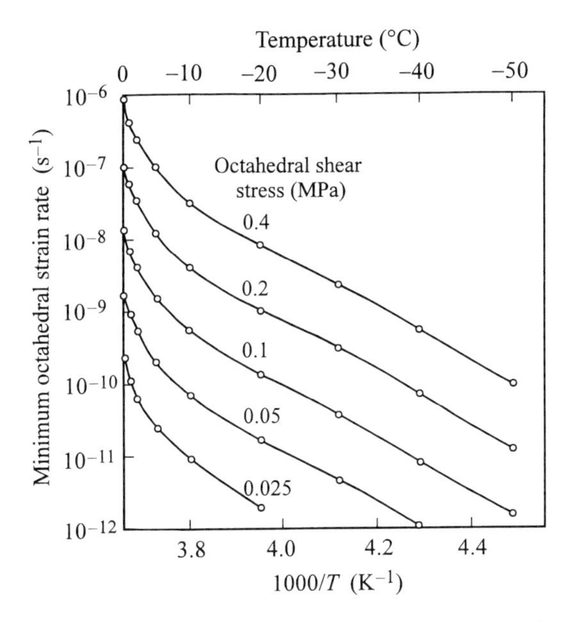

Fig. 8.15 Arrhenius plots of the minimum octahedral shear strain rate of granular polycrystalline ice in uniaxial creep tests at various stresses. Reproduced from *Cold Regions Science and Technology*, **16**, W.F. Budd and T.H. Jacka, A review of ice rheology for ice sheet modelling, pp. 107–44, Copyright 1989, with permission from Elsevier Science.

at this stage of the creep curve. A different form of creep was observed by Goldsby and Kohlstedt (1997) in extremely fine grained ice (diameters ~3–90 μm), which had to be kept and deformed at low temperatures to prevent spontaneous grain growth. The creep rate increased with decreasing grain size, and the stress exponent n was less than 3. On the basis of microscopic evidence they interpret these observations as due to grain boundary sliding.

In §8.2.1 the creep of monocrystals of ice was shown to be not much affected by a confining hydrostatic pressure. Jones and Chew (1983) examined this question in experiments on polycrystalline ice up to 60 MPa at −9.6 °C. The creep rate initially fell with increasing pressure but then increased, probably because of increasing proximity to the melting point (see Table 2.2). The overall conclusion seems to be that within the range of ductile deformation experiments and away from the melting point the creep rate of ice is not much affected by confining stresses up to 100 times the uniaxial or shear stresses causing the creep. Under pressures greater than about 200 MPa ice Ih transforms to other phases as explained in Chapter 11. The plasticity of ice II, ice III, ice V, and ice VI has been studied by Durham *et al.* (1987, 1997) and is relevant to the moons of the outer planets (§12.7.1).

8.3.4 *Tertiary creep*

S. Steinemann (1958) showed that the deformation of ice at high strains was associated with dynamic recrystallization, a phenomenon which also occurs in the deformation of many metals at high homologous temperatures (Sellars 1978). Duval (1981) and Jacka and Maccagnan (1984) determined the distribution of grain sizes and orientations (the 'fabric') as a function of strain during creep tests, and showed that the minimum in the creep rate coincides with the start of dynamic recrystallization. Beyond this point new grains are formed that are more favourably oriented for basal slip. At strains of more than about 10% steady-state tertiary creep is established at a rate $\dot{\varepsilon}_{max}$ and the fabric is controlled completely by the deformation. If the original grain size is larger than the dynamic equilibrium value new grains are nucleated at the grain boundaries, but if the original grain size is too small new or existing favourably oriented grains will expand, destroying those around them. The grain size during dynamic recrystallization is a function primarily of the stress as shown in Fig. 8.16, and the functional dependence is consistent with theoretical predictions (Jacka and Li 1994).

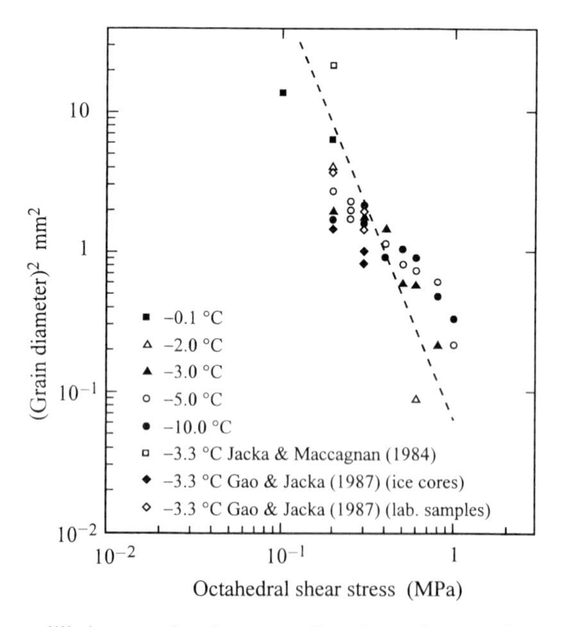

Fig. 8.16 The equilibrium grain size as a function of stress in polycrystalline ice deformed in uniaxial compression under conditions of dynamic recrystallization (steady-state tertiary creep). The line drawn has gradient $= -3$ and is based on a simple theoretical model (Jacka and Li 1994). Reproduced by courtesy of the International Glaciological Society from *Annals of Glaciology*, Vol. 20, 1994, pp. 13–18, Fig. 3.

For a given kind of deformation the ratio of the strain rate for steady-state tertiary creep $\dot{\varepsilon}_{max}$ to the minimum strain rate $\dot{\varepsilon}_{min}$ is found to be a constant independent of both stress and temperature. In uniaxial compression $\dot{\varepsilon}_{max}/\dot{\varepsilon}_{min} \approx 3$, and for shear the ratio is ≈ 8 (Budd and Jacka 1989). This shows that the flow processes are similar in both secondary and tertiary creep, with the difference between them arising from the more favourable fabric established in the tertiary stage. In the case of shear the preferred fabric has all grains with their *c*-axes parallel, but in compression the *c*-axes are distributed over a cone, and this is how shear deformation can give a greater enhancement than compression. If ice deformed into the tertiary stage is subsequently tested in a different orientation it is much harder, but eventually recrystallization changes its fabric to one appropriate to the new deformation. It is very important to recognize that ice flowing in an ice sheet has highly anisotropic properties developed from its recent history.

We cannot discuss here the large and highly complex subject of the modelling of the flow of ice sheets, but two recent attempts to model the physical processes occurring in the ice during tertiary flow are Castelnau *et al.* (1996) and Wenk *et al.* (1997).

8.3.5 *Primary creep*

We turn finally to the transient or 'primary' creep which occurs in the portion BC of Fig. 8.11. Following Glen (1955) and P. Barnes *et al.* (1971) the transient component is often fitted by an Andrade time dependence as $t^{1/3}$, so that the total strain in the region BC is given by

$$\varepsilon(t) = \beta t^{1/3} + \kappa t, \tag{8.8}$$

where β and κ are constants representing the primary and secondary components of the creep rate. In the early stages of the deformation ($\varepsilon \sim 0.01\%$) the primary creep strain at a given time is directly proportional to the stress. On this basis and making assumptions about the similarity of the creep curves, Budd and Jacka (1989) show that the power of 1/3 occurring here is the reciprocal of the value of $n = 3$ in equation (8.6).

Experiments at very small strains ($< \sim 0.01\%$) show that the creep strain is largely recoverable on unloading (Jellinek and Brill 1956; Sinha 1978*b*; Duval 1978; Duval *et al.* 1983), and these effects are therefore sometimes referred to as 'delayed elastic' or 'anelastic' deformation. They are closely related to the phenomena studied by internal friction (e.g. Perez *et al.* 1979). Budd and Jacka (1989) quote unpublished evidence that as much as 0.2% of the primary creep may be recoverable. The simplest model of primary creep is that it is a result of the dislocation glide building up internal stresses within the grains. On unloading some of these dislocations can glide back under the internal stress. However, the interpretation by Sinha (1984) is that it involves grain boundary sliding which is a precursor to the formation of cracks.

8.4 Brittle fracture of polycrystalline ice

8.4.1 *Fracture in tension*

For ice at temperatures above $-20\,^{\circ}\mathrm{C}$ fully ductile creep without cracking is only observed at strain rates less than about $10^{-7}\,\mathrm{s}^{-1}$ ($\sim$1% per day) and this is the case denoted as (a) in Fig. 8.1. Higher deformation rates are usually studied in constant strain rate tests, and the properties of ice in tensile tests of this kind are described by Schulson (1987). For strain rates faster than $\sim 10^{-6}\,\mathrm{s}^{-1}$ the ice fractures with little if any prior plastic deformation, as in the case (e) in Fig. 8.1. The fracture is clean across the specimen, and it seems that the first crack to be formed propagates immediately leading to complete failure. At intermediate strain rates $\sim 10^{-7}\,\mathrm{s}^{-1}$ there is some plastic deformation (up to about 0.05%) during which many small stable cracks are formed until one is produced which is large enough to propagate, leading to failure. This case is represented by (c) in Fig. 8.1 and may be referred to as ductile fracture. The ductile fracture represented by (b) with stable cracks present during appreciable ductile flow is more typical of a compression test.

Figure 8.17 shows the failure stresses for specimens of different grain sizes (diameters $d = 1$–10 mm) deformed at strain rates of 10^{-3} and $10^{-7}\,\mathrm{s}^{-1}$ at $-10\,^{\circ}\mathrm{C}$. For the faster strain rate (and others above $10^{-6}\,\mathrm{s}^{-1}$) the tensile failure stress $\sigma_{\mathrm{T}}^{\mathrm{N}}$ is that for the nucleation of a crack, and it can be seen to satisfy the equation

$$\sigma_{\mathrm{T}}^{\mathrm{N}} = \sigma_0 + kd^{-1/2}, \tag{8.9}$$

where σ_0 and k are constants. This equation is of a form commonly observed in metals, and in that context it is interpreted as the stress required to form a

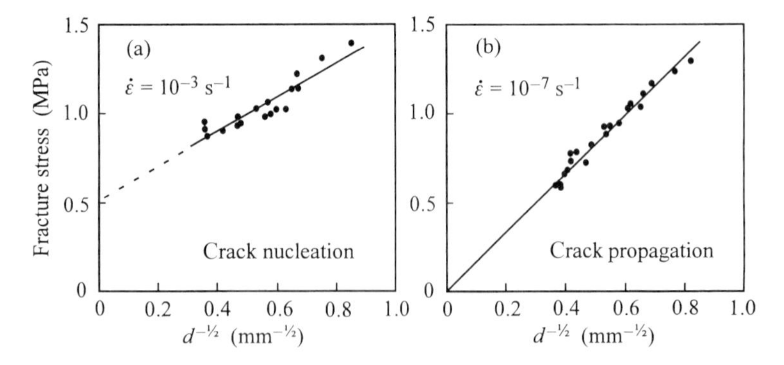

Fig. 8.17 The tensile fracture stress at $-10\,^{\circ}\mathrm{C}$ of specimens of granular polycrystalline ice of different grain size d, plotted as functions of $d^{-1/2}$, (a) for high strain rates where crack nucleation is rate limiting, and (b) for low strain rates where the limiting factor is crack propagation. The line drawn in (b) is that calculated from equation (8.10) with the parameters given in the text. (From Schulson 1987.)

dislocation pile-up sufficient to nucleate a crack at a grain boundary. However, in ice it is unlikely that dislocations can move sufficiently rapidly to do this, and evidence from compression tests to be described in the next section points towards a mechanism involving grain boundary sliding.

For lower strain rates and smaller grain sizes the critical factor is crack propagation rather than crack nucleation. In these cases many cracks are observed which do not lead to fracture. The theory for the propagation of an embedded disc-like crack of radius a lying perpendicular to a tensile stress yields a critical stress (T.L. Anderson 1995)

$$\sigma_T^P = \frac{\pi^{1/2}}{2} K_{Ic} a^{-1/2}, \tag{8.10}$$

where K_{Ic} is a property of the ice known as the 'fracture toughness' for a mode I crack (i.e. a crack opening under tension). In independent experiments on notched specimens fractured at high strain rates Nixon and Schulson (1987) found that for ice at $-10\,°C$ $K_{Ic} \approx 0.08$ MPa m$^{1/2}$. The diameters of the largest residual cracks in the specimens shown in Fig. 8.17(b) were about 3.7 times the grain diameter, and using these data equation (8.10) gives the solid line in the figure; this is in good agreement with the experimental points. Schulson *et al.* (1989) performed high-strain-rate tensile tests on specimens that had been slowly pre-strained to introduce cracks of measured size. Where fracture was initiated from these cracks, the results were in agreement with equation (8.10) with $K_{Ic} = 0.089$ MPa m$^{1/2}$. However, in some cases fracture was initiated at a freshly nucleated crack rather than at one of those introduced previously; this shows that even quite large cracks can become blunt or healed within a few minutes.

If we put $a = \alpha d$, where α is a constant of order unity, equations (8.9) and (8.10) yield the critical stresses for crack nucleation and propagation in material of grain size d. The larger of $\sigma_T^N(d)$ and $\sigma_T^P(d)$ determines the stress at which fracture will occur. In the range of values appropriate to ice at intermediate strain rates, nucleation is the critical process for large grain size. For small grains, however, the lines describing equations (8.9) and (8.10) cross so that the stress for propagation becomes greater than that for nucleation. Only small cracks of the order of the grain size are nucleated, and the critical stress is that necessary to make these small cracks grow.

8.4.2 *Fracture in compression*

The brittle fracture of ice under a compressive stress is different from that which occurs in tension, and this topic has been reviewed by Schulson (1997). Initially, just as in tension, there is a very small amount of plastic deformation leading to the nucleation of small cracks with dimensions similar to the grain size. These start to appear at $\frac{1}{3}$ to $\frac{1}{2}$ of the final fracture stress and increase in number as the stress increases. They can be observed most clearly in type S2 columnar ice (described in §12.1) with the stress applied perpendicular to the columns so as to produce an effectively two-dimensional system (Gold 1972; Cannon *et al.* 1990; Schulson *et al.* 1991). Figure 8.18 is an example of cracks formed in such a sample; the columns are perpendicular to the figure and the specimen was compressed in the vertical direction with a smaller confining stress perpendicular to this to stabilize the cracking. The initial cracks form at places like 'a', 'b', and 'c' on grain boundaries lying at approximately 45° to the stress. The development of these cracks is illustrated in Fig. 8.19, where AB represents the initial crack. In fracture mechanics such cracks are referred to as 'mode II' cracks. They cannot open under the compressive stress, but sliding occurs between their faces and this leads to the opening of the 'wing cracks' AC and BD. These start off perpendicular to AB and then swing round into the plane parallel to the compressive stress. This process has been modelled by introducing cracks into completely brittle materials by Nemat-Nasser and Horii (1982) and by Ashby and Hallam (1986).

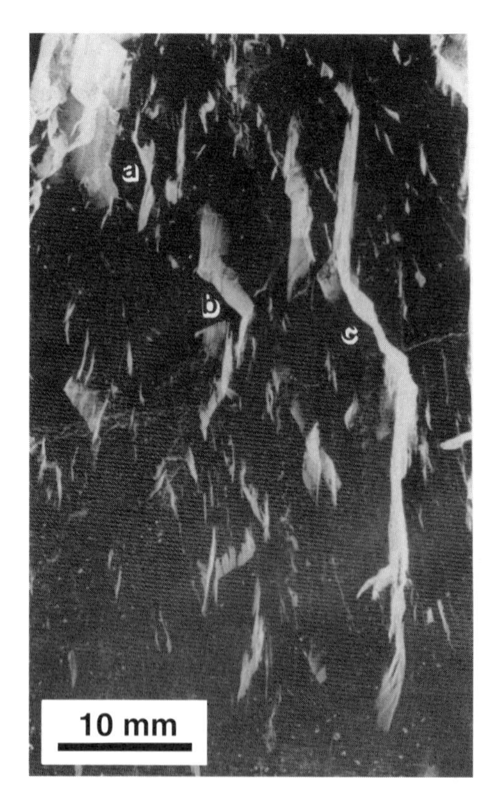

Fig. 8.18 Photograph of wing cracks in S2 columnar ice of column diameter ~8 mm compressed vertically across the columns to the point of failure. A smaller (5%) compressive stress was also applied transversely. Temperature = −10 °C, strain rate $= 4 \times 10^{-3}\,s^{-1}$. Unpublished photograph provided by D. Iliescu and E.M. Schulson of Dartmouth College, Hanover, NH, 1998.

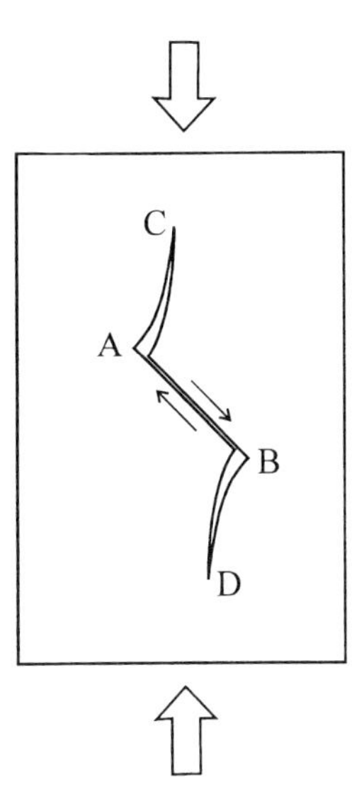

Fig. 8.19 Diagram illustrating the formation of wing cracks in a specimen under compressive stress.

At high strain rates the wing cracks propagate leading to failure, but, if the strain rate is below that for the ductile-to-brittle transition indicated in Fig. 8.1, the cracks are stabilized by plastic flow in the regions of the tips C and D. Significant plastic deformation can then be achieved with the formation of a large number of small cracks, until these finally link together and the specimen collapses. This is the case illustrated in Fig. 8.1(b). For both brittle and ductile failure the specimen appears quite cloudy due to the cracks formed prior to final failure. The mode of fracture in the brittle range depends on the constraints at the ends of the specimen and the effect of any lateral confining pressure. A specimen that is free of any transverse constraint will probably fail by splitting along its length, but one with the ends fixed to rigid plattens is more likely to fail by shear across a plane oblique to the axis.

Columnar ice is of practical importance because floating ice sheets have this form and compression across the columns is the kind of deformation involved in an impact with a fixed structure. The same processes also occur in the compression of granular (T1) polycrystalline ice except that the cracks are now more confined in the plane perpendicular to that of Figs 8.18 and 8.19 (Schulson 1990).

A theoretical model of the formation of wing cracks was developed by Ashby and Hallam (1986) and has been applied to ice by Schulson (1990). Treating the critical stage as the propagation of the wing cracks, it predicts that in ice of grain size d the stress for fracture σ_C^P is given by the equation

$$\sigma_C^P = \frac{ZK_{Ic}d^{-1/2}}{(1-\mu)}, \tag{8.11}$$

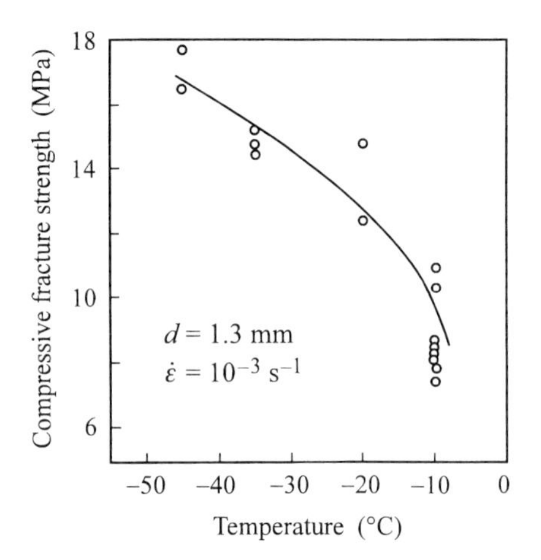

Fig. 8.20 Temperature dependence of the fracture strength σ_C^P of granular polycrystalline ice under uniaxial compression. Reproduced from *Acta Metallurgica et Materialia*, **30**, E.M. Schulson, The brittle compressive fracture of ice, pp. 1963–76, Copyright 1990, with permission from Elsevier Science.

where K_{Ic} is the fracture toughness, μ is the coefficient of friction between the faces of the original crack, and Z is a numerical constant independent of temperature and strain rate. The dependence of σ_C^P on grain size is consistent with experiment, though with large scatter in the data. Figure 8.20 shows the temperature dependence of σ_C^P in granular (T1) ice. This dependence is stronger than that expected from the temperature dependence of K_{Ic}, and it has been attributed to the effect of the friction between the faces of the sliding crack. There is no direct information about the friction of ice on ice under these conditions, but experiments involving the sliding of two blocks over one another (D. E. Jones *et al.* 1991) do show marked effects of temperature.

According to the model of Schulson (1990) ice will be ductile, even in the presence of cracking, if the strain rate is low enough for plastic flow in a zone around the crack tip to prevent the propagation of the crack. He assumes that the ductile-to-brittle transition occurs when the size of this plastic zone is a certain fraction f of the grain size, and that plastic deformation in this microscopic zone is governed by equation (8.6), although this equation is strictly applicable to the macroscopic deformation of polycrystalline ice. On this basis the critical strain rate for the ductile-to-brittle transition is given by the equation

$$\dot{\varepsilon}_{D/B} = \frac{4ZBK_{Ic}^3}{3\pi f(1-\mu)d^{3/2}}, \tag{8.12}$$

in which B is the constant in the flow law (eqn (8.6)) with $n = 3$. The dependence on $d^{-3/2}$ is consistent with the observations, with a typical value of $\dot{\varepsilon}_{D/B}$ being $10^{-4}\,\text{s}^{-1}$ at $d \sim 5$ mm. The transition described is that between two kinds of

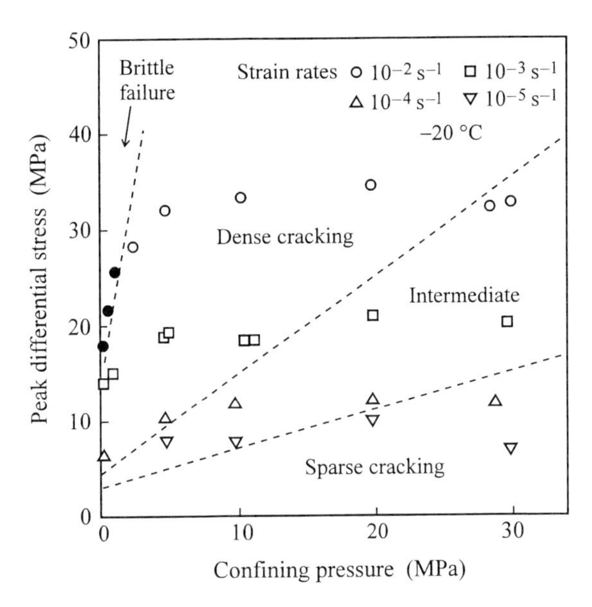

Fig. 8.21 Summary of results of constant-strain-rate compression tests on granular polycrystalline ice under confining pressures at −20 °C by Rist and Murrell (1994). The differential stress is the total uniaxial compressive stress minus the hydrostatic pressure. The solid points denote brittle shear fracture and the open points denote ductile failure with differing degrees of microcracking. Strain rate is indicated by the shape of the points. Reproduced by courtesy of the International Glaciological Society from *Journal of Glaciology*, Vol. 40, No. 135, 1994, pp. 305–18, Fig. 6.

macroscopic failure. As explained earlier the critical strain rate for ductile flow without any cracking is much less at $\sim 10^{-7}\,\mathrm{s}^{-1}$.

Although the criterion for fracture depends on crack propagation, an interesting problem concerns the mechanism of crack nucleation. The cracks that lead to brittle fracture are nucleated at the fastest strain rates, and there is insufficient time for dislocations, which move only slowly in ice, to form the pile-ups commonly associated with equation (8.9). At lower strain rates dislocation glide does of course lead to deformation within grains and so to internal stresses that may nucleate cracks at grain boundaries. It may also relax stresses at crack tips and prevent crack propagation as already explained. However, returning to the question of crack nucleation at the highest strain rates, Picu and Gupta (1995*a*) and Elvin and Sunder (1996) have shown that incompatibility of the elastic strains in differently oriented anisotropic grains is insufficient to nucleate cracks at triple grain boundary junctions, and they propose that cracks are nucleated by grain boundary sliding. In experiments on columnar ice at −10 °C and $\dot{\varepsilon} = 10^{-3}\,\mathrm{s}^{-1}$, Picu and Gupta (1995*b*) observed 'pockets of decohesion' on the grain boundary facets, which could subsequently spread over the whole boundary. The sliding that nucleates the cracks is then the same process as that which produces the wing cracks at the ends of the boundary.

In practical situations ice is subjected to complex stress systems often with considerable lateral confinement. Laboratory experiments and theoretical models have therefore been extended to treat ice under multi-axial compressive loading (e.g. Weiss and Schulson 1995; Schulson and Buck 1995; Gratz and Schulson 1997). Rist and Murrell (1994) performed a comprehensive series of experiments in which granular polycrystalline ice specimens were deformed in compression under confining hydrostatic pressures of up to 30 MPa. The relevant quantity here is the differential stress, which is the total uniaxial stress minus the hydrostatic pressure, and Fig. 8.21 summarizes the results of compression tests, which yielded stress–strain curves similar to those in Fig. 8.12. The peak differential stress and the type of fracture depend on both the strain rate and the confining pressure. A large hydrostatic pressure, such as would be present at some depth in an ice sheet, effectively eliminates macroscopic cracking.

At the other extreme, when a mass of ice suffers an impact with another object the deformation takes the form of crushing, which involves extensive fracture at high strain rates (Jordaan and Timco 1988). Experiments involving the fracture of large ice masses are reviewed by Sanderson (1988). A perhaps surprising observation is that the cracking of ice generates radio-frequency electromagnetic waves, and these can be used to detect cracking both in the laboratory and in the field (Fifolt *et al.* 1993; Petrenko 1993*a*).

8.5 Summary

For most potentially ductile solid materials there is a critical stress, flow stress, or elastic limit below which the material is elastic and above which plastic deformation occurs. For ice this is not so. Plastic flow of ice can occur at any stress at a strain *rate* which, ignoring the small additional effect of primary transient creep, increases as the nth power of the stress. For polycrystalline ice $n \approx 3$. If an attempt is made to deform the ice too rapidly, which requires in practice the application of too large a stress, brittle failure will occur.

These properties depend on two aspects of the physical processes involved. The first is that the plastic properties of single crystals are highly anisotropic, with easy deformation occurring by slip only on the basal planes. In a polycrystalline aggregate deformation of the grains by such slip leads either to the build-up of huge stresses within the grains or to loss of cohesion at the grain boundaries. The second important aspect of the physical properties is that both dislocation glide and the processes which can relax the stresses built up in polycrystals are thermally activated. These processes and hence ductile flow can only occur at a certain rate. If this rate is exceeded failure occurs by the non-thermally-rate-limited process of cracking, and this leads sooner or later to brittle failure. Ice therefore exhibits a ductile-to-brittle transition at a critical strain rate, which depends on the temperature. The two extremes of the mechanical behaviour of ice are typified by the ductile flow of an ice sheet under a confining hydrostatic pressure, and the crushing of an ice floe under the impact of an ice breaker.

The dominance of thermally activated processes in the mechanical properties of ice is related to our interest in ice at temperatures unusually close to its melting point T_m. The range of most experiments is above about $-20\,°C$, which is greater than $0.9T_m$, although creep has been observed to occur quite normally down to $-70\,°C$. These temperatures are well above those considered in the study of most crystalline materials. Even so ice is remarkable in retaining brittle properties right up to its melting point.

At the molecular level the process of slip on the basal planes depends on the ability of dislocations both to glide and to multiply on these planes. As seen in Chapter 7, this seems likely to depend on the way in which dislocations dissociate into partial dislocations on these planes, and it also depends on there being a mechanism to overcome the barrier presented by proton disorder. The crystal structure is such that it is very difficult for slip with a component in the [0001] direction to take place. In a polycrystal the processes that occur to retain compatibility between the grains are primarily grain boundary migration, grain boundary sliding, and recrystallization. It is especially important that ice which is highly deformed by creep develops a grain fabric which is determined by its recent deformation history, and is appropriate to the continuation of this deformation.

In brittle fracture it is necessary to distinguish between the nucleation of cracks with dimensions typical of the grain size and the propagation of these cracks leading to failure. At intermediate strain rates many cracks are formed which accommodate quite stably some of the incompatibility between the slip in the grains. The further growth of these cracks depends on the stress, their orientation and whether the strain rate is low enough to allow plastic deformation to occur at the crack tip.

9
Optical and electronic properties

9.1 Introduction

In this chapter we interpret 'optical' as referring not just to visible light but to a wide range of the electromagnetic spectrum. Figure 9.1 displays this spectrum on a logarithmic scale spanning 13 decades of frequency, and including the corresponding free-space wavelengths and other units of interest. The chapter deals with the propagation of these waves though ice, and the variety of ways in which they interact with the structure and individual molecules of the ice. Energy is dissipated to the ice most strongly when the frequency of the waves matches frequencies or time scales associated with processes in the ice, and the processes of interest are marked in the figure. At low frequencies the response of ice to an alternating electric field arises from the Debye dispersion which has already been discussed in detail in Chapters 4 and 5. In the infrared region of the spectrum waves can couple to the modes of vibration of the molecules or the lattice, providing evidence about the spectrum of these modes as described in Chapter 3.

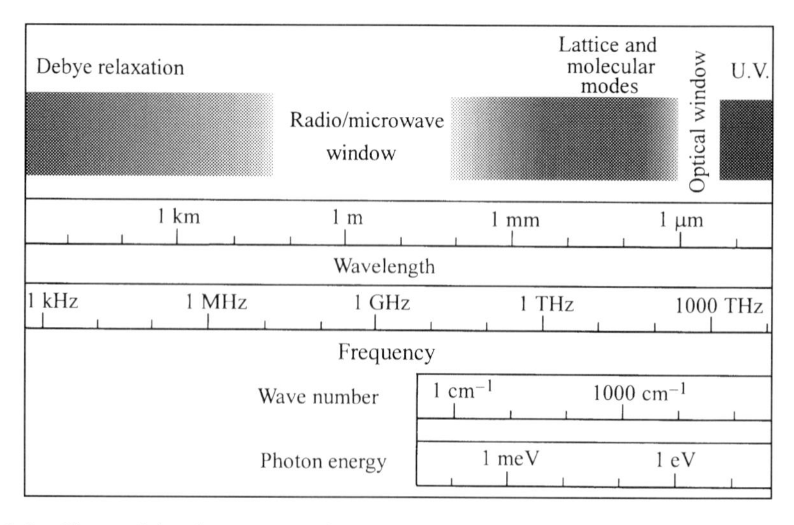

Fig. 9.1 Chart of the electromagnetic spectrum showing the regions of absorption and transparency of ice.

As in any material, ultraviolet radiation can excite electronic transitions. In these three regions of the spectrum there is strong absorption of radiation, but between them in the radio and visible parts of the spectrum there are 'windows' in which the ice is almost transparent. A survey of the properties of ice as understood from its electromagnetic spectrum has been written by Johari (1981).

This chapter deals with the propagation of electromagnetic waves of frequencies greater than 1 MHz, with particular emphasis on the windows where the ice is transparent. It concludes with a brief account of the electronic structure, which relates to the absorption of ultraviolet radiation. Photoconductivity, which results from such ultraviolet absorption, has been described in §6.10.

9.2 Propagation of electromagnetic waves in ice

An attenuated plane electromagnetic wave of amplitude E_0 and angular frequency ω travelling in the $+x$ direction with velocity v is described by the equation

$$E(x,t) = E_0 \exp\left(-\frac{\alpha x}{2}\right) \exp\left[\mathrm{i}\omega\left(t - \frac{x}{v}\right)\right]. \tag{9.1}$$

This wave has the property that the intensity $I(x)$ decays as

$$I(x) = I_0 \exp(-\alpha x), \tag{9.2}$$

and α is defined as the absorption coefficient, with units of m^{-1}. Alternatively the wave can be written as

$$E(x,t) = E_0 \exp\left[\mathrm{i}\omega\left(t - \frac{nx}{c}\right)\right], \tag{9.3}$$

where c is the velocity of electromagnetic waves in free space and $n = n' - \mathrm{i}n''$ is the complex refractive index. Writing equation (9.3) as

$$E(x,t) = E_0 \exp\left(-\frac{\omega n''}{c} x\right) \exp\left[\mathrm{i}\omega\left(t - \frac{n' x}{c}\right)\right], \tag{9.4}$$

we see that

$$v = \frac{c}{n'}, \tag{9.5}$$

and

$$\frac{\alpha}{2} = \frac{\omega n''}{c}, \tag{9.6}$$

or

$$\alpha = \frac{4\pi n''}{\lambda_0}, \tag{9.7}$$

where λ_0 is the wavelength of waves of this frequency in free space. The quantity n'' may be interpreted physically by noting that in propagating through a distance λ_0 the intensity falls by a factor of $e^{-4\pi n''}$.

According to Maxwell's equations the velocity of an electromagnetic wave in a non-magnetic material of (relative) permittivity ε is

$$v = c\varepsilon^{-1/2}, \tag{9.8}$$

and the refractive index, defined as c/v, is then

$$n = \varepsilon^{1/2}. \tag{9.9}$$

This is valid even when $n = n' - in''$ and $\varepsilon = \varepsilon' - i\varepsilon''$ are complex quantities, and it is simple to show that

$$\varepsilon' = n'^2 - n''^2, \tag{9.10}$$

and

$$\varepsilon'' = 2n'n''. \tag{9.11}$$

In the literature on this subject the absorption properties of ice may be given in terms of α, n'', or ε''. In the absence of significant absorption the measured refractive index n' can be written just as n.

For waves incident at normal incidence on a plane surface the reflection coefficient, defined as the fraction of the incident intensity that is reflected, is given by (Ditchburn 1963; Johari 1981)

$$R = \frac{(n' - 1)^2 + n''^2}{(n' + 1)^2 + n''^2}. \tag{9.12}$$

This approaches 1 when n'' is large, and, when the absorption is too large to measure directly, the absorption spectrum has to be deduced from measurements of the reflectivity using this equation. However, the scattering or reflection of light from a surface is also strongly dependent on the state of the material, as can be appreciated by comparing the appearance of solid ice and fresh snow.

Experiments over the band of frequencies from 10^6 to 10^{16} Hz involve a wide range of techniques performed on different specimens at different temperatures. Warren (1984) has made a critical compilation of the available data standardized at $-7\,^\circ$C, and more recently Matsuoka *et al.* (1996) have reviewed measurements of n'' in the radio and microwave region. Values of n' and n'' taken from these

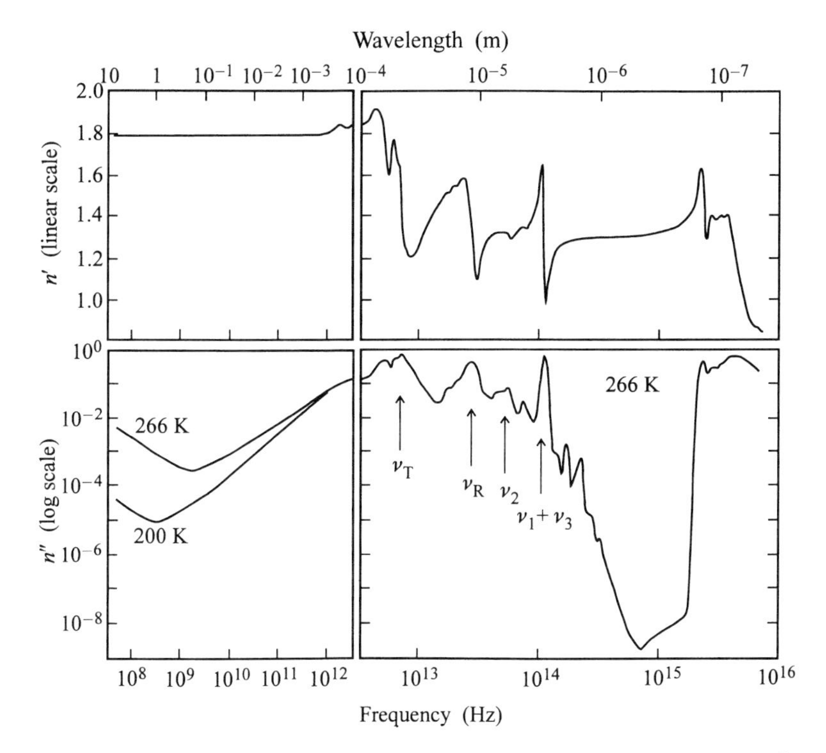

Fig. 9.2 Frequency dependence of the real part n' and the imaginary part n'' of the refractive index of ice at 266 K and at 200 K (low frequency n'' only). Reproduced with changes from S.G. Warren, *Applied Optics*, **23**, 1206–25 (1984), and including revised data of Matsuoka *et al.* (1996).

reviews are plotted in Fig. 9.2 for a temperature of 266 K. The temperature dependence is significant primarily for n'' in the radio-frequency region, and in this case values are also given at 200 K. This figure is intended to provide an overall impression of the optical properties of ice. To assess the evidence concerning a particular portion of the figure the reader should refer to the original sources and specific references given later in this chapter.

To appreciate the extent to which electromagnetic waves of different frequencies penetrate ice it is helpful to compare values of the absorption coefficients, or better the distances α^{-1} over which the intensity is reduced by a factor of e. Values of α^{-1} derived using equation (9.7) from the data in Fig. 9.2 are given in Table 9.1. These show the extreme transparency of ice in the optical and radio windows as compared with its opacity at the peaks of the infrared and ultraviolet absorption.

The Debye dispersion indicated in Fig. 9.1 lies off the scale to the low-frequency end of Fig. 9.2. The behaviour of n' in this range follows the square root

Table 9.1 Attenuation of electromagnetic waves in ice at $-7\,^\circ C$

Frequency (Hz)	Wavelength λ_0 (m)	n''	α^{-1} (m)	
1×10^8	3	3.3×10^{-3}	70	
1×10^9	0.3	3×10^{-4}	80	Radio min. of n''
1×10^{10}	0.03	8×10^{-4}	3.0	
1×10^{11}	0.003	7×10^{-3}	0.034	
1×10^{12}	3×10^{-4}	1.5×10^{-2}	0.0016	
6×10^{12}	5×10^{-5}	6×10^{-1}	7×10^{-6}	ν_T peak
1×10^{14}	3×10^{-6}	6×10^{-1}	4×10^{-7}	$\nu_1 + \nu_3$ peaks
6×10^{14}	5×10^{-7}	2×10^{-9}	20	Mid. visible†
1.5×10^{15}	2×10^{-7}	1.3×10^{-8}	1.2	
2.2×10^{15}	1.36×10^{-7}	4×10^{-1}	3×10^{-8}	First uv max.

†Absorption ~10 times less in deep Antarctic ice (Fig. 9.3).
Derived from data in Fig. 9.2.

of the permittivity ε' in Fig. 4.5, rising to a low frequency value of $n' \approx 10$. In the radio/microwave region above the Debye dispersion the permittivity is $\varepsilon_\infty = 3.16$, which corresponds to a value of $n' = 1.79$ that is almost independent of frequency, and for which Matsuoka *et al.* (1997) observed a small birefringence (see §5.3.1).

For frequencies $\omega \gg \tau_D^{-1}$ the theory of the Debye dispersion (eqn (4.7b)) gives

$$\varepsilon'' = \chi'' = \frac{\chi_s}{\omega \tau_D} \ll 1, \tag{9.13}$$

Using equations (9.10) and (9.11) with $\varepsilon' = \varepsilon_\infty$, we then find

$$n'' = \frac{\chi_s}{2\omega \tau_D \varepsilon_\infty^{1/2}}, \tag{9.14}$$

and from equation (9.6) the distance in which the waves are attenuated by a factor of e is

$$\alpha^{-1} = \frac{\tau_D c \varepsilon_\infty^{1/2}}{\chi_s} = \frac{c \varepsilon_0 \varepsilon_\infty^{1/2}}{\sigma_\infty}. \tag{9.15}$$

The transmission of the waves through ice is then independent of frequency, but dependent on temperature through the conductivity σ_∞.

Using values for pure ice at 265 K from §5.3.2 equation (9.15) gives $\alpha^{-1} = 250$ m, which should be compared with the value of 70 m obtained from microwave measurements at 100 MHz (Table 9.1). With increasing frequency the absorption coefficient remains constant up to the minimum of n'' at about 1 GHz, and it then rises in the tail of the infrared absorption band. These properties determine the frequency band used for radio echo sounding in ice sheets (see §12.5.2).

9.3 Infrared range

The strong absorption in the infrared region arises from the excitation of the various modes of molecular and lattice vibration described in §3.4, and the main peaks in Fig. 9.2 are labelled to correspond to those in Fig. 3.5. Over the whole infrared absorption band n' falls from 1.79 to 1.30, corresponding to a change in ε' from 3.16 to 1.69. The difference in these permittivities represents the polarization due to the full set of molecular and lattice modes. In the microwave region these motions all occur in phase with the electric field, but at optical frequencies no such motions are possible.

In the near infrared (high frequency) region the complex structure of the absorption spectrum arises from combinations of the basic lattice and molecular modes and is characteristic of water ice. This spectrum is used to identify the presence of ice, for example in astronomical situations (§12.7). Precise measurements of n' between 1.4 and 7.8 μm are described by Gosse *et al.* (1995); the temperature dependence of α between 1.0 and 2.7 μm has been determined and its interpretation discussed by Grundy and Schmitt (1998). The tail of the infrared absorption extends right into the red end of the visible spectrum as seen in the next section.

9.4 Visible optical range—birefringence

In the visible and near ultraviolet range of frequencies good quality ice is highly transparent to electromagnetic radiation and its optical properties can be easily studied. Ice was identified by Brewster (1814, 1818) as a 'positive uniaxial doubly-refracting' material. The uniaxial birefringence is now understood as a consequence of the hexagonal crystal symmetry. It means that for light propagating perpendicular to the *c*- or 'optic' axis the refractive index is different for waves polarized with the **E**-field parallel or perpendicular to the *c*-axis. For light travelling along the *c*-axis all polarizations are equivalent and the refractive index is that corresponding to **E** being perpendicular to the *c*-axis. Such waves or rays are referred to as 'ordinary', and those with **E** parallel to the *c*-axis are called 'extraordinary'. In ice the refractive index for the extraordinary waves n^{E} is larger than the refractive index n^{O} for the ordinary waves, and the material is then said to have *positive* birefringence.

The refractive indices as determined by Ehringhaus (1917) for a number of wavelengths in the visible spectrum are given in Table 9.2. The increase with increasing frequency is that normally observed in transparent materials, and is what would be expected for the tails of the infrared and ultraviolet absorption bands in Fig. 9.2. Ehringhaus made measurements at −3.6, −65, and ∼−110 °C. Assuming a linear dependence as

$$n(T) = n(T_0)(1 - a(T - T_0)), \tag{9.16}$$

Table 9.2 Refractive indices of ice in the visible range at −3.6 °C

Wavelength λ_0 (nm)	n^{O}	n^{E}
405	1.3185	1.3200
436	1.3161	1.3176
492	1.3128	1.3143
546	1.3105	1.3119
589	1.3091	1.3105
624	1.3082	1.3096
691	1.3067	1.3081

Selected data from Ehringhaus (1917). Accuracy $\sim \pm 0.0001$.

his results at −3.6 and −65 °C give $a = 2.7 \times 10^{-5}\,\mathrm{K}^{-1}$. The value of a falls at lower temperatures.

Examination of the frequency dependence of n'' in Fig. 9.2 shows that it has a minimum in the visible range at 640 THz (470 nm) and rises steeply towards the red end of the spectrum. Measurements of the absorption coefficient α in this region by Grenfell and Perovich (1981) and Perovich and Govoni (1991) are plotted in Fig. 9.3. The preferential absorption towards the red end of the visible spectrum accounts for the fact that light scattered from deep within a large volume of ice appears blue, and this is the reason for the vivid blue-green colour seen in crevasses etc. (Bohren 1983). The absorption coefficient of liquid water is generally similar though differing in detail (Smith and Baker 1981), and the light transmitted by water also appears greenish. As pointed out by Braun and Smirnov (1993), the molecular absorption bands ν_1 and ν_3 for D_2O are displaced to lower frequencies (§3.4.4), and one would not therefore expect a crevasse in a glacier formed from D_2O ice to appear blue!

In the process of developing a detector for cosmic muons and neutrinos deep in the Antarctic ice sheet the AMANDA collaboration has studied the propagation of short (~12 ns) pulses of light through the ice at depths of 0.8 to 1.0 km (Askebjer *et al.* 1995, 1997). From the shape of pulses which had travelled a nominal distance of 20 m through the ice, it was possible to distinguish between direct propagation and light that had been multiply scattered by air bubbles. They were therefore able to determine the true absorption coefficient for clear ice, and the results are plotted as the solid triangles in Fig. 9.3. The values follow the extrapolation of the infrared absorption into the visible region, and towards the blue end of the spectrum they fall an order of magnitude below the results for laboratory ice. Price and Bergström (1997) have interpreted the measured absorption in laboratory ice between 200 and 500 nm, which varies approximately as λ^{-4}, as due to Rayleigh scattering by nanoscale defects in the ice. Further data can be expected from measurements deeper in the Antarctic

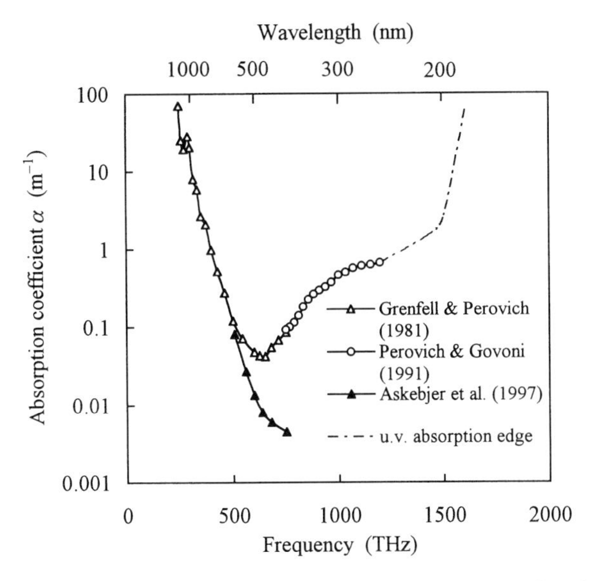

Fig. 9.3 Measurements of the absorption coefficient of ice in the visible and near visible portion of the spectrum. For explanation see text.

ice, where the air bubbles have been converted to clathrates (§12.5). In perfect ice the absorption in the near ultraviolet should be lower than anything observed so far.

The birefringence of ice is extremely useful for determining the orientations of the *c*-axes in crystals and observing the grain structure in thin sections of polycrystalline ice. The methods used are similar to those used in mineralogy (e.g. Hartshorne and Stuart 1970), but because the birefringence of ice is weak they can be used on larger samples, and there is no need for a microscope. Consider a thin monocrystalline specimen with its *c*-axis lying in its plane, and placed between a pair of crossed polarizing filters. If the *c*-axis is parallel to the **E**-vector of the light transmitted by the polarizer, only the extraordinary wave will propagate through the crystal, and as this will not be transmitted by the analyser the specimen will appear dark. Similarly the specimen will appear dark if it is rotated through 90° in its plane so that the *c*-axis is perpendicular to the **E**-vector of the polarizer. If, however, the plate is rotated so that the *c*-axis is at 45° to the plane of polarization, the incident wave must be resolved into ordinary and extraordinary components. These travel at different speeds, emerging with a phase difference $\phi = 2\pi\delta nd/\lambda_0$, where $\delta n = n^{\mathrm{E}} - n^{\mathrm{O}}$, d is the thickness of the specimen, and λ_0 is the wavelength in free space. If $\phi = \pi$, the plane of polarization is rotated through 90° and the wave will be transmitted by the analyser so that the specimen appears bright. If, however, $\phi = 2\pi$ the plane of polarization is the same as that of the original wave and the specimen appears dark. For light in the middle of the visible spectrum the 'first order extinction', where $\phi = 2\pi$, occurs for $d \approx 0.4$ mm. In white light a specimen of the order 1 mm thick will appear coloured due to the elimination of some of the wavelengths present, except when it is oriented with its c-axis parallel or perpendicular to the **E**-vector of the polarizers, in which case it will be dark.

A monocrystalline plate cut perpendicular to the c-axis and viewed between crossed polarizing filters remains dark when rotated in its own plane. When the c-axis is at an angle to the plane of the specimen, rotation in the plane of the specimen reveals the component of the c-axis in that plane. For specimens more than about 1 mm thick the colours become weak due to the simultaneous extinction of several wavelengths within the visible band, but the dark extinction positions can still be determined. A poor quality crystal can be recognized if it does not appear uniformly dark in the extinction position. In thin polycrystalline specimens the individual grains appear different colours, and monochrome examples are Figs 8.8 and 12.1. The orientations of the c-axes cannot be inferred from the colours (or the shades of grey in the monochrome figures). To find the orientations of the components of the c-axes of the individual grains in the plane of the specimen, it is necessary to rotate the specimen in its plane and determine the extinction position for each grain.

Another birefringent property which is of practical use in work with ice is the conoscopic figure. This is produced when a thick single crystal is viewed between crossed polarizers with an optical system focused on infinity. A typical conoscopic figure for a crystal of ice viewed along the c-axis is illustrated in monochrome in Fig. 9.4(b), in which each point in the figure corresponds to light travelling in a particular direction through the crystal. The dark arms of the cross ('isogyres') are parallel to the planes of polarization of the polarizers, for which only one wave is propagated in the crystal. In other directions coloured rings are seen, depending on the degree of birefringence produced. In the case of ice a conoscopic figure can be observed by holding a thick ($>\sim 1$ cm) specimen between crossed polarizers close to the eye, and looking through (not at) it towards a diffuse source of light as shown in Fig. 9.4(a). If the c axis is roughly along the line of sight the isogyres can usually be seen, and in favourable conditions for a thick enough specimen some of the colours will also be visible. For practical purposes the important feature of this conoscopic figure is that as the specimen is tilted the centre of the cross moves with the c-axis (allowing where necessary for refraction), and the approximate orientation of the optic axis can thus be identified for a large specimen before cutting it up. This technique can also resolve the ambiguity of 90° that arises in determining the optic axis from the extinction direction observed in a single plane.

All transparent materials exhibit a stress-induced birefringence, or the photoelastic effect. In the case of ice it has been measured by Ravi-Chandar *et al.* (1994) for light travelling along the c-axis while a compressive stress was applied perpendicular to this. In this orientation there is of course no intrinsic birefringence. The effect is stronger than for glass, but for realistic stresses the birefringence is

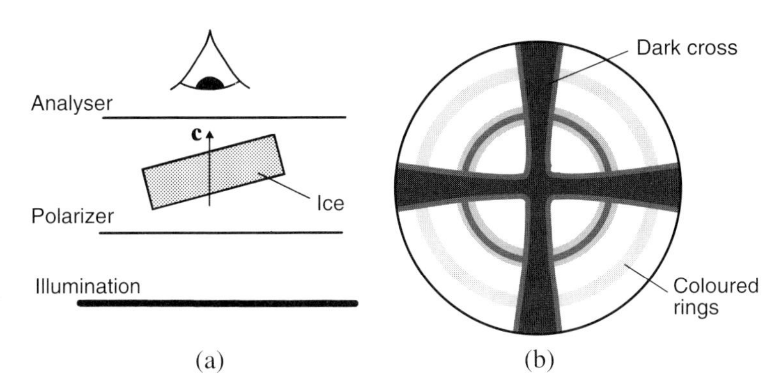

Fig. 9.4 Conoscopic figure produced in ice viewed along the *c*-axis. (a) Simple arrangement for viewing the figure with the unaided eye looking through the ice and focused on infinity. (b) Monochrome impression of the conoscopic figure.

two orders of magnitude smaller than the intrinsic birefringence that would be observed in a perpendicular orientation.

9.5 Ultraviolet range

As seen in Fig. 9.2 there is a very sharp absorption edge in the ultraviolet spectrum of ice at about 1900 THz (160 nm). This represents a photon energy of 7.8 eV, which is the minimum required to excite transitions between the electronic energy levels in the ice. For photons of energy larger than this and up to the soft X-ray region, ice is almost completely opaque. Transmission experiments are impossible, and the optical properties must be determined from the reflectivity (eqn (9.12)). Figure 9.5 shows the reflectivity of a specimen of ice Ih as measured by Seki *et al.* (1981). Measurements on vapour-deposited amorphous ice were similar, but the features were less well resolved. The values of n' and n'' shown for this region of the spectrum in Fig. 9.2 were determined from the reflectivity by a Kramers–Kronig analysis. This involves integration across the whole spectral range, and thus requires some assumption about the soft X-ray region as well (Warren 1984).

A closely related experimental technique is photoelectron spectroscopy, in which the specimen is illuminated with ultraviolet or soft X-radiation and the energy distribution of the emitted electrons is recorded. Figure 9.6 shows the results for excitation with Al K_α X-rays (photon energy 1.49 keV) reported by Shibaguchi *et al.* (1977), and compared with similar measurements on water vapour (Siegbahn 1974) with the energy scale displaced to make the curves match. It is clear that although the energy levels are broader in ice, they are still essentially the same as in the free molecule. Photoemission spectra for ultraviolet excitation (Shibaguchi *et al.* 1977; Abbati *et al.* 1979) reveal the upper group of three bands in Fig. 9.6, but they are less well resolved.

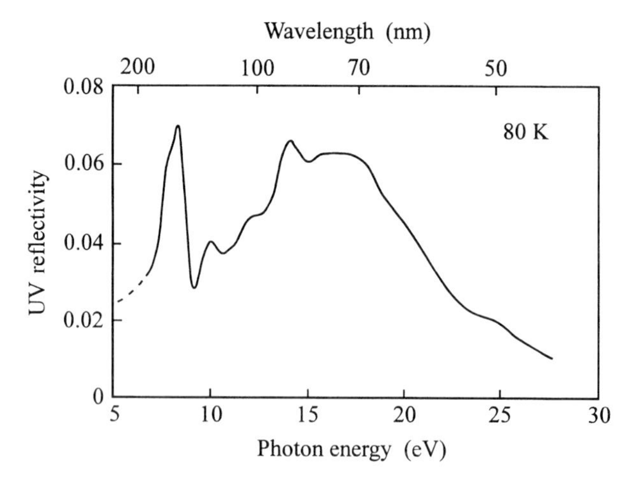

Fig. 9.5 Reflectivity of ice in the region of strong ultraviolet absorption as determined for a freshly cleaved surface *in vacuo* at 80 K with the **E**-vector parallel to the *c*-axis. (From Seki *et al.* 1981.)

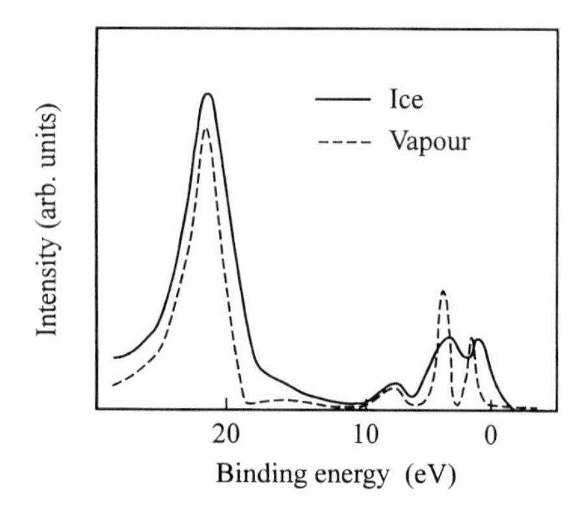

Fig. 9.6 Photoelectron spectrum of ice illuminated with Al K_α X-radiation. The spectrum from water vapour (Siegbahn 1974) is shown for comparison with the energy axis displaced to match. (From Shibaguchi *et al.* 1977.)

9.6 Electronic structure

The spatial distribution of electron density, averaged as in the half-hydrogen model (§2.2.1), can be calculated by Fourier synthesis of the observed X-ray diffraction data. Such calculations by Sakata *et al.* (1992) show high densities centred around the oxygen sites linked along the hydrogen bonds as shown in Fig. 9.7. The valence electrons appear partially delocalized in the structure, and features specific to hydrogen bonding are not apparent at this level.

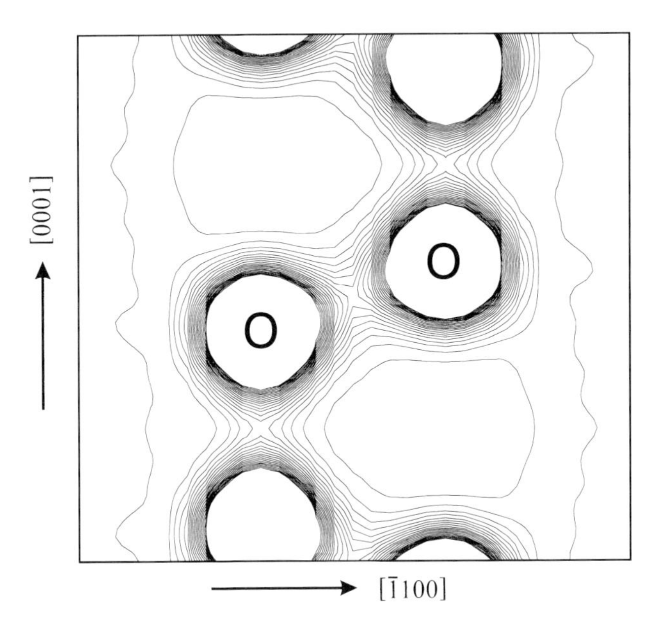

Fig. 9.7 The spatial electron density in ice as determined by X-ray diffraction. Contours are drawn from 0.05 to 1.0 at intervals of 0.05 electron per Å^3. The high density close to the oxygen nuclei in the regions marked O is excluded. (From Sakata *et al.* 1992.)

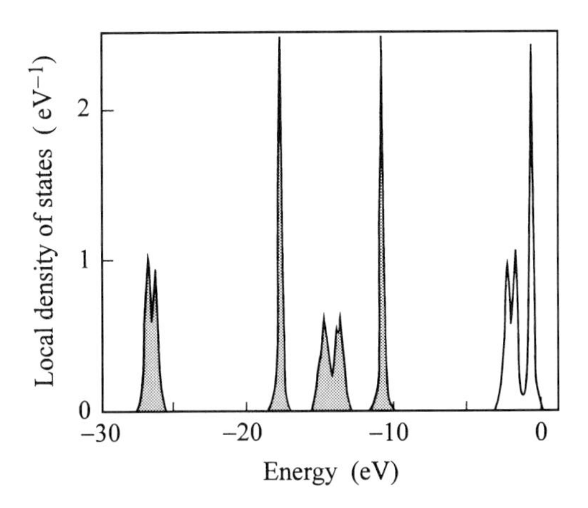

Fig. 9.8 Calculated energy density of states for electrons in ice Ih. The occupied levels are shaded. Reproduced from V.F. Petrenko and I.A. Ryzhkin, *Physical Review Letters*, **71**, 2626–9, Copyright 1993, American Physical Society.

Because of the proton disorder in ice Ih, early calculations of the electron band structure by Pastori Parravicini and Resca (1973) were performed for ordered ice Ic. More recently Petrenko and Ryzhkin (1993) have calculated the electron energy spectrum for proton-disordered ice Ih, using the tight-binding approximation for 1s and $2sp^3$ orbitals. Their results are shown in Fig. 9.8, in which the zero of energy is the vacuum level, and the occupied levels are shown shaded. The band gap of 7.8 eV has been fitted to the experimental absorption edge in Fig. 9.5. It is close to the value of 7.5 eV required for the ultraviolet excitation of a free molecule (§1.3). The number of energy bands in Fig. 9.8 corresponds to the number of peaks in the photoemission spectrum, but the spacings still need refinement. At this level of approximation the theory is not sensitive to proton disorder or to the difference between hexagonal or cubic symmetry.

10
The surface of ice

10.1 Introduction

The surface of a solid, whether it be the interface with a vacuum, the vapour, a liquid, or even another solid, is a region with properties different from those of the bulk material. The fundamental source of this difference is that atoms or molecules at a free surface only experience bonding forces to other molecules from one side, and at other interfaces there is a similar imbalance. This results in displacements of atoms from their normal sites, changes in the energies and force constants, and consequent effects on the layers below. Surface science is a large and expanding field of pure and applied research. Ice is of special interest in surface science, and, judging from the numbers of papers in the 1996 International Symposium on the Physics and Chemistry of Ice the surface properties have become the largest field of current research within ice physics (Glen 1997). The whole topic is relevant to such practical issues as frost heave (§12.6), friction and adhesion (Chapter 13), or surface catalysis by ice particles of reactions in the upper atmosphere (§12.3.3).

Ice is the only substance which most of us commonly experience under conditions close to its melting point. Just below 0 °C snow readily forms a compact snowball, but cold dry ice remains powdery. Long ago in studies of the melting and 'regelation' (or what we would now call 'sintering') of ice Faraday (1850, 1860) and Tyndall (1858) recognized that there was something special about the surface of ice. The free surface was in some sense liquid-like, and if two pieces of ice were brought together the material between them became solid. Such pre-melting phenomena are now recognized as a common property of many materials (van der Veen *et al.* 1988), and the application of these ideas to ice is reviewed by Dash *et al.* (1995). However, as explained in the review by Petrenko (1994*a*), experiments on ice reveal surface properties that extend to lower temperatures and to greater depths than can be accounted for simply by pre-melting.

The interpretation of these effects is not yet well developed, and this chapter deals with what is probably the most complex and uncertain topic in this book. We start with an account of the experimental evidence about the structure and physical properties of the surface with particular emphasis on the unusual properties near the melting point, and we finally turn to the models that have been proposed to account for these properties. As is well known in surface science, the properties of the surface will be dependent on the nature of the adjacent medium, not just whether it is another solid, but for a free surface whether it is in vacuum, water vapour or air. Properties are also likely to be critically dependent on surface contamination, and hence on how the surface has been formed. These are not the principal matters of interest, but they may give rise to unrecognized complications in some experiments.

10.2 Surface structure

10.2.1 *Structural studies at low temperatures*

Probably the most powerful techniques for studying the atomic structure of the surface layers of crystalline solids are scanning tunnelling microscopy (STM) and low-energy electron diffraction (LEED). These and other surface diffraction techniques show that in the case of silicon, for example, the surface is 'reconstructed' to eliminate dangling bonds (Haneman 1987), but these methods are not well suited to the study of ice and yield only limited information. Using STM Morgenstern *et al.* (1997) showed that at 140 K H_2O molecules form a bilayer on a Pt(111) surface with a structure resembling a (0001) bilayer of ice Ih, but this cannot be taken as typical of bulk ice. The electrical conductivity of ice is so low it has not so far been possible to use this technique on a layer more typical of the bulk.

Vapour-deposited (0001) layers of crystalline ice Ih (or possibly {111} layers of ice Ic) grown on Pt(111) surfaces at 125–140 K have been studied by LEED (Materer *et al.* 1997) and by helium atom diffraction (Braun *et al.* 1998). These experiments show that the (0001) surface has a 'full-bilayer termination', which means that molecules in the outer layer are bound to three molecules in the layer below and have just one dangling bond pointing outward. This is the termination of the upper surface of the structure shown in Fig. 2.3, and corresponds to a termination on a plane of the shuffle set in Fig. 7.2. Braun *et al.* also concluded that there was greatly enhanced vibrational motion of the molecules on the outer surface, which is consistent with the LEED observations. The formation of similar crystalline multilayers on Ru(001) has been described by Brown and George (1996). Using optical sum-frequency vibrational spectroscopy Su *et al.* (1998) have shown that ice multilayers deposited on Pt(111) are at least partially ordered ferroelectrically perpendicular to the surface. This is to be expected because, if there is a preferential orientation for the bonding of molecules to the Pt surface, the ice rules require that this polarization will propagate into the bulk (see also §10.4.3).

Fletcher (1992) has speculated about the low-temperature structure of the surface layer, including the possibility of ordered arrangements of the dangling bonds, but we must remember that even bulk ice will transform to an ordered structure at low temperatures (see §11.2) if there is sufficient proton mobility for this to happen.

In §3.4.4 it was shown that localized O—H bonds in D_2O or O—D bonds in H_2O yielded sharp features in the infrared absorption spectrum of ice. Similarly molecules at the surface of nanocrystals of ice, which have a large surface/volume ratio, produce distinct absorption lines. Detailed investigations coupled with computer modelling have identified three characteristic surface molecules: those with a dangling O—H bond, those with a dangling bond not occupied by H, and four-co-ordinated molecules that are distorted from the usual tetrahedral symmetry (Rowland and Devlin 1991; Rowland *et al.* 1995). The modes in the absorption spectrum are also sensitive to the adsorption of gas molecules on the surface of the ice, and provide a powerful tool for the study of adsorption processes (Buch *et al.* 1996; Delzeit *et al.* 1996). The interpretation of these extensive investigations of the surface bonding in ice at low temperatures is described by Devlin and Buch (1995, 1997). It appears, as expected, that a free surface with dangling bonds is unstable and that molecules are displaced in ways which partially eliminate the dangling bonds, but this reconstruction is disordered. Such disordered reconstruction may not be distinguishable from the apparent greatly enhanced molecular vibrations reported in the experiments of Materer *et al.* and Braun *et al.* described above.

These low-temperature experiments indicate nothing anomalous about the surface properties of ice, but unfortunately they cannot be extended to temperatures close to the melting point where evaporation and self diffusion become important, and to which we will devote our attention for the remainder of the chapter.

10.2.2 *X-ray diffraction*

X-ray diffraction is the one crystallographic technique which is not restricted to low temperatures and can also be set up to study diffraction from the near-surface region alone. To do so we make use of the fact that X-rays incident on a material at a glancing angle less than the critical angle (typically 1–2 mrad for ice) undergo total external reflection. An evanescent wave then penetrates the surface to a depth dependent on the angle of incidence, and in the case of ice this depth can be varied between about 5 and 1000 nm. This evanescent wave is diffracted by appropriately oriented planes in the crystal to produce Bragg reflections that emerge from the surface. Kouchi *et al.* (1987) first reported that these surface-diffracted beams disappeared as the temperature was raised above $-2\,^{\circ}\mathrm{C}$. The effect was studied in some detail by Dosch *et al.* (1995) using the highly collimated X-radiation from a synchrotron source and very carefully prepared specimens maintained under their saturated vapour pressure so that they neither grew nor sublimed during the measurements. They confirmed that the long-range order giving rise to Bragg reflections was indeed lost in a layer close to the surface, and by varying the angle of incidence they determined the thickness L of the non-crystalline layer as a function of temperature for the $\{0001\}$, $\{10\bar{1}0\}$, and $\{11\bar{2}0\}$ surfaces. Their results are plotted in Fig. 10.1, and over most of the range can be fitted by the equation

$$L = A \ln |T_s/(T_m - T)|, \qquad (10.1)$$

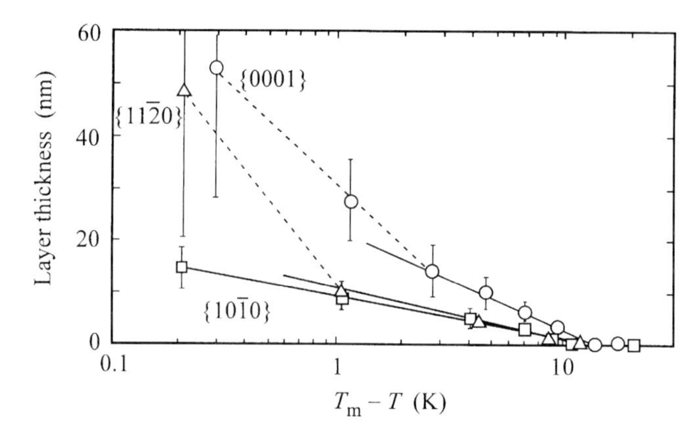

Fig. 10.1 Thickness of the oxygen-disordered layer on the surface of ice crystals of various orientations, as determined by glancing-angle X-ray diffraction. Solid lines are fits to equation (10.1). Reproduced from *Surface Science*, **327**, H. Dosch, A. Leid, and J.H. Bilgram, Glancing-angle X-ray scattering studies of the premelting of ice surfaces, pp. 145–64, Copyright 1995, with permission from Elsevier Science.

where T_m is the melting temperature, $T_s \approx -13\,^\circ$C, and $A \approx 8.4$ nm and 4.0 nm for basal and prismatic surfaces respectively. The results show that there is a significant non-crystalline or 'quasi-liquid' layer at all temperatures above about $-13\,^\circ$C. Figure 10.1 shows some evidence that L increases more rapidly than represented by equation (10.1) within $\sim 1\,^\circ$C from the melting point. The results were reproducible on heating or cooling, apart from some complications arising from changes in the roughness of the surface.

In a further experiment Dosch *et al.* (1996) studied the 0004 reflection from a $\{10\bar{1}0\}$ surface. This is a very weak reflection, because the contribution to the structure factor from the oxygen atoms is zero and the intensity therefore comes only from the hydrogen atoms. For diffraction from bulk ice this is a normal Bragg reflection with angular width of the order of 0.01° limited by the instrumental resolution. However, for diffraction from the surface layer with a penetration depth of 250 Å the line width is 1.5° and independent of temperature from -20 to $-0.4\,^\circ$C. The stronger reflections involving scattering by the oxygen atoms are sharp even when the 0004 reflection is broad. The integrated intensity of the surface 0004 reflection is equal to that from an equivalent volume of bulk ice from -20 to $-10\,^\circ$C, but diminishes at higher temperatures as the non-crystalline layer grows. These observations show that there is something anomalous about the protonic subsystem in the ice in a region below the quasi-liquid layer, even though the oxygen lattice is fully crystalline. In their paper Dosch *et al.* refer to this as a region of 'rotational Bjerrum disorder', but a satisfactory model of the proton disorder in this region has not yet been developed. We note that this region

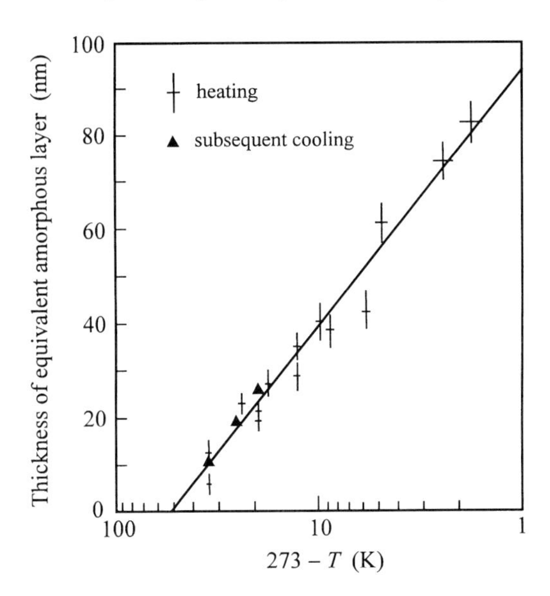

Fig. 10.2 Thickness of the 'equivalent amorphous layer' required to account for the proton back-scattering from the (0001) surface of an ice crystal. (From Golecki and Jaccard 1978.)

persists at temperatures down to at least −20 °C, at which there was no observable quasi-liquid layer.

10.2.3 *Proton channelling*

Golecki and Jaccard (1978) studied the back-scattering of 100 keV protons from the (0001) surface of ice between −130 and −2 °C. Such scattering depends on the extent to which the incident particles can travel along open channels in the structure (Gemmell 1974). Below −60 °C the back-scattering was only weakly temperature dependent, and could be attributed to the normal thermal motions of the atoms perpendicular to the [0001] direction. At higher temperatures the back-scattering increased more rapidly indicating additional blocking of the channels. The data were modelled in terms of an equivalent amorphous layer with a thickness that varied with temperature as shown in Fig. 10.2. This equivalent amorphous layer is much thicker than the disordered layer observed by Dosch *et al.* (1995). However, Golecki and Jaccard preferred to interpret their observations not in terms of an amorphous layer, but as showing anomalously large amplitudes of molecular vibration near the surface.

10.3 Optical ellipsometry and microscopy

If a beam of plane-polarized light is reflected at non-normal incidence from a material of refractive index n_2 covered with a thin layer of thickness L and different refractive index n_l, the light becomes elliptically polarized, and its state of polarization can be used to determine n_l and L. Using this technique Beaglehole and Nason (1980) detected a 'liquid-like' layer on the prismatic surfaces of ice above −5 °C. A more refined experiment was performed by Furukawa *et al.* (1987), who studied the plane facets of 'negative crystals' formed inside a specimen of ice by inserting a hypodermic needle and pumping out the water vapour, though the surfaces were subsequently opened up to air. Surface layers were observed on both basal and prismatic facets, and the data were fitted to values of n_l and L. The refractive index of the layers was 1.330, which is closer to that of water (1.3327) than that of ice (1.3079). The thickness L of the layers varied with temperature as shown in Fig. 10.3. Although there was scatter from sample to sample the difference between the faces was significant.

Other experiments using similar optical techniques have been reported by Beaglehole and Wilson (1993) and Elbaum *et al.* (1993), but the results of such measurements are not consistent and depend on the condition of the surface. Elbaum *et al.* (1993) found that for vapour-deposited crystals in the absence of air neither basal nor prismatic facets undergo complete surface melting, and at the triple point droplets were formed which did not wet the remainder of the surface.

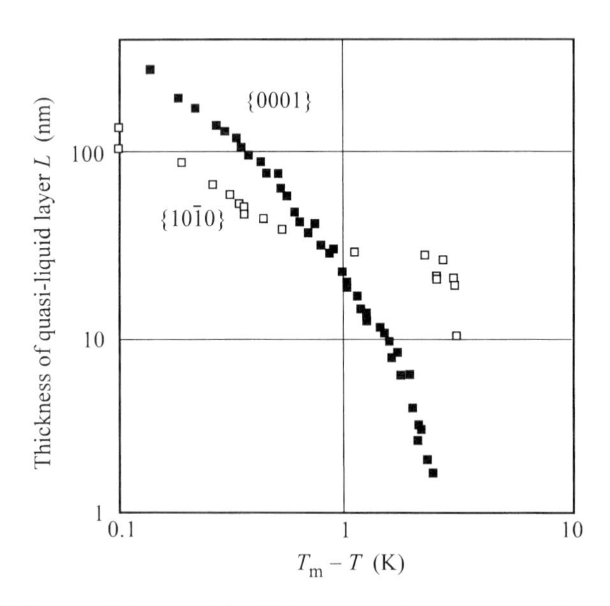

Fig. 10.3 Thickness of the quasi-liquid layers on {0001} and {10$\bar{1}$0} ice surfaces as determined by optical ellipsometry by Furukawa *et al.* (1987) and plotted in this form by Furukawa and Nada (1997). Reproduced with permission from *Journal of Physical Chemistry*, **B101**, 6167–70. Copyright 1997 American Chemical Society.

However, if air was allowed into the chamber macroscopically thick liquid layers formed on basal facets. Their measurements showed thick layers forming within only tenths of a degree from 0 °C. Beaglehole and Wilson (1994) performed optical measurements for an ice–glass interface. For smooth clean glass no interfacial layer could be detected, but pre-melting was observed at rough or contaminated interfaces. There was no pre-melting if the glass was coated with a hydrophobic film.

The conclusion of these measurements is that a layer with density close to that of liquid water is frequently present on the surface of ice within not more than 5 °C of the melting point, and the nature of this pre-melted layer is critically dependent on the conditions of the experiment.

10.4 Electrical properties of the surface

Imagine two electrodes placed on the surface of a specimen as shown in Fig. 10.4. If the bulk conductivity σ_{bulk} of the material (at some fixed frequency) is known, the bulk contribution to the conductance G_{b} between the electrodes can be calculated. The actual conductance G may be larger than this by an amount G_{s} attributable to the surface of the material lying between the electrodes, such that

$$G = G_{\mathrm{b}} + G_{\mathrm{s}}. \tag{10.2}$$

In practice measurements of G_{s} have been restricted to conditions where it is much larger than G_{b}. The value of G_{s} depends on the dimensions marked on the figure, and we can define a specific conductance of the surface

$$g_{\mathrm{s}} = \frac{G_{\mathrm{s}} d}{w}. \tag{10.3}$$

This quantity resembles σ_{bulk} for the bulk, but its units are siemens (S or Ω^{-1}) not S m^{-1}. It will be frequency dependent, and we can define both the d.c. or low-frequency limit g_{ss} and the high-frequency limit $g_{\mathrm{s}\infty}$.

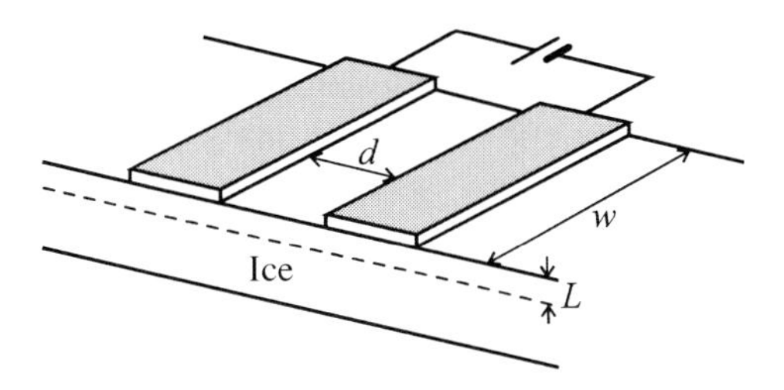

Fig. 10.4 Geometry of electrodes on an ice surface for the definition of the surface conductance occurring within a layer of thickness L.

In practice surface conduction will occur in some layer of effective thickness L as marked in Fig. 10.4, within which there is an additional volume conductivity σ_{surf}, such that

$$g_s = \sigma_{\text{surf}} L. \tag{10.4}$$

If for simplicity we consider only one kind of charge carrier of charge e, mobility μ, and concentration per unit volume n,

$$\sigma_{\text{bulk}} + \sigma_{\text{surf}} = ne\mu, \tag{10.5}$$

and then σ_{surf} can be seen to arise from an increase in n near the surface or a change in μ or both.

At any surface in thermal equilibrium one expects there to be some electrical charge on the surface, consisting in ice of an excess or deficit of protons on the dangling bonds, and this charge λ_s per unit area will be compensated by a layer of oppositely charged point defects of thickness L. The total number of these excess defects per unit area is ΔnL, and from equations (10.4) and (10.5) this contributes directly to g_s.

10.4.1 *Surface conductivity*

Many experiments have shown that it is important to take account of the effect of the surface in electrical measurements on ice above -20 to $-30\,°\text{C}$. Figure 10.5 shows measurements of the d.c. surface conductance g_{ss} on various monocrystalline samples of ice by Maeno and Nishimura (1978). A typical value for their 'pure' ice at $-10\,°\text{C}$ is 3×10^{-11} S with an activation energy below $-5\,°\text{C}$ of 1.2 eV. Above $-5\,°\text{C}$ g_{ss} rises more rapidly towards the melting point, probably corresponding to the effect of the quasi-liquid layer observed in structural and optical experiments. The surface conductance was changed by roughening the surface or by pumping air and water vapour away from it.

Many similar measurements have been reported with values of g_{ss} lying in the whole range between the data for pure ice and the HF-doped ices in Fig. 10.5, with higher values normally associated with lower activation energies (e.g. Ruepp and Käss 1969; Camp *et al.* 1969; Maidique *et al.* 1971; Maeno 1973). The differences are presumably associated with specimen purity and the state of the surface. As for the steady-state conductivity of bulk ice considered in §5.3.3, the lowest values of g_{ss} in Fig. 10.5 can only be taken as upper limits on the intrinsic surface conductance of ice.

The frequency dependence of the surface conductance was studied for polycrystalline ice by Caranti and Illingworth (1983*a*), with the results shown in Fig. 10.6. The values for both bulk and surface conductivities show that their specimens were not particularly pure. They found that the surface current remained almost in phase with the p.d. across the whole frequency range, and that the ratio between the high- and low-frequency limits was not very large. This

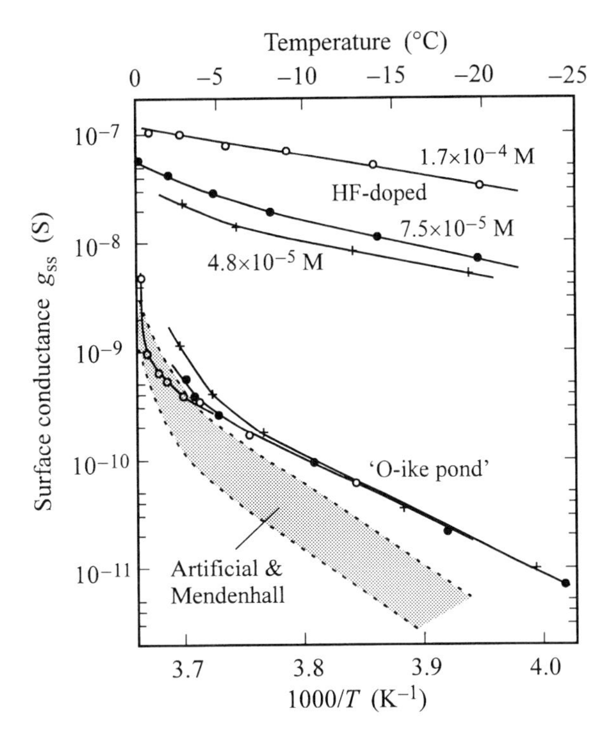

Fig. 10.5 The d.c. surface conductance g_{ss} of various specimens of ice as determined by Maeno and Nishimura (1978). The artificial ice and ice from the Mendenhall Glacier were nominally pure. Almost as pure were the samples from the O-ike pond in Antarctica. HF-doped samples have concentrations as determined after melting. Reproduced by courtesy of the International Glaciological Society from *Journal of Glaciology*, **21**(85), pp. 193–205, 1978, Fig. 2.

means that the surface behaves like a normal conductor, with no strong Debye relaxation such as is observed in bulk ice. If the Jaccard theory of §4.5 is applicable to the surface, this corresponds to both Bjerrum and ionic partial conductivities being comparable in magnitude. Alternatively, the surface region is so disordered that the strict requirements of the ice rules are relaxed, and one can just consider a single type of dominant charge carrier. One further possibility is that there is indeed a Debye relaxation in the surface but at a frequency above that of current experimental measurements.

In §5.3.3 it was explained that there can be no Hall effect in bulk ice, but Caranti and Lamfri (1987) claim to have observed the effect for surface currents. They used an alternating magnetic field at 50 Hz and a current at 107 Hz, detecting the Hall voltage at 157 Hz. They did not determine the sign of the charge carriers, but assuming only one type of carrier they deduced a mobility of about

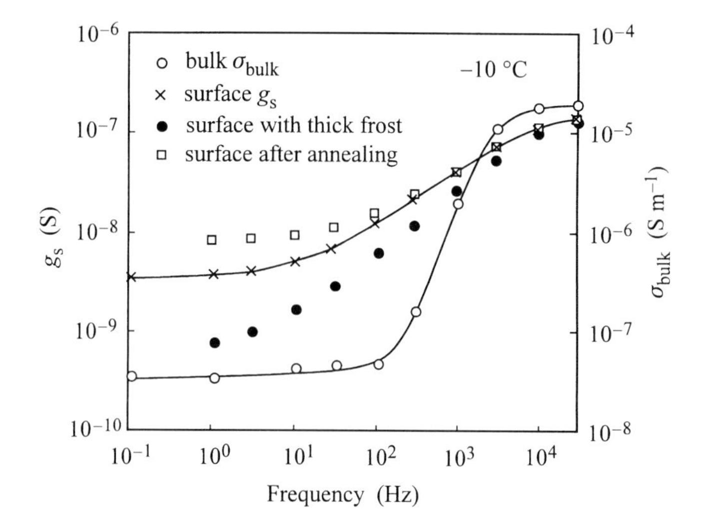

Fig. 10.6 Frequency dependence of the surface conductance g_s of 'pure' polycrystalline ice at −10 °C as determined by Caranti and Illingworth (1983*a*). The bulk conductivity σ_{bulk} is shown for comparison. Reproduced with permission from *Journal of Physical Chemistry*, **87**, 4078–83. Copyright 1983 American Chemical Society.

3×10^{-4} m^2 V^{-1} s^{-1} at −10 °C which decreased rapidly with rising temperature. Such a mobility is orders of magnitude larger than that of anything possible in the bulk. The result must be treated with caution, but even the existence of the effect is an indication of the very different nature of the surface.

Chrzanowski (1988) describes a related experiment in which he measured the in-plane d.c. electrical conduction of thin films of ice deposited at 220 K on glass or silica plates. The current flowing for a given p.d. increased rapidly with thickness up to 0.5 μm and more slowly thereafter, providing evidence for enhanced conductivity in regions within ~0.2 μm of the free or dielectric interface. Such regions are thicker than anything we have envisaged so far.

10.4.2 *Interpretation and further experiments*

In attempting to interpret the observations of the surface conductance of ice we will focus attention on results for 'pure' ice below −5 °C, so that possible premelting phenomena can be ignored. We first observe that if we substitute values for g_s and corresponding estimates of σ_s for bulk ice into equation (10.4) we obtain values of L greater than or equal to the dimensions of the specimen. This means that the results really do correspond to an additional conductivity in the surface. If we use instead the value of σ_s for pure liquid water (5 μS m^{-1}) we obtain

$L \sim 1$–$10\,\mu$m, which is several orders of magnitude thicker than any conceivable surface layer. The surface conductance cannot therefore be accounted for by surface melting, and we are looking for greatly enhanced conduction in an essentially crystalline subsurface layer.

As already stated at the beginning of §10.4, the surface of ice is expected to carry an electric charge, and experiments to be described will show that it is positive in sign. For a clean surface on pure ice such a charge would be present if more than 50% of the molecules in the outer layer of the crystal had protons on the outward pointing dangling bonds. This would result from a polarization **P** of the ice directed normal to the surface. There can be no polarization deep in the interior of the ice, and **P** must therefore decay with depth x below the surface as in Fig. 10.7. The local charge density $\rho(x)$ is equal to $-\nabla \cdot \mathbf{P}$, and this is also shown in Fig. 10.7. It consists of a delta function corresponding to a charge λ_s per unit area on the surface and an equal and opposite charge in a layer below the surface. This space charge will be composed of OH^- ions or L-defects, but if we suppose that there are only OH^- ions of concentration $n(x)$, then

$$\lambda_s = \int_0^\infty e_\pm n(x)\,\mathrm{d}x. \tag{10.6}$$

If these ions have mobility μ_- the conductivity at a depth x in the surface layer is $\sigma_s(x) = n(x)e_\pm\mu_-$, and the conductance of the surface layer, measured parallel to the surface, is

$$g_s = \int_0^\infty \sigma_s(x)\,\mathrm{d}x = \lambda_s\mu_-. \tag{10.7}$$

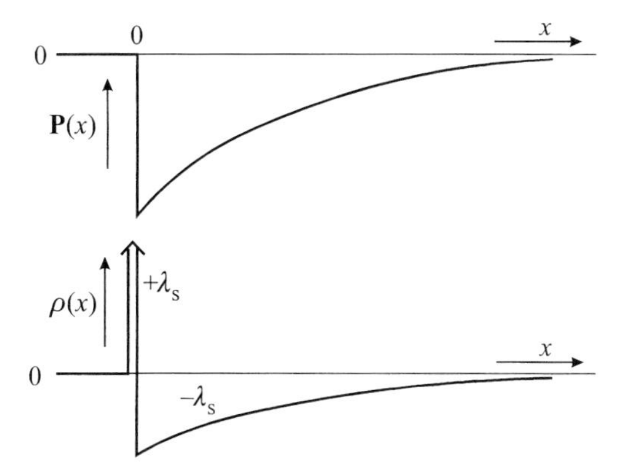

Fig. 10.7 Polarization $\mathbf{P}(x)$ and space charge density $\rho(x)$ in the charge double layer at a crystal surface. The actual surface charge $+\lambda_s$ per unit area is represented as a delta function at $x = 0$.

If we take $g_s = 3 \times 10^{-11}$ S and assume the bulk ionic mobility $\mu_- = 3 \times 10^{-8}\,\mathrm{m^2\,V^{-1}\,s^{-1}}$ from Table 6.4, this equation gives $\lambda_s = 10^{-3}\,\mathrm{C\,m^{-2}}$. The density of molecules on a (0001) surface of ice is $N_s = 5.7 \times 10^{18}\,\mathrm{m^{-2}}$, and thus the saturation surface charge is $\lambda_{max} = \frac{1}{2} e_{DL} N_s = 0.17\,\mathrm{C\,m^{-2}}$. The required charge is less than 1% of this, so that the model for the surface conduction is quite plausible. The actual surface charge is difficult to measure, but one experiment that sets a lower limit is electrification by sliding friction, from which Petrenko and Colbeck (1995) set a lower limit on λ_s of the order of $10^{-4}\,\mathrm{C\,m^{-2}}$, and we will describe in §10.6 an experiment indicating a value of at least $0.01\,\mathrm{C\,m^{-2}}$.

In developing this idea we should not forget the effect of the ice rules, which require that the steady state conductivity is limited by the smaller of the partial conductivities $\sigma_\pm$ and σ_{DL} defined by equations (4.48) and (4.49). In accounting for the d.c. surface conductance in terms of an enhanced concentration of ions we have to assume that the rate of molecular reorientation by Bjerrum defects is also enhanced. This would displace the Debye relaxation in the subsurface layer to frequencies higher than that in the bulk and above the range of current measurements of surface properties. Our model of the subsurface region then becomes one in which molecules rotate rather more easily than in the bulk, but ionic states still transfer from one molecule to another by proton hopping; they do not diffuse as entities as they do in the liquid.

The model has not so far involved the thickness L of the subsurface layer. Simple electrostatics shows that L is related to the potential difference across the surface layer V_s by the relation

$$\lambda_s \approx \frac{\varepsilon_0 \varepsilon_\infty V_s}{L}. \tag{10.8}$$

λ_s and V_s have to be calculated by solving the differential equations for the space charge region that were introduced in §4.6. The solutions quoted there assumed that $eV \ll k_B T$ and yielded the characteristic screening lengths κ_1^{-1} and κ_2^{-1} given in equations (4.64), but for the surface charge densities we are considering here this approximation is far from being valid. A better approximation was used by Petrenko and Maeno (1987), but a more complete analysis of the surface layer for $eV > k_B T$ and including the effect of the ice rules is that of Petrenko and Ryzhkin (1997). They show that, for $\lambda_s = 0.1\lambda_{max}$ and other parameters appropriate for $-10\,°\mathrm{C}$, 95% of the screening charge lies within 1.6 nm of the surface, though the potential falls more slowly than this. Because of their larger charge the concentration of ionic defects near the surface is enhanced relative to that in the bulk by a larger factor than that for Bjerrum defects.

The above interpretation of the surface conduction is supported by an experiment by Khusnatdinov *et al.* (1997), who used the field effect transistor technique described in §5.6.2 to study the ice–SiO_2 interface at $-10\,°\mathrm{C}$. By applying both positive and negative potentials to the silicon gate electrode they could introduce measurable negative and positive space charges in the subsurface

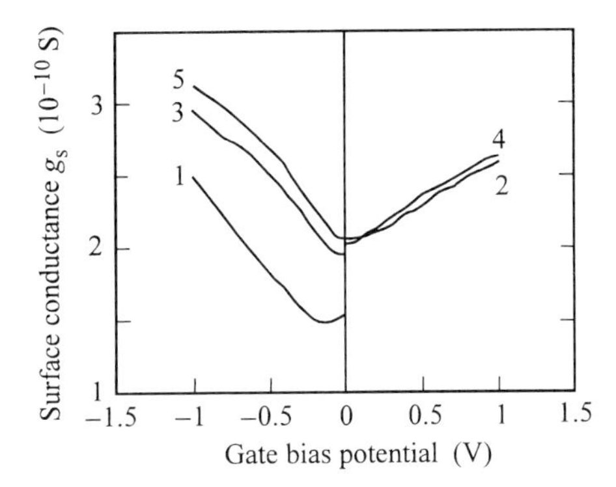

Fig. 10.8 Observations of surface conductance g_s at 10 Hz and −10 °C as functions of gate bias potential in experiments with an ice field effect transistor similar to that in Fig. 5.15. The measurements were taken in the order indicated by the numbering of the curves (Khusnatdinov *et al.* 1997). Reproduced with permission from *Journal of Physical Chemistry*, **B101**, 6212–14. Copyright 1997 American Chemical Society.

layer of the ice. Measurement of the surface capacitance set an upper limit on L of 2 nm. Figure 10.8 shows how the conductance g_s between the gold source and drain electrodes varied with the potential applied to the gate. The different slopes for positive and negative potentials show that the positive and negative charge carriers have different mobilities. Knowing λ_s and using equation (10.7) they calculated mobilities of $(1.2 \pm 0.3) \times 10^{-7}$ and $(3 \pm 0.5) \times 10^{-7}$ $m^2 V^{-1} s^{-1}$ for the positive and negative ions respectively. These are somewhat larger mobilities than those estimated for the bulk in Table 6.4, and reflect different behaviour in the subsurface layer. The conductance in Fig. 10.8 passes through a minimum for a gate potential which may differ from zero by a few tenths of a volt. This potential, which is not reproducible between specimens or on cycling the potential, is taken to represent the natural state of the surface, and corresponds to a charge density on the ice–solid interface of the order of $\pm 10^{-4}$ C m^{-2}.

Finally we consider the effect which the charges at the surface may have on the crystal structure. The electrostatic pressure at the top of the subsurface layer is of the order of $\lambda_s^2/\varepsilon_0\varepsilon_\infty$, and for charges of $0.1\lambda_{max}$ this is sufficient to depress the melting point of ice by a few degrees, so that it could be relevant to the formation of the quasi-liquid region at the highest temperatures (Petrenko and Ryzhkin 1997).

10.4.3 *Surface potential and surface charge*

Various experiments have studied the contact potentials between metals and ice, and the transfer of charge when contact is made. These experiments involve the transfer of electronic charge which has to be converted by electrochemical processes in the ice. Buser and Jaccard (1978) observed the charge transfer on contact with various metals and inferred a work function for electrons in ice of 4.4 eV. Mazzega *et al.* (1976) used the Kelvin vibrating capacitor technique to determine the contact potential between ice and gold, obtaining a similar value for the work function and finding that the contact potential changes by about 0.1 V in the temperature range from $-30\,^{\circ}C$ to the melting point. They attributed this change to the development of a potential difference across the surface layer within the ice, with the free surface becoming rapidly more negative as the temperature approached the melting point. This sign is opposite to what would be expected from experiments we have already described, and the whole topic of surface potentials, which must include both the free surface and the ice–metal contact, is very hard to interpret.

The experiments of Su *et al.* (1998) on ice multilayers deposited on Pt(111), which have been referred to in §10.2.1, led to the conclusion that there is some ferroelectric polarization of the ice at the interface to the metal, and that this polarization decays into the bulk over a characteristic distance of ~30 monolayers. The magnitude and direction of the polarization were not determined, and the measurements were confined to films annealed at 137 K; the use of higher temperatures is limited by sublimation of the ice.

Surface potentials have also been observed in the growth of crystals from the liquid (Caranti and Illingworth 1983*b*) or from the vapour (Takahashi 1970, 1973).

10.5 Nuclear magnetic resonance

In §2.5.1 and §6.6 we showed how nuclear magnetic resonance (n.m.r.) can provide information about the location and motion of the protons in ice. When n.m.r. experiments are performed on finely powdered ice a narrow line is observed superimposed on the normal broad line, and this narrow line must be attributed to protons located close to the surface (Clifford 1967; Kvlividze *et al.* 1974). The most recent observations are those of Mizuno and Hanafusa (1987), and Fig. 10.9 shows their n.m.r. spectra for finely powdered ice at temperatures from -100 to $-10\,^{\circ}C$ together with that for liquid water at $+5\,^{\circ}C$. The spectrum of bulk ice, which is about 60 times broader, is not seen on the scale of this figure. The narrow surface line was only observed for particles of diameter less than 150 μm, which provide sufficient surface area for it to appear above the noise level. The temperature dependence of the intensity of the line is related to the fraction of protons contributing to it, but it is difficult to draw conclusions about the temperature dependence of this fraction because sintering occurs during the course of the experiment, especially at the higher temperatures. Similar observations were obtained on powders that had been slightly melted and re-frozen to contain very small bubbles. Barer *et al.* (1977) observed a similar narrow line in

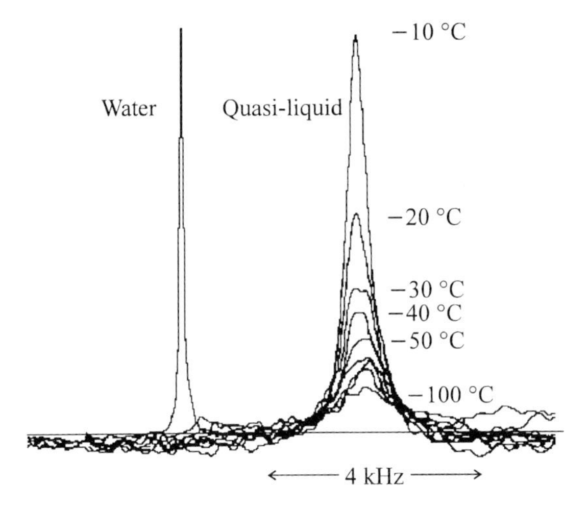

Fig. 10.9 Nuclear magnetic resonance spectrum from the quasi-liquid layer on frozen water droplets of diameter less than 150 μm at temperatures from −100 to −10 °C, together with the single narrower line for liquid water at +5 °C. The radio frequency for these measurements was 99.5 MHz. The line arising from bulk ice is too broad to be seen on the scale of these plots. (From Mizuno and Hanafusa 1987.)

the n.m.r. spectrum of ice containing finely divided silica particles, and, assuming that the surface and bulk line intensities are proportional to the numbers of molecules contributing to them, they concluded that the thickness of the surface layer in this case was 11 Å at −12 °C rising to 52 Å at −2 °C.

The fact that the surface n.m.r. line is narrow is evidence that the protons are moving from one position to another, thus experiencing an averaged local magnetic field. Figure 10.10 shows the spin–lattice relaxation time T_1 as a function of temperature for both the powdered and the re-frozen ice of Mizuno and Hanafusa (1987). This passes through a minimum at −35 °C, which implies that at this temperature the radio-frequency ω and the correlation time for proton hopping τ_c satisfy the condition $\omega\tau_c \approx 0.6$. (In §6.6 we noted that such a minimum could not be attained in bulk ice; its observation here means that the interpretation for the surface can be made more reliably than for the bulk). At −35 °C the correlation time τ_c in the surface region was therefore shown to be about 10^{-9} s, which is closer to the value in liquid water (10^{-12} s) than to that in bulk ice (10^{-4} s). The temperature dependence of τ_c can also be estimated, but this involves assumptions and the results differ for the powdered and re-frozen samples. The conclusion from these experiments is that even at −100 °C the protons, or whole molecules, in at least the surface monolayer move from one site to another at a rate orders of magnitude faster than in bulk ice. As the temperature rises the thickness of the region of high mobility increases substantially, long before there is any evidence of structural melting.

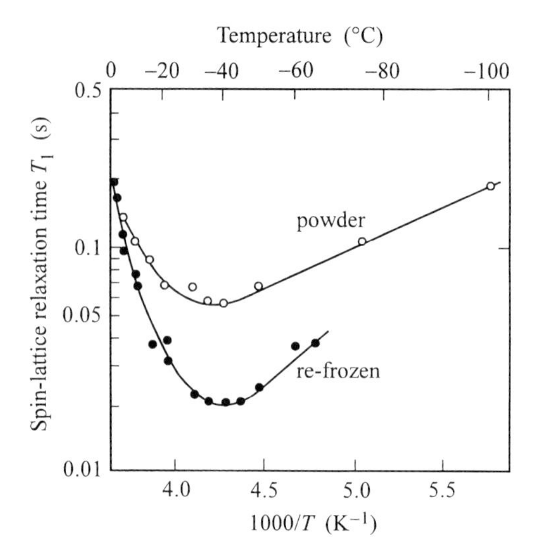

Fig. 10.10 Temperature dependence of the spin–lattice relaxation time T_1 in nuclear magnetic resonance measurements at 100 MHz on the quasi-liquid layer in finely powdered ice and in the same material after re-freezing to contain very small bubbles. (From Mizuno and Hanafusa 1987.)

10.6 Scanning force microscopy

The scanning force microscope, SFM, otherwise known as the atomic force microscope, AFM, is an instrument in which a probe tip is mounted on a cantilever held above a surface and the force of attraction or repulsion is measured (Sarid 1994). In the scanning mode the tip is servo-controlled to keep the force constant and so determine the profile of the surface. Alternatively, force–distance curves can be obtained as the tip is moved towards and away from a fixed point on the surface. The resolution in the plane of the surface is typically 5–100 nm, which is much less than the atomic resolution of the STM referred to in §10.2.1, but the SFM is suitable for use at higher temperatures and in air or under inert liquids. Its resolution perpendicular to the surface is typically 0.1 Å.

In the scanning mode Nickolayev and Petrenko (1995) observed relatively sharp features of height ~50 nm on an ice–air interface at −25 °C. At −11 °C the surface became smooth on a scale ten times smaller than this, and at −5 °C it developed a waviness of amplitude 100 nm that changed with time over periods of the order of 60 s.

Force curves have been studied by Slaughterbeck *et al.* (1996) on vapour-deposited films *in vacuo* and by Petrenko (1997) on the (0001) face of single crystals in air or an inert liquid. With the surface and tip immersed in decane, Petrenko found that, after the first contact of an insulating tip with the ice, in subsequent approaches of the tip to the ice there was a smoothly increasing repulsion prior to actual contact. He attributed this to a surface charge of at least $0.01\ \mathrm{C\ m^{-2}}$ over the temperature range -1 to $-20\,^{\circ}\mathrm{C}$. For free surfaces above $-13\,^{\circ}\mathrm{C}$ both Slaughterbeck *et al.* (1996) and Petrenko (1997) found evidence for a liquid-like layer which draws the tip down and then exhibits capillary attraction as the tip is withdrawn. Contact between the tip and the surface was observed to damage the surface of the ice, but such damage heals quite rapidly. There was no evidence for pressure melting by the tip below $-1\,^{\circ}\mathrm{C}$. While the SFM shows clear evidence of high surface mobility of molecules above about $-13\,^{\circ}\mathrm{C}$ and is in principle promising, it will be some time before the technique is developed to yield really quantitative information about the ice surface.

10.7 Surface energy

An important thermodynamic parameter of the interface between two phases is the free energy per unit area γ. In the case of the interface between a liquid and its vapour this is the familiar surface tension γ_{lv}, equal to the work which must be performed isothermally to create unit area of liquid surface. In the case of a solid–vapour interface γ_{sv} is half the 'work of adhesion', or the work required per unit area to split the material into two parts, without dissipating energy in plastic deformation but allowing any restructuring of the surface (see for example Israelachvili 1991). There is also an interfacial free energy γ_{sl} for the solid–liquid interface.

At the triple point all phases co-exist in equilibrium, and on a flat solid interface the angle of contact θ defined as in Fig. 10.11 is given by the condition

$$\gamma_{\mathrm{sv}} = \gamma_{\mathrm{sl}} + \gamma_{\mathrm{lv}} \cos\theta. \tag{10.9}$$

The situation shown is only possible if $\gamma_{\mathrm{sv}} < \gamma_{\mathrm{sl}} + \gamma_{\mathrm{lv}}$. If this is not so, the liquid will completely wet the surface. At temperatures below the triple point it has been proposed that, if γ_{sv} for a theoretically 'dry' surface were greater than $\gamma_{\mathrm{sl}} + \gamma_{\mathrm{lv}}$, a layer of liquid would always be present on the surface of the solid. The thickness of this layer would be determined by balancing the free energy required to melt the layer against the free energy gained by forming the two interfaces. This way of thinking about the surface is controversial, and we discuss the model in §10.9.1. At present we are concerned with the experimental evidence about the surface energies. In this context γ_{sv} is the free energy of the whole surface at a specified temperature inclusive of any special structure it may have.

Knight (1966, 1971) made several attempts to determine the contact angle for water droplets on ice close to $0\,^{\circ}\mathrm{C}$, obtaining values between 0° and 12°. Elbaum *et al.* (1993) observed that water did not wet ice in pure vapour, but did so if air were present. Makkonen (1997) argued that in considering the condition $\gamma_{\mathrm{sv}} \gtrless \gamma_{\mathrm{sl}} + \gamma_{\mathrm{lv}}$ it is essential to use a value of γ_{sv} for a 'dry' surface, i.e. one that is at

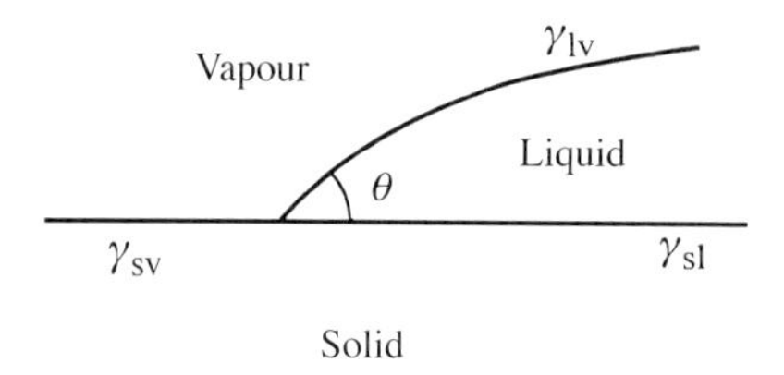

Fig. 10.11 Illustration of the meaning of the contact angle θ of a liquid on a solid surface.

a sufficiently low temperature that no special surface layer is present. He was able to observe water drops at $+85\,°C$ present on ice at $-25\,°C$ with the ice surface remaining flat. The angle of contact in the water was then $\theta = 37°$. Adjusting for the differences in temperature he showed that γ_{sv} could not be greater than $\gamma_{sl} + \gamma_{lv}$. By measuring the angles at grooves where grain boundaries cut the water–ice interface in a temperature gradient Hardy (1977) deduced that $\gamma_{sl} = 29\,\text{mJ m}^{-2}$.

van Oss *et al.* (1992) have determined the surface energy of ice within the context of modern surface chemistry (Good and Chaudhury 1991; Good *et al.* 1991). The analysis involved separating the surface free energies into Lifshitz–van der Waals and Lewis acid–base components, measurements of the contact angles of various liquids on ice and experiments to observe the exclusion or incorporation of various colloidal particles in an advancing freezing interface. They finally conclude that at $0\,°C$

$$\gamma_{sv} = 69.2\ \text{mJ m}^{-2} \qquad \gamma_{lv} = 75.8\ \text{mJ m}^{-2},$$

and other parameters of these surfaces are given in their paper. Only $26.9\,\text{mJ m}^{-2}$ of the value of γ_{sv} arises from the Lifshitz–van der Waals interaction. From their data they deduce that

$$\gamma_{sl} \approx 0.04\ \text{mJ m}^{-2} \qquad \text{and} \qquad \theta = 24.2°.$$

These values presumably include any special equilibrium features of the surface. They supersede earlier figures for γ_{sv}, and there should definitely be a finite contact angle. These results provide no driving force for the formation of a liquid layer on ice. Although γ_{sl} is too small for its numerical value to be meaningful, it must be positive to be consistent with the observation of Faraday (1860) that when two pieces of ice are brought into contact under water at the freezing point they stick together.

The surface free energy $\gamma_{sv} = 69.2\,\text{mJ m}^{-2}$ is equivalent to 0.076 eV per bond across a (0001) surface. From §2.6 the lattice energy of ice at 0 K is 0.306 eV per bond, and as cleaving the crystal exposes two surfaces this sets an upper limit of 0.153 eV per bond on the surface energy. The actual value of γ_{sv} must be less than this due to surface relaxation and entropic contributions to the free energy, so these results are entirely compatible.

10.8 Review of experimental evidence

The many experiments described show beyond any doubt that at high temperatures the surface of ice has properties that are very different from those of the bulk. Figure 10.12 indicates the temperature ranges over which such surface properties have been observed by different techniques. The differences may be largely determined by the nature and sensitivity of the techniques. In n.m.r., for example, one observes a new sharp line forming in the spectrum, whereas the surface electrical conductance has to be measured above the background arising from the bulk, and in X-ray diffraction one is looking for the diminution of an existing Bragg reflection. Nevertheless the temperature dependencies of some effects are certainly different. Optical measurements, X-ray diffraction, and electrical conductivity all show rapid changes within about 1 °C from the melting point which can be labelled as 'pre-melting' phenomena, but they also show effects extending to lower temperatures with more gradual temperature dependencies. Where the techniques yield direct evidence of layer thickness, these are summarized in Fig. 10.13. At −10 °C the effects shown are mainly confined to the first four double layers of molecules on a (0001) face (i.e. twice the *c* lattice parameter), though there may be effects such as the subsurface space charge layer extending deeper.

The experimental observations are basically of two kinds: structural and dynamic. The X-ray experiments show a breakdown of the long-range order of the oxygen lattice. This is probably correlated with optical observations of a

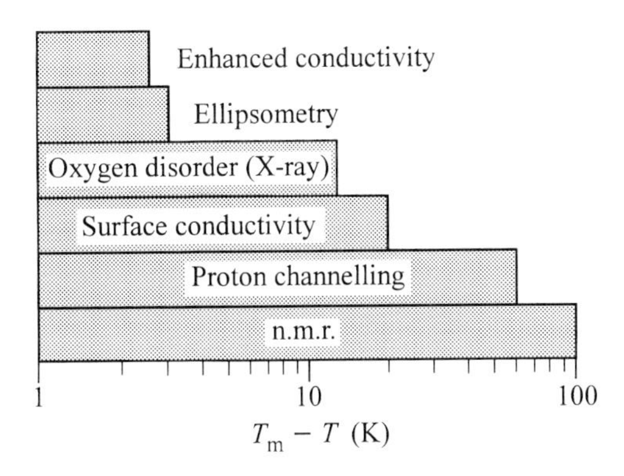

Fig. 10.12 Chart showing on a logarithmic scale the temperature range below the melting point T_m over which different techniques reveal special properties characteristic of the surface layer on ice.

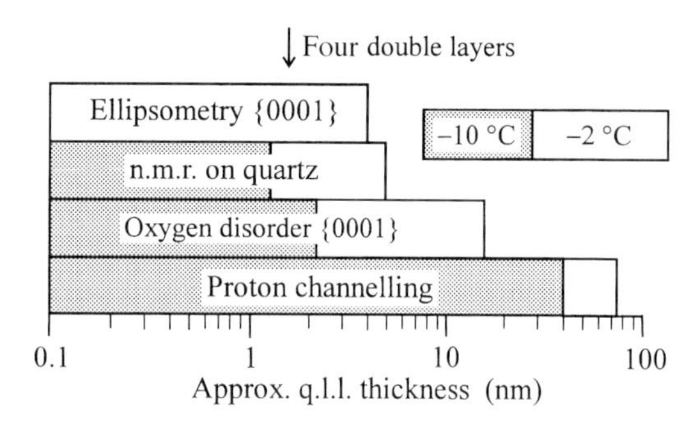

Fig. 10.13 For experiments which can be interpreted as yielding a thickness of the quasi-liquid surface layer, this chart shows the approximate thicknesses deduced at −10 and −2 °C.

region of higher refractive index, and by inference a density close to that of liquid water. However, the X-ray technique appears to be more sensitive to a gradual breakdown of crystallographic order than are the optical measurements, which have to be fitted to a layer of quite distinct refractive index. Structural changes will also affect the channelling measurements, and it is puzzling that the current interpretation of the channelling experiments requires disorder at greater depths and lower temperatures than other structural observations.

The n.m.r. experiments reveal that a fraction of the protons have a greatly increased rate of motion from place to place. This could arise either from molecular diffusion or from protons hopping between molecules as they do in the motion of protonic defects in bulk ice. The conductivity near the surface is also greatly increased, and this has to be due to increased numbers or mobilities of charge carriers in the subsurface layer. It cannot be accounted for by structural melting in which the molecules remain intact. The surface conductance measurements do not yield a thickness for the subsurface layer, which can only be inferred from a model of the surface and subsurface charge distributions.

For many purposes it is simpler to consider the free surface, but we have also reported surface effects at the interface with other materials. These effects may be more stable, and have yielded particularly informative results in the case of the field effect transistor and for n.m.r. where the finely divided material does not sinter during the course of the measurements.

Finally, we must be very cautious in interpreting these data because surface experiments are notoriously sensitive to surface contamination. Elbaum *et al.* (1993), for example, found that optical observations of surface wetting were completely changed by exposure to air, and there is also the danger of accumulation of impurities from within the ice on a surface subjected to sublimation. Progress in this field may be dependent on better control of experimental conditions, and on facing the challenge of performing more than one kind of measurement on the same surface!

We have not in this chapter considered the role of the surface in the growth of ice crystals with the diverse forms to be described in §12.3. These features appear to be related to the dynamics of crystal growth, which would be seriously affected by the presence of a disordered surface. One might expect surface pre-melting to smooth out local crystallographic character, but in contrast Kuroda and Lacmann (1982) attempted to interpret the diverse habits of vapour-grown crystals in terms of the properties of quasi-liquid layers on the different facets. At the present time our understanding of the nature of the surface is too uncertain to be able to trust such models of what are very subtle effects.

10.9 Theoretical models

There is no satisfactory model for the special properties of the surface of ice observed at temperatures approaching the melting point, but several quite distinct models have been proposed which help to advance our understanding of this intriguing topic. The basis of these models will be outlined in the following sections.

10.9.1 *The liquid-like layer*

As mentioned in §10.7, if the free energy of the ice–vapour interface γ_{sv} is greater than the combined free energies of the ice–liquid and the liquid–vapour interfaces ($\gamma_{sl} + \gamma_{lv}$), the total free energy of the surface could be reduced by covering the surface with a thin layer of liquid. This hypothesis was applied to the case of ice by Lacmann and Stranski (1972) and has been developed by Dash *et al.* (1995), who review its application to other materials as well. The increase in the free energy per unit area of a surface by the formation of a liquid layer of thickness d is written as

$$\Delta G = \rho_l \Delta\mu\, d + \Delta\gamma f(d). \tag{10.10}$$

Here $\Delta\mu$ is the difference in free energy per unit mass between the liquid and the solid, ρ_l is the density of the liquid, $\Delta\gamma = \gamma_{lv} + \gamma_{sl} - \gamma_{sv}$, and $f(d)$ is a function varying between 0 and 1 as d changes from 0 to ∞. The function $f(d)$ takes account of the fact that the two interfaces will not be independent of one another. Ignoring pressure effects which can be shown to be small,

$$\Delta\mu = \frac{q_m(T_m - T)}{T_m}, \tag{10.11}$$

where q_m is the latent heat of melting and T_m is the melting temperature. Different assumptions about $f(d)$ yield different functional forms for $d(T)$. A common assumption based on van der Waals interactions is

$$f(d) = \frac{d^2}{d^2 + \sigma^2}, \tag{10.12}$$

where σ is a constant of the order of the molecular spacing, and this assumption gives (for $d \gg \sigma$)

$$d = \left(-\frac{2\sigma^2 \Delta\gamma}{\rho_1 q_m} \frac{T_m}{T_m - T} \right)^{1/3}. \tag{10.13}$$

This model describes formally correct thermodynamics, but the weight of the experimental evidence in §10.7 is that for ice $\Delta\gamma$ is not negative. A liquid layer should therefore not be formed on a clean free surface. Further, the factor $f(d)$ expresses the fact that the interfaces of the liquid layer with the solid and the vapour cannot be treated as independent and the layer itself will not have the properties of the normal liquid. Knight (1996*b*) has criticized the model on the grounds that it is not appropriate to treat the properties of a single non-homogeneous boundary layer as consisting of a homogeneous medium separated by two sharp interfaces, and indeed the use of $f(d)$ shows that the model is not really what it purports to be. In our view the adoption of this model to describe the free surface of ice, with the use of the words 'liquid' or 'liquid-like', is misleading and does not advance our understanding of the nature of the surface. We prefer the phrase 'quasi-liquid' as referring to the experimental evidence for structural disorder without implying that the layer has the bulk or interface properties of water. The model may of course have some validity for other materials or even possibly for the interface between ice and other solids.

10.9.2 *Subsurface pressure melting*

It is generally accepted that when a solid is split into two parts the molecules on the newly exposed surface will experience inward forces due to the attraction by their neighbours not being balanced by attraction from the part that has been removed. The surface molecules therefore move inwards, and this relaxation reduces the surface free energy γ_{sv} below half the work of cohesion. These effects exert a pressure on the layers of molecules below, and it has been proposed by Fukuta (1987) and Makkonen (1997) that in the case of ice this results in pressure melting of the subsurface layers. Using dT/dp for the normal ice–liquid phase boundary (§2.1.1) and models of the intermolecular forces that do not include hydrogen bonding, they estimate the magnitude of the effect; it is not negligible.

10.9.3 *Molecular dynamics simulations*

To take the above idea further it is useful to perform molecular simulations, and a molecular dynamics simulation using the TIP4P potential has been described by Furukawa and Nada (1997). They used a model of ice containing 720 molecules with one free surface and periodic boundary conditions on surfaces perpendicular to this. Sets of molecular positions obtained in such simulations at 265 K are shown in Fig. 10.14, and it is quite clear that the outer two bilayers are greatly

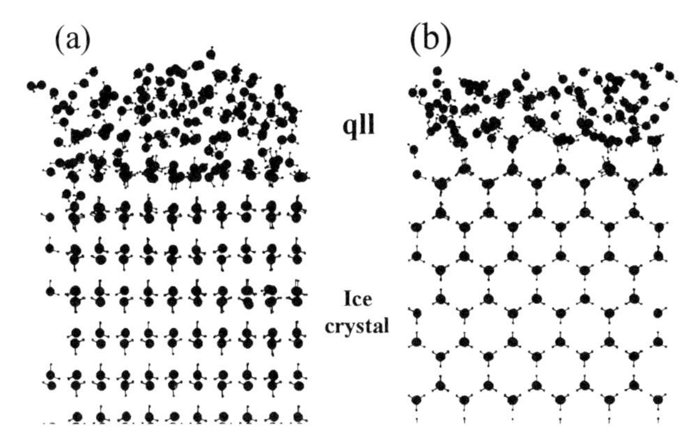

Fig. 10.14 Actual positions of molecules at a given instant from molecular dynamics simulations of (a) the {0001} and (b) the {10$\bar{1}$0} surface of ice at 265 K by Furukawa and Nada (1997). Reproduced with permission from *Journal of Physical Chemistry*, **B101**, 6167–70. Copyright 1997 American Chemical Society.

disordered. They found that the thickness of the disordered layer varied from 18 Å at 272 K to 2 Å at 250 K on a {0001} surface and 14 to 4 Å at the same temperatures on a {10$\bar{1}$0} surface. There was appreciable diffusion within the surface layers.

Kroes (1992) describes a similar simulation using the same potential, and gives a fuller analysis of the results. In this case the effects seem to occur at lower temperatures than expected but he does not simulate actual melting to establish a reference point. A particularly interesting feature is that at low temperatures the surface molecules tend to rotate to minimize the number of dangling bonds, and as the temperature rises this leads to a constantly changing disorder in the outer two bilayers. The outer layers become partially polarized with the dipole moments pointing inwards, even though the model is set up to allow no polarization at the inner boundary. The model thus predicts a negative surface charge, with a compensating positive space charge distributed within the subsurface disordered region.

Similar molecular dynamics simulations are reported by Devlin and Buch (1995) and Buch *et al.* (1996), who found that if an ordered structure was set up at a low temperature and given a simulated anneal at 200 K the surface structure became disordered and it remained disordered on cooling. Such modelling indicates two different things. The first is that it is energetically favourable for the surface to be reconstructed to eliminate at least some of the dangling bonds. Within the boundary conditions imposed in the modelling this reconstruction is disordered (see §10.2.1). The second conclusion from the modelling is that there is dynamic disorder in the surface approaching the melting point, but this does not imply that the structure is like the liquid. Realistic modelling within only a few degrees of the melting transition will be very difficult, and may be affected by the boundary conditions imposed at the inner boundary to crystalline material.

10.9.4 *Dispersion force modelling*

All molecules interact via dispersion or Lifshitz–van der Waals forces, and a detailed theoretical model exists to describe the interaction between the surfaces of different media in terms of the dependence of their permittivities on frequency over the whole range of atomic and molecular resonances (Dzyaloshinskii *et al.* 1961). Wilen *et al.* (1995) applied this theory to the interface between ice and a variety of substrates, looking for the formation of some liquid-like layer, but no significant effects were predicted for temperatures below −0.1 °C. They conclude that additional contributions to the intermolecular interactions are necessary to account for the special properties of the ice surface, and this is not surprising. In this connection we should note that the Lifshitz–van der Waals contribution to the free energy of the ice–vapour interface calculated by van Oss *et al.* (1992) and referred to in §10.7 was less than half the total free energy.

10.9.5 *Surface charge model*

Fletcher (1962, 1968) proposed that as a result of electrostatic dipole and quadrupole interactions the water molecules on the outer surface of ice or liquid water might be partially ordered with an excess of OH bonds pointing in to or out of the surface. This produces a surface charge which must be screened by oppositely charged point defects as in Fig. 10.7. This is the origin of the surface charge model already considered in §10.4.2, but the reasons originally proposed for the surface polarization are not valid (Fletcher 1973), and the model does not therefore explain why the surface charge is formed. The as yet unidentified driving force has to be quite strong to provide sufficient free energy to offset the electrostatic energy of the double layer and the energy of formation of the point defects that make up the space charge. As we have seen, the subsurface charge seems necessary to account for the high electrical conductance of the surface, and there is no reason to believe that the presence or thickness of this layer is related to the disorder on the oxygen lattice.

The latest calculations on the space charge region by Petrenko and Ryzhkin (1997) show the importance of allowing the surface potential to be much larger than $k_B T/e$, and this results in very large increases in the concentrations of charge carriers near the surface. They suggest that in extreme cases the electrostatic pressure at the surface could result in pressure melting, which would make the oxygen disorder a consequence of the charged double layer rather than vice versa.

10.10 Conclusions

First of all it may be necessary to distinguish certain pre-melting phenomena that are observed within about 1 °C of the melting point. These can yield very thick surface layers and may be particularly sensitive to contamination and external conditions like the presence of air. Much more careful experimentation will be required before drawing conclusions about this region of temperature.

At lower temperatures the special properties of the ice surface are well established and include molecular disorder, greatly enhanced diffusivity, surface charges and conductivity. Melting while retaining intact molecules cannot account for the electrical conductance, and the electrical effects and structural disorder are not necessarily correlated. There is currently no satisfactory explanation of the primary cause of these effects. The disorder appears naturally in molecular simulations, but as these models take neutral molecules as the basic unit for modelling they have no way of relating to the electrical properties.

11
The other phases of ice

11.1 Introduction

The water substance exhibits a fascinating range of solid phases, and all of these are referred to as forms of 'ice'. Most of these phases are produced by the application of high pressures, which result in denser packings of molecules than in ice Ih. The first high-pressure phases were discovered almost a century ago by Tammann (1900) in a programme to study the pressure–volume–temperature relationships of various materials, and he named these phases 'ice II' and 'ice III'. His discovery was extended in experiments by Bridgman (1912), which reached 2 GPa and led to the discovery of ices V and VI. Ice IV was not assigned by Bridgman until 1935, the designation IV having originally been 'left blank because of uncertainty with regard to some unstable forms, the existence of which was suspected by Tammann'. Later it became possible to determine the structures of the ice phases by X-ray and neutron diffraction, and there has been a continual extension of the range of pressures over which ice has been studied. The formation of these phases can be vividly demonstrated using a diamond-anvil pressure cell on an optical microscope (Whatley and Van Valkenburg 1966). A drop of water in the cell is compressed to a suitable volume and fine adjustment of the pressure is achieved by changing the temperature at effectively constant volume. In this way faceted crystals of various phases can be observed to grow from the liquid phase. Examples of ices IV and V from the video recordings of Chou and Haselton (1998) are shown in Fig. 11.1; similar observations of ice VII with *in situ* X-ray diffraction of the same crystal were reported by Yamamoto (1980).

These phases of ice have been labelled with the Roman numerals I–XII in the approximate order in which they were produced experimentally, with the result that the numbers indicate nothing about the structures or the relations of the phases to one another. Each phase (except ices IV, IX, and XII) is stable over a certain range of temperature and pressure, but it is a feature of the ice system that many phases are metastable well outside their regions of stability and others have no region of absolute stability at all. Several phases can be quenched in liquid nitrogen and the pressure released without detransformation. Provided the sample is then kept at a sufficiently low temperature, its structure and properties can be studied outside the pressure cell, although there can be significant differences between such observations and ones made under pressure (Besson *et al.* 1997). The metastability of these 'recovered' phases results from the difficulty of rearranging the water molecules in the solid at low temperatures.

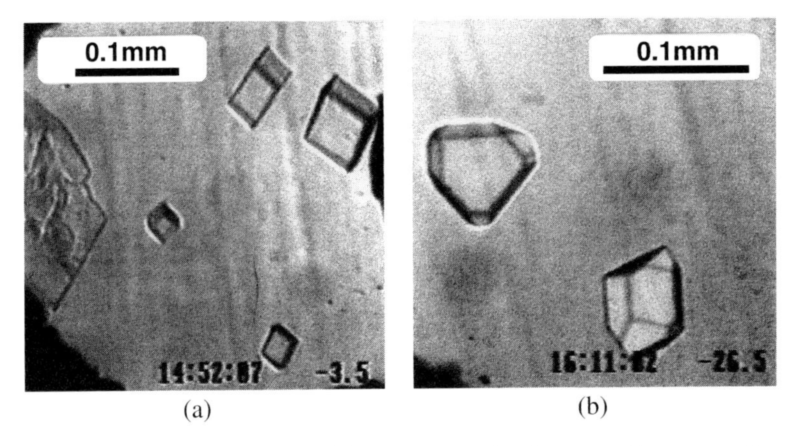

Fig. 11.1 Crystals of (a) ice IV and (b) ice V growing from the liquid in a diamond-anvil pressure cell. (From Chou and Haselton 1998.)

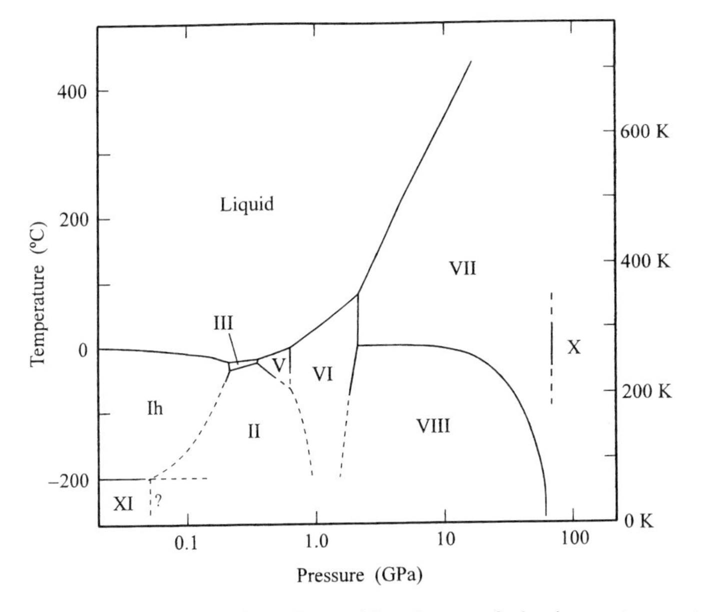

Fig. 11.2 Phase diagram showing the stable phases of the ice–water system on a logarithmic scale of pressure. The central region is shown expanded on a linear scale in Fig. 11.10.

Figure 11.2 shows the regions of stability of the crystalline phases of ice on the pressure–temperature phase diagram. This diagram has been plotted on a logarithmic scale of pressure because the range involved is now too large to use a linear scale as in the often used diagram compiled by Whalley (1969). An expanded portion of the central region using a linear scale is given later as Fig. 11.10. Figure 11.2 has not been extended to include features involving equilibrium with the vapour; these are shown in Fig. 2.1. The solid lines in the phase diagram represent experimental measurements, and the broken lines represent extrapolated or inferred phase boundaries. The key parameters in the diagram are the triple points where three phases are in equilibrium with one another, and these are listed in Table 11.1. The values given for H_2O and D_2O differ slightly, but although D_2O has to be used for some neutron diffraction experiments, we will not be concerned with these differences in this chapter. Internationally adopted empirical equations for the melting curves of the phases of ice are given by Wagner *et al.* (1994).

Several phases are known which are not included in Fig. 11.2 because they have no region of stability. These are ice IV, IX, and XII (which are discussed in §11.6 and shown in Fig. 11.10), ice Ic, and the amorphous phases. In addition there have over the years been a number of predictions based on intuition or theoretical modelling. Some have subsequently been realized experimentally and others not, and occasionally this has led to confusion over the naming of phases. We take the view that the series of Roman numerals should only be used for crystalline phases that are experimentally well established, for example by crystallography or spectroscopy. Hypothetical or predicted structures need a different nomenclature (e.g. by space group or author) so that in the future it will be possible to say, for example, that ice XV has the structure 'abc', or that a predicted phase 'pqr' has now been found and designated 'ice XVI'.

The problem of nomenclature can be illustrated by considering the case of ice XII. Sirota and Zhapparov (1994) describe pressure–volume–temperature measurements on D_2O ice leading to a phase diagram rather different from that generally accepted and including a new phase below about 110 K at 1–2 GPa. They named this 'ice XII', but in spite of attempts it has not been reproduced or

Table 11.1 Triple points of the stable phases of ice

	H_2O		D_2O	
	p (MPa)	T (°C)	p (MPa)	T (°C)
L–Ih–III	209	−22.3	220	−18.8
L–III–V	350	−17.5	348	−14.5
L–V–VI	632	0.1	629	2.4
L–VI–VII	2210	81.6	2060	78[a]
Ih–II–III	213	−34.7	225	−31.0
II–III–V	344	−24.3	347	−21.5
VI–VII–VIII	2100	≈0[a,b]	1950	≈0[a,b]
Ih–XI–vap.	0	−201[c]	0	−197[d]

Compiled from Bridgman (1935), Engelhardt and Whalley (1972), and Wagner *et al.* (1994), together with [a]Pistorius *et al.* (1968), [b]Johari *et al.* (1974), [c]Tajima *et al.* (1984), and [d]Matsuo *et al.* (1986).

characterized by other measurements. Next Svishchev and Kusalik (1996) found a new structure, which they named 'ice XII', in a molecular dynamic simulation of the freezing of water under pressure in an electric field, and most recently Lobban *et al.* (1998) have determined the crystal structure of a metastable phase formed within the region of stability of ice V. In conformity with the criteria given above they propose that this should be called 'ice XII'. We concur with their view and adopt it in this chapter. If a phase matching the observations of Sirota and Zhapparov is properly characterized, it will be 'ice XIII' or whatever is the next number then available. The predicted structure of Svishchev and Kusalik will only acquire a Roman numeral if and when it is actually identified in a real experimental situation. It can be hoped that an appropriate international body may produce some guidance on the naming of such phases, but this has not so far been done.

In considering the phases of ice there are two guiding themes: the fourfold co-ordination of the water molecules and hydrogen disorder. Except at the very highest pressures (ice X and beyond) the water molecules remain intact and retain their fourfold co-ordination as in ice Ih, donating two hydrogen bonds and accepting two others, even if the bond angles are distorted. Ice Ih is a very open structure and the high-pressure phases adopt denser forms of packing. The Pauling disorder is a dominant consideration in the structure and properties of ice Ih, and it was recognized from the start that, according to the third law of thermodynamics, this phase could not be the thermal equilibrium structure at low temperatures. However, in spite of many experiments and theoretical speculations about its structure the ordered phase of ice Ih was only identified in 1972, and by the time it was named in 1984 it had to be called ice XI. The possibility of ordered and disordered forms arises in a number of the structures: ordered ice VII is ice VIII, and ordered ice III is ice IX.

A very important feature of the phase diagram is the slope $\mathrm{d}T/\mathrm{d}p$ of the line representing the equilibrium between two phases. At every point along this line the Gibbs free energies of the two phases must be the same, and this leads to the Clausius–Clapeyron equation for the gradient of the phase boundary:

$$\frac{\mathrm{d}p}{\mathrm{d}T} = \frac{\Delta S}{\Delta V}, \tag{11.1}$$

where ΔS is the difference in entropy and ΔV is the difference in volume between equal amounts of the two phases. Along the liquidus curve the entropy of the liquid is always greater than that of the ice because of the latent heat of melting. Ice Ih contracts on melting, making ΔV negative and $\mathrm{d}T/\mathrm{d}p$ also negative as seen in Fig. 11.2. For all the other phases that have a boundary with the liquid the ice is denser than the liquid and the phase boundary line slopes the other way. Within the solid phases at constant temperature the volume always decreases with increasing pressure. Between two disordered phases ΔS is almost zero so that the phase boundary in Fig. 11.2 is nearly vertical, and this is the case for the boundary between ice Ih and ice III. However, Tammann (1900) found that the ice Ih–ice II

Table 11.2 Structures of the crystalline phases of water

Parameters are at the temperature T and pressure p stated. All densities given are for H_2O

Ice	Crystal system	Space group	Proton order?	Molecules per cell	T (K)	p (GPa)	Density (Mg m^{-3})	Cell parameters (Å)	Reference
Ih	Hex.	$P6_3/mmc$	N	4	250	0	0.920	$a = 4.518$, $c = 7.356$	Röttger *et al.* (1994)
Ic	Cubic	Fd3m	N	8	78	0	0.931	$a = 6.358$	Kuhs *et al.* (1987)
II	Rhomb.	$R\bar{3}$	Y	12	123	$0^\dagger$	1.170	$a = 7.78$, $\alpha = 113.1°$	Kamb (1964)
III	Tetrag.	$P4_12_12$	N	12	250	0.28	1.165	$a = 6.666$, $c = 6.936$	Londono *et al.* (1993)
IV	Rhomb.	$R\bar{3}c$	N	16	110	$0^\dagger$	1.272	$a = 7.60$, $\alpha = 70.1°$	Engelhardt and Kamb (1981)
					260	0.50	1.292		Lobban *et al.* (1998)
V	Monocl.	A2/a	N	28	98	$0^\dagger$	1.231	$a = 9.22$, $b = 7.54$, $c = 10.35$, $\beta = 109.2°$	Kamb *et al.* (1967)
					223	0.53	1.283		Kamb *et al.* (1967)
VI	Tetrag.	$P4_2/nmc$	N	10	225	1.1	1.373	$a = 6.181$, $c = 5.698$	Kuhs *et al.* (1984)
VII	Cubic	Pn3m	N	2	295	2.4	1.599	$a = 3.344$	Kuhs *et al.* (1984)
VIII	Tetrag.	$I4_1/amd$	Y	8	10	2.4	1.628	$a = 4.656$, $c = 6.775$	Kuhs *et al.* (1984)
IX	Tetrag.	$P4_12_12$	Y	12	165	0.28	1.194	$a = 6.692$, $c = 6.715$	Londono *et al.* (1993)
X	Cubic	Pn3m	n/a	2	300	62	2.79	$a = 2.78$	Hemley *et al.* (1987)
XI	Ortho.	$Cmc2_1$	Y	8	5	0	0.934	$a = 4.465$, $b = 7.858$, $c = 7.292$	Line and Whitworth (1996)
XII	Tetrag.	$I\bar{4}2d$	N	12	260	0.50	1.292	$a = 8.304$, $c = 4.024$	Lobban *et al.* (1998)

† Samples recovered at low temperature.

boundary had a positive slope, and according to the measurements of Bridgman (1912) $dT/dp = 1.07 \times 10^{-6}\,K\,Pa^{-1}$ and $\Delta V = -3.6 \times 10^{-6}\,m^3\,mol^{-1}$. From equation (11.1) the entropy of ice II is therefore less than that of ice Ih by about $3.4\,J\,K^{-1}\,mol^{-1}$. The vibrational entropies of the molecules will be similar, as they were for ice Ih and ice III. The fact that this calculated entropy difference is close to the Pauling entropy of ice Ih (§2.3) therefore shows that ice II is ordered. It is interesting that this feature was known long before Pauling's model for ice Ih was introduced. By similar arguments we can see from the ice VI–ice VIII phase boundary that ice VI is disordered and ice VIII is not. Ice VII is then disordered and the broken line between ice VI and ice II is drawn by inference.

In this chapter the emphasis will be on the structures of the phases, and they will be described not in numerical order but in a sequence chosen for the development of this theme. A summary of the structural parameters of the crystalline phases is given in Table 11.2. Little will be said about the techniques of producing the ices or about the crystallographic evidence for the structures; for such matters the reader should consult earlier reviews (e.g. Hobbs 1974) or the original papers. Towards the end of the chapter there is a brief account of the important and closely related clathrate hydrates, and finally some common features concerned with lattice vibrations and the properties of the hydrogen bond are described.

11.2 Ice XI—The ordered form of ice Ih

The temperature of the phase transition of ice Ih to a proton-ordered structure is 72 K (76 K in D_2O), but this transition never takes place in pure ice. It has only been observed in ice suitably doped with alkali hydroxides, and the reason for this is clear from Fig. 5.10. The process of ordering the protons in ice involves the reorientation of bonds or molecules, but this can only occur through the motion of protonic point defects, exactly as in the case of dielectric relaxation. In pure ice the relaxation rate at 72 K would be far too slow for the transformation to be observed, but doping with hydroxides speeds up this process by the introduction of OH^- ions which remain sufficiently mobile down to 72 K. KOH is the most effective and extensively used dopant in such experiments.

The phase transition was first reported by Kawada (1972), who observed that the dielectric permittivity of KOH-doped ice became very small below about 70 K. He also reported evidence of a latent heat, seen as a plateau in the temperature–time graph as the doped ice was warmed through this temperature. More precise calorimetric experiments were performed by Tajima *et al.* (1984), who subsequently named the ordered phase 'Ice XI' (Matsuo *et al.* 1986). Their experiments involved holding the doped ice a few degrees below the transition temperature for several days, cooling to a lower temperature, and then measuring the heat capacity during gradual warming. A typical set of observations is shown in Fig. 11.3, in which the peak at 72 K represents the latent heat of disordering of the protons. The entropy change associated with the phase transition depended

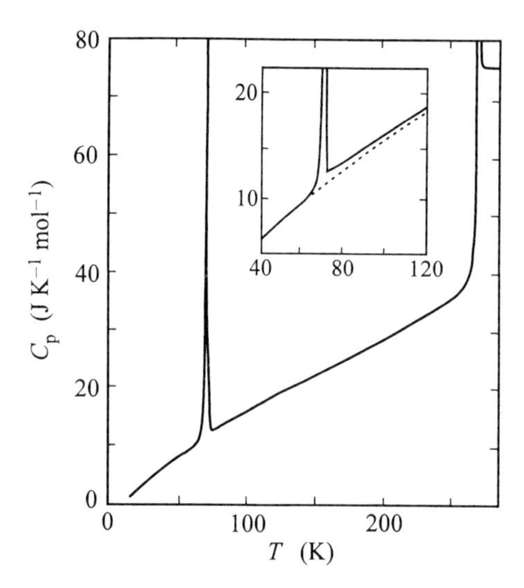

Fig. 11.3 Measurements of the molar heat capacity C_p of ice frozen from 0.01 M KOH solution and annealed below 72 K. The peak at 72 K marks the transition of ice XI to ice Ih. The inset is an enlargement of the region near the transition; note in particular the increased heat capacity above the transition temperature. Reproduced from *Journal of Physics and Chemistry of Solids*, **45**, Y. Tajima, T. Matsuo, and H. Suga, Calorimetric study of phase transition in hexagonal ice doped with alkali hydroxides, pp. 1135–44, Copyright 1984, with permission from Elsevier Science.

on the level of doping and on the time and temperature of the annealing below the transition temperature. This entropy change is a measure of the degree of transformation achieved; in these experiments it was never more than 68% of the Pauling entropy of ice Ih, but later experiments have observed changes of up to 82% (Suga *et al.* 1992).

The likely pattern of ordering in ice Ih (i.e. the lowest energy state compatible with the ice rules, and assuming no rearrangement of the oxygen atoms) has been a matter of much speculation ever since the issue was raised by Bernal and Fowler (1933) (see §2.2.1). All simple cases have been enumerated by Howe (1987); there are 17 structures possible within an 8-molecule cell and many more if larger cells are considered. To locate the protons in the ordered ice it is necessary to use neutron diffraction. The first such experiment was reported by Leadbetter *et al.* (1985), but they achieved only a very small degree of transformation. Further experiments have presented technical difficulties, but it is now established that ice XI has the orthorhombic structure shown in Fig. 11.4 (Line and Whitworth 1996; Jackson *et al.* 1997). The unit cell is one which can also be used to describe ice Ih and is shown in Fig. 2.7. The space group is $Cmc2_1$. All bonds parallel to the c-axis are oriented in the same direction, which means that the ordering is

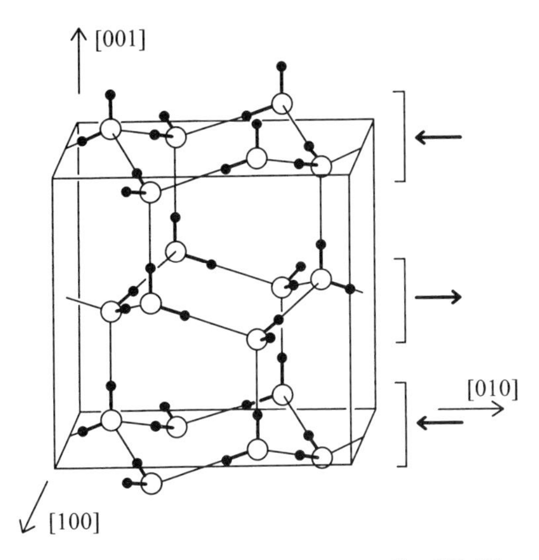

Fig. 11.4 The orthorhombic $Cmc2_1$ crystal structure of ice XI. The arrows on the right indicate the directions of small displacements of the layers. (From Howe and Whitworth 1989*a*.)

ferroelectric in character. The (001) layers of puckered hexagonal rings lying perpendicular to this direction are polarized along the *b*-direction, but alternate layers are polarized in opposite senses. The symmetry allows these layers to be displaced in this *b*-direction, and the neutron diffraction data are consistent with displacements of about 0.02*b* in the directions shown in Fig.11.4. This structure, which was proposed by Kamb (1973), has a primitive unit cell of only four molecules and is the simplest possible ordered structure of ice Ih. The ferroelectric type of ordering is very significant, and has been confirmed on a macroscopic scale by thermally stimulated depolarization experiments (Jackson and Whitworth 1995). In these experiments the crystal was cooled through the transition and then annealed in an electric field. On subsequent warming without the field there was a large release of charge at the transition temperature for polarizing fields along the *c*-direction but very little effect for fields applied perpendicular to this. For this type of ordering there are in principle two order parameters, one for ordering the bonds parallel to the *c*-axis and the other for ordering in the planes perpendicular to this, but all experiments indicate only a single first order transition.

High-resolution neutron diffraction experiments on both monocrystalline and powder samples show that the transformed ice is a two-phase mixture of ice XI and untransformed ice Ih. The lattice parameters of these phases together with those of the same ice in the detransformed state and of pure ice are shown in Table 11.3. It can be seen that *c* is smaller for ice XI than for ice Ih, and that *c* has

Table 11.3 Lattice parameters of KOD-doped D_2O ice XI and ice Ih for orthorhombic cell measured at 70 K

	a (Å)	$b/\sqrt{3}$ (Å)	c (Å)
Components of transformed ice			
Ice XI	4.464	4.537	7.291
Ice Ih	4.499		7.308
Detransformed doped ice	4.4964		7.3173
Pure (untransformable) ice	4.4966		7.3222

From Line and Whitworth (1996).

also been reduced in the ice Ih that remains in the transformed mixture. This contraction in the Ih component must be an elastic distortion consequent on the contraction of the ice XI component, and the two phases must therefore be intimately mixed. To understand the parameters a and b given in the table we note that for ice Ih the hexagonal symmetry requires that in this unit cell $b = a\sqrt{3}$. The table therefore lists only a for ice Ih and gives a and $b/\sqrt{3}$ for ice XI. When ice XI forms within the hexagonal lattice of ice Ih there are six possible domain orientations. To avoid a net polarization there will be equal proportions oriented in opposite directions along the c-axis, and for each of these there are three possible directions of the orthorhombic b-axis within the hexagonal cell. These domains exert stresses on one another and there are potential incompatibilities in linking them together. For example, it is not possible to place two oppositely polarized domains end to end in the c-direction without seriously violating the ice rules on the boundary. Such problems may explain why some ice Ih always remains; it has not yet been possible to form a specimen consisting of a single domain of ice XI.

Yamamuro *et al.* (1987) found that the gradient of the ice Ih–ice XI boundary in the phase diagram was $0.015 \pm 0.001\ \mathrm{K\ MPa^{-1}}$. Neutron diffraction measurements give the change in volume ΔV as $0.15 \pm 0.03\%$, and substituting in the Clausius–Clapeyron equation (11.1) this gives $\Delta S = 1.9\ \mathrm{J\ K^{-1}\ mol^{-1}}$ compared with the full Pauling entropy of $3.4\ \mathrm{J\ K^{-1}\ mol^{-1}}$. This entropy change relates to only the ice XI component of the transformed mixture, but the volume change is subject to error due to the stress arising from the ice Ih, and it is only possible to conclude that the transition corresponds to the loss of much but not all of the entropy of disorder. The failure to recover the full Pauling entropy in calorimetric experiments will be due in part to the presence of ice Ih in the transformed material and in part to incomplete ordering within the ice XI phase itself. Incomplete ordering is expected because the last remaining OH^- ions would have to travel over exactly the right paths to remove the last remnants of disorder. The relationship between the entropy and the fraction of correctly ordered bonds was derived by Howe and Whitworth (1987).

In Fig. 11.3 it can be seen that there is a small excess heat capacity between 72 and about 100 K. This corresponds to about 5% of the Pauling entropy, and shows that some partial short-range ordering occurs in doped ice Ih as it is cooled towards the transition. Comparison of the *c*-lattice parameter of doped ice with that of pure ice shows a similar effect (Line and Whitworth 1996). Even in pure ice the relaxation time above 100 K is short enough for slow partial ordering to occur. This was observed in the heat capacity measurements of Haida *et al.* (1974), who describe the freezing out of the ordering below 100 K as a glass transition. At one time, before the KOH-induced transformation had been discovered, effects at around 100 K were misinterpreted as evidence of the expected ordering transition itself. (See for example the discussion by Onsager 1967.)

The pattern of ordering that occurs in ice Ih must depend on the nature of the intermolecular interactions, and Minagawa (1990, 1992) has developed a model of ordering for simple empirically defined forms of interaction. His model also predicts the magnitude and sign of Δ in the Curie–Weiss temperature dependence of the dielectric susceptibility, $\chi_s \propto 1/(T - \Delta)$. It shows, as would be expected, that an interaction which leads to a positive Δ for fields along a particular crystallographic direction results in ferroelectric ordering parallel to that direction. Antiferroelectric ordering is associated with negative values of Δ. Experiments above 72 K show that Δ is positive for fields along the *c*-axis, both in pure ice (§5.3.2) and in ice doped with KOH (§5.4.3.4; Oguro and Whitworth 1991). The observed ferroelectric ordering is consistent with this.

The temperature at which the transition occurs is related to the energy ΔE gained in ordering the protons by the simple equation $\Delta E = T\Delta S$, where ΔS is the full Pauling entropy. The known values of T and ΔS give $\Delta E = 0.0025$ eV per molecule. An important objective in the theory of ice is to predict which ordered structure has the lowest energy, and out of the many possible structures attention has been focused on two. These are the observed ferroelectric $Cmc2_1$ structure and the simplest antiferroelectric structure, which has space group $Pna2_1$. Barkema and de Boer (1993), using Nagle's unit model (§4.3.2) for treating the electrostatic energy terms, have shown that, with the oxygen atoms kept on site, this antiferroelectric ordering gives a lower energy than ferroelectric ordering. Lekner (1998) describes a more exact calculation of the energy of a set of point charges placed on the O and H sites in the lattice, and shows that the structure with the lowest electrostatic energy is antiferroelectric. The absolute accuracy of the *ab initio* calculations referred to in §2.4 is not sufficient to distinguish between the possible ordered structures (Casassa *et al.* 1997). Buch *et al.* (1998) have shown that the introduction of an appropriate polarizability into the empirical pair potentials (see §1.4) may favour the $Cmc2_1$ ordered structure, but the feature that must be included in the theory to drive the ordering to be ferroelectric, against the electrostatic preference for antiferroelectric order, is not yet clearly established. Some authors (e.g. Barkema and de Boer 1993; Casassa *et al.* 1997) suggest that the type of order produced in experiments on KOH-doped ice is actually determined by the KOH. However, this cannot account for the fact that well above the transition temperature Δ is positive for fields along the c-axis even in pure ice (§5.3.2), and this is indicative of intermolecular interactions that would produce ferroelectric ordering at low temperatures. It seems likely that the KOH acts purely as a catalyst of this transformation, and this is a presupposition in giving the name 'ice XI' to the phase so produced.

A consequence of the molecules being ordered is that ice XI does not have the large dielectric permittivity that is characteristic of ice Ih. As already mentioned the transition to ice XI was first recognized by the disappearance of the Debye relaxation below about 70 K. Dielectric measurements are in practice the easiest way to monitor whether a sample of doped ice will transform, and the preparation and properties of transformable single crystals were first studied by this means (Kawada *et al.* 1989; Oguro and Whitworth 1991; Kawada and Tutiya 1997). Kawada (1989*b*) has investigated what features of the dielectric properties above the transition are a requirement for the transformation to occur, and Zaretskii *et al.* (1991) have used such measurements to show that the most effective transformation is obtained if the ice is first cooled to around 60–62 K to nucleate the ice XI and then annealed nearer the transition temperature for the new phase to grow. However, dielectric measurements do not give a simple relaxation spectrum in the region of the transition, the time constants become very long as the transformation proceeds, and it is not possible to interpret the permittivity directly in terms of the fraction of ice transformed.

There have not been many studies of other physical properties of the ice XI phase, but we note that ordering does sharpen up features in the lattice vibration spectrum (Li *et al.* 1995). It also affects the processes which limit thermal conductivity, and Andersson and Suga (1994) have observed a 10–20% *increase* of thermal conductivity in transformed ice, which is of course not 100% ice XI. The nature of ice XI should in principle tell us a great deal about the nature of the interactions between molecules within the ice Ih structure, but it is a difficult phase to produce or study, and its structure is not yet understood theoretically.

11.3 Ices VII and VIII

The phase ice VII and its ordered form ice VIII occupy a large area of the phase diagram. Ice VII was first identified by Bridgman (1937), and of all the high-pressure phases of ice these two have the simplest high-density packing of H_2O molecules. The earliest X-ray crystallographic studies of Bertie *et al.* (1964) on recovered samples and of Kamb and Davis (1964) under pressure revealed a body-centred cubic arrangement of oxygen atoms. These experiments must in fact have been on ice VIII, but the existence of that phase was not known at the time, and the arrangement of oxygen atoms is essentially the same in both phases. Proper structural studies on ice VII were subsequently carried out by Walrafen *et al.* (1982) using X-rays and Kuhs *et al.* (1984) using neutrons. In this structure, which is shown in Fig. 11.5, each oxygen atom has eight nearest neighbours but is

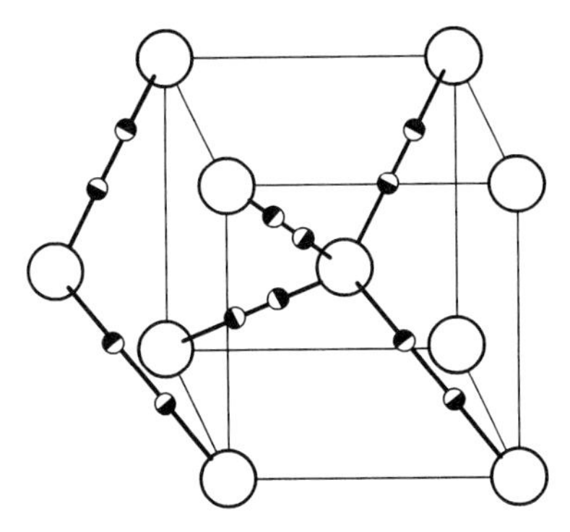

Fig. 11.5 The cubic crystal structure of ice VII, showing the hydrogen bonds of one sublattice within the unit cell and bonds to one molecule of the other sublattice in an adjacent cell. The half-filled circles represent positions for the protons according to the half-hydrogen model, as in Figs 2.6 and 2.7.

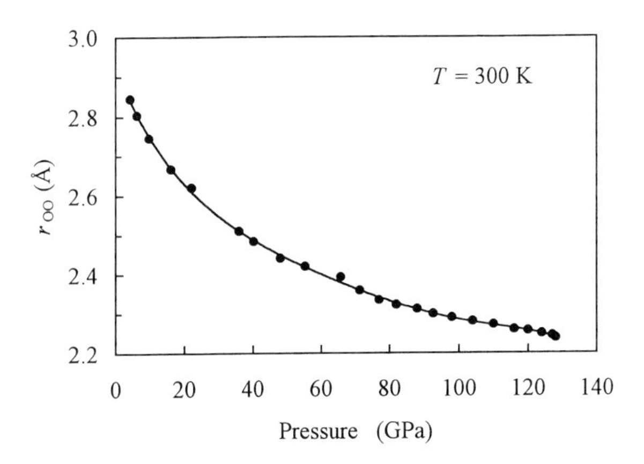

Fig. 11.6 Pressure dependence of the O—O bond length r_{OO} in ice VII to very high pressures. (Calculated from the data of Hemley *et al.* 1987.)

tetrahedrally linked by hydrogen bonds to only four of them. There are two interpenetrating but independent sublattices, each with the structure of ice Ic, the cubic equivalent of ice Ih. It is not geometrically possible to form such an interpenetrating structure based on the hexagonal lattice of ice Ih. As can be seen in Fig. 11.5, each oxygen atom has four non-bonded neighbours at the same inter-site distance as those to which it is linked by hydrogen bonds. The effect of the non-bonded oxygen atoms is to push the structure apart, so that even at 2.3 GPa the oxygen–oxygen distance r_{OO} of 2.92 Å is slightly larger than the value of 2.76 Å in ice Ih at zero pressure. Hemley *et al.* (1987) have determined the lattice parameter of ice VII up to 128 GPa, and Fig. 11.6 shows how the molecules are squeezed closer together as the pressure is increased. At 8 GPa r_{OO} is the same as in ice Ih, and the density of ice VII at this pressure is therefore double that of ice Ih. As seen in Fig. 11.2 such compressed ice would melt at over 300 °C!

The neutron diffraction data show that the hydrogen atoms are disordered subject to the ice rules, and they are therefore shown in Fig. 11.5 using the half-hydrogen symbol used in Fig. 2.6. This disorder will give the same Pauling entropy per molecule as in ice Ih (§2.3). As discussed for ice Ih in §2.5.2, there is evidence of associated displacements of the oxygen atoms off their sites to maintain reasonable molecular geometry within the lattice (Kuhs *et al.* 1984; Nelmes *et al.* 1998*b*).

Because of the hydrogen disorder the dielectric permittivity of ice VII has a Debye relaxation similar to that in ice Ih. The theory of Chapter 4 should be exactly applicable, including the use of $G = 3.0$ in equation (4.22). Johari *et al.* (1974) made measurements of the dielectric properties of ice VII as a function of temperature at 2.33 GPa and found that the low-frequency permittivity could be described by the empirical equation $\varepsilon_s = 111.4 - 0.213t$, where t is the temperature in °C. If this is converted to the more appropriate Curie–Weiss form

$$\varepsilon_s - \varepsilon_\infty = \frac{A}{T - \Delta} \tag{11.2}$$

(using $\varepsilon_\infty = 4.75$ from measurements on ice VIII) we find $A = 4.48 \times 10^4$ K and $\Delta = -418$ K. In mean-field theory a large negative Δ implies an antiferroelectric interaction between the molecular dipoles, but A is still given by equation (4.22). Using this equation and the appropriate density, we obtain the effective molecular dipole moment $p = 10 \times 10^{-30}$ C m, which is very close to the value of 9.69×10^{-30} C m deduced for ice Ih in §5.4.1. We may therefore conclude that the mechanism of polarization in ice VII at pressures around 2.3 GPa is by the motion of Bjerrum (not ionic) defects within the sublattices. The Debye relaxation time obtained by Johari *et al.* (1974) was 1 μs at 270 K with an activation energy of 0.58 ± 0.02 eV. Comparison with Table 5.1 shows that the activation energy is almost identical with that in ice Ih, but the rate is an order of magnitude faster.

Whalley *et al.* (1966) showed that the Debye relaxation disappeared below 0 °C and so discovered the ordered phase which they named 'ice VIII'. This transformation between ice VII and ice VIII occurs at a temperature where the relaxation time is of the order of a microsecond, so that there is no difficulty in achieving transformation, such as there is in the case of ice XI. Nevertheless, the transition exhibits some hysteresis (Pruzan *et al.* 1992), which makes an accurate determination of the transition temperature difficult. The arrangement of the oxygen atoms in ice VIII remains approximately body-centred cubic, and neutron diffraction experiments on D_2O (Kuhs *et al.* 1984; Nelmes *et al.* 1993) have revealed that the deuterons on the two sublattices are ferroelectrically ordered in

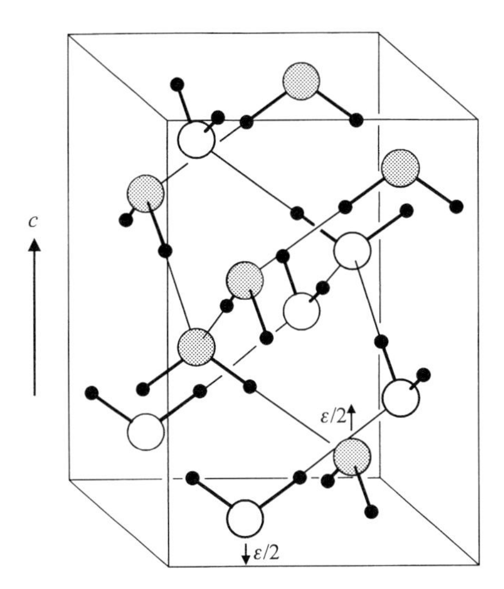

Fig. 11.7 The tetragonal crystal structure of ice VIII. The oxygen atoms of the two interpenetrating but unconnected sublattices are shown open and shaded. As compared with ice VII these sublattices are displaced in the *c*-direction by the amounts labelled $\varepsilon/2$.

opposite senses along one of the $\langle 001 \rangle$ directions, giving the overall antiferroelectric ordering anticipated from the sign Δ of in ice VII. One unit cell of the resulting structure is shown in Fig. 11.7, with the oxygen atoms of the two sublattices shown open or shaded. The lattice is tetragonal with the fourfold *c*-axis along the direction of the ordering and the *a*- and *b*-axes along $\langle 1\bar{1}0 \rangle$ directions of the ice VII lattice. The reduction of symmetry allows a slight distortion of the cubic lattice parameters and a relative displacement of the two sublattices in the *c*-direction by a distance $\varepsilon = 0.2$ Å. The displacements marked $\varepsilon/2$ in Fig. 11.7 are such that the non-bonded O—O distance becomes less than the O—D $\cdots$ O bond length.

The structural parameters of D_2O ice VIII up to 20 GPa have been determined using neutron diffraction (Besson *et al.* 1994; Nelmes *et al.* 1998*a*). In contrast with the oxygen–oxygen distance which falls as in Fig. 11.6, the O—D bond length remains remarkably stable at approximately 0.972 Å, rising by barely more than the experimental scatter to not more than 0.980 Å at 20 GPa. This is very close to the value in ice Ih (§2.5.1 and Fig. 2.11). It shows that the molecules remain intact, but the energy barrier between the sites for the protons on a given bond becomes narrower and consequently lower as the pressure rises. The structure and properties of ice VIII have been modelled in some detail using *ab initio* methods by Ojamäe *et al.* (1994). They found that the dependence of the O—H bond length on pressure is small up to 10 GPa, in agreement with experiment, but that it increases more rapidly towards 30 GPa.

Structural studies become increasingly difficult at higher pressures, but optical measurements on small samples in a diamond-anvil cell remain possible. Pruzan *et al.* (1992, 1993) found that the Raman spectrum between 300 and 600 cm^{-1} changes from three sharp lines in ice VIII to two broad lines in ice VII, and in this way they followed the boundary line in the phase diagram with increasing pressure with the remarkable result shown in Fig. 11.2. The line curves downwards to meet the pressure axis perpendicularly at 62 GPa (72 GPa for D_2O), and such behaviour was predicted theoretically by Schweizer and Stillinger (1984). The conclusion from the Clausius–Clapeyron equation must be that ice VII has lost its entropy of disorder at such pressures. As the oxygen atoms are squeezed closer together the energy barrier between the proton sites decreases and the protons become delocalized between the two sites. The loss of entropy is a gradual process within the ice VII. Pruzan *et al.* (1993) comment that at these pressures ice VII is analogous to hydrogen-bonded ferroelectrics like KH_2PO_4, the properties of which are understood in terms of the co-operative motion of protons on bonds rather than the orientation of discrete molecules. Under these conditions the hydrogen bond is becoming symmetrical, but we defer consideration of this topic until §11.7.

11.4 Ice VI

The crystal structure of ice VI was determined by Kamb (1965) using samples recovered at atmospheric pressure in liquid nitrogen. As in ice VII there are two interpenetrating but independent networks of hydrogen-bonded molecules, and Kamb refers to such structures as 'self-clathrates'. The lower density but more complicated structure of ice VI avoids the close non-bonded contacts between oxygen atoms that are an inevitable feature of ice VII. The structure is illustrated in projection in Fig. 11.8, but its construction can be visualized more vividly from the disassembled diagrams in Kamb's paper. The symmetry is tetragonal, and the structure is formed from chains built up of hydrogen-bonded molecules lying parallel to the fourfold [001] axis. The chains centred on the corners of the unit cell in Fig. 11.8 are linked together by hydrogen bonds parallel to [100] and [010]. A similar chain, shown shaded in the figure, fits inside this open structure at the centre of the cell, and this chain is linked to other such chains by bonds passing through the cell faces between the bonds of the first network. There are no hydrogen bonds between molecules in the two subsystems. The symmetry of this structure and the properties of the phase diagram indicate that the hydrogen positions are disordered.

The above structure has been confirmed by neutron diffraction on a D_2O sample under pressure by Kuhs *et al.* (1984) giving the lattice parameters in Table 11.2. There are several non-equivalent hydrogen bonds and considerable distortions of bond angles. The inter-site distances between hydrogen-bonded oxygens range from 2.72 to 2.79 Å at 225 K and 1.1 GPa, which is much less than

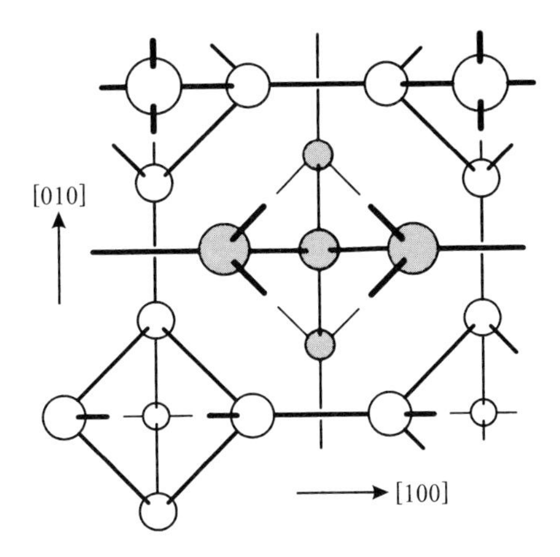

Fig. 11.8 The crystal structure of ice VI viewed along the tetragonal [001] direction. Only oxygen atoms are shown. The larger circles and thicker lines represent atoms and bonds nearer the front of the unit cell. The oxygen atoms at the front bottom corners of the cell are omitted to expose the atoms at the back.

the closest non-bonded O–O contact at about 3.4 Å. Further data for both H_2O and D_2O ice VI are given by Kuhs *et al.* (1989).

As expected for a disordered structure, the dielectric properties of ice VI exhibit a Debye relaxation. Johari and Whalley (1976) followed this relaxation down to 128 K, where the relaxation time had risen to about 300 s following a simple Arrhenius law. The activation energy was 0.47 eV, and there was no indication of a change of slope to an extrinsic region as occurs in ice Ih (Fig. 5.3). The permittivity obeyed equation (11.2) with $A = 32\,300$ K and $\Delta = 47$ K, but this polycrystalline average is not very informative. Unless the extrapolated phase boundaries with ice II and ice VIII meet above 0 K, there should be an ordered form of ice VI, but the transition will be at a temperature where it is difficult for ordering to occur. Evidence for such a phase or for partial ordering in ice VI is doubtful and inconsistent (Kamb 1973; Johari and Whalley 1979; Kuhs *et al.* 1984).

11.5 Ice II

Ice II has a truly ordered structure that is stable to relatively high temperatures and for which there is no corresponding disordered phase. It is formed by compressing ice Ih at −60 to −80 °C (or by decompression of ice V at −30 °C), and if heated it transforms to ice III which has a totally different arrangement of oxygen atoms. Ice II is not easily formed on cooling ice III, which remains metastable and finally orders to ice IX as explained in the next section. The evidence that ice II is ordered comes firstly from the slope of the phase boundary to ice Ih (see §11.1) and secondly from the absence of a Debye relaxation in the dielectric permittivity (Wilson *et al.* 1965). The crystal structure has been determined by Kamb *et al.* (1971) using neutron diffraction on monocrystalline D_2O samples recovered at low temperature. Their structure is fully consistent with an earlier X-ray determination by Kamb (1964) in which the hydrogen positions were inferred from the displacements of the oxygen atoms off sites appropriate to a similar arrangement of higher symmetry. The ordered structures so obtained are believed to be applicable to ice II in its range of stability because of the independent evidence for ordering under these conditions.

The unit cell is rhombohedral with the parameters given in Table 11.2, but the structure can also be described in a larger hexagonal cell of 36 molecules with $a = 12.97$ Å and $c = 6.25$ Å. Figure 11.9 is a stereo-pair of this structure viewed at a small angle to the hexagonal c-axis. It contains puckered hexagonal rings, one of which can be seen lying perpendicular to the c-axis in the centre of the unit cell. These hexagonal rings are linked to one another in a complicated way which achieves a higher density than that in ice Ih. There are four non-equivalent hydrogen bonds, and the molecules are in two different situations with D—O—D angles of 103.2° and 107.6°. A feature of the order is the sharpness of the infrared absorption lines from the OD or OH stretch mode of HOD in an H_2O or D_2O

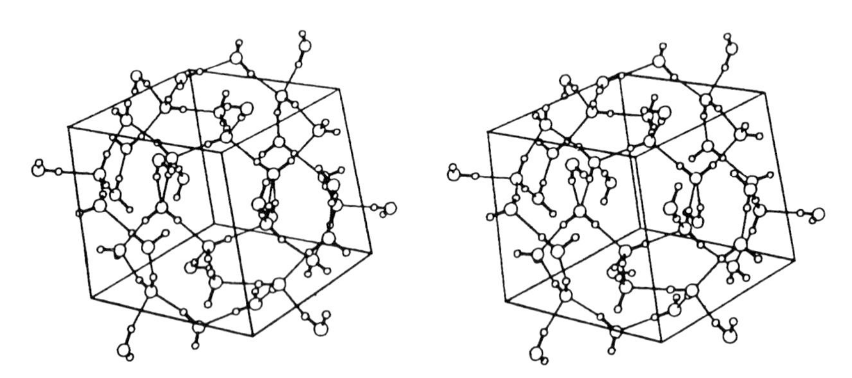

Fig. 11.9 Pair of stereo images of the structure of ice II viewed at an angle of about 30° to the hexagonal c-axis. The contents of one rhombohedral unit cell and portions of adjacent cells are shown. (From Kamb *et al.* 1971.) Readers not already familiar with viewing such stereo pairs should hold the page close to the face so that one image is in front of each eye but out of focus. Then move the page slowly away while trying to keep the eyes relaxed and looking towards the distance. Three images will be seen and the central one of these becomes the stereo-image, but avoid trying to focus on it until its three-dimensional character has become clear. If this cannot be made to work, try viewing through two 10 cm converging lenses, one in front of each eye. It is worth the effort!

matrix (§3.4.4). In ice II these lines are split into four, corresponding to the four types of bond (Bertie and Whalley 1964). In disordered structures the lines are broader. For an interpretation of these lines see Knuts *et al.* (1993).

Handa *et al.* (1988) have made calorimetric measurements of the transformation of ice II to ice Ic and then to ice Ih on heating at zero pressure. The first step is endothermic because of the entropy of disorder in ice Ic but the second stage is exothermic. They find that the enthalpy of ice II is less than that of ice Ih by $18\,\mathrm{J\,mol^{-1}}$ (a very small amount ≈ 0.18 meV per molecule), and they predict that ice II would therefore just be stable relative to ice Ih at zero temperature and pressure. This would be compatible with a possible extrapolation of the ice Ih–ice II boundary in the phase diagram, ignoring ice XI. However, the enthalpy change in the ice Ih–ice XI transition is $240\,\mathrm{J\,mol^{-1}}$, making ice XI stable relative to ice II. There should in principle be a vertical boundary on the phase diagram between ice XI and ice II at some undetermined pressure, and this is indicated by the broken line marked '?' in Fig. 11.2. It seems very unlikely that such a phase transition could be achieved in practice.

The structure of ice II is very similar to that of helium hydrate. This is an inclusion compound in which up to two helium atoms per unit cell are incorporated at sites between the hexagonal rings of water molecules (Londono *et al.* 1992). The helium stabilizes the ice II structure and increases the range of stability right up to the liquidus line on the phase diagram. The compound has similarities with the clathrate hydrates described in §11.10.

In studying crystallographic features of the ice phases little attention is paid to the process by which transformations occur within the solid state. An exception is the transformation between ice Ih and ice II which has been investigated by Kirby *et al.* (1992). Under pressure platelets of ice II form which profoundly affect the local stress distribution; the associated release of latent heat may also be relevant.

11.6 Ices III, IV, V, IX, and XII

In the central region of the phase diagram there are a number of phases with complex structures which are stable over narrow ranges of temperature and pressure or which exist only as metastable phases. This portion of Fig. 11.2 is shown expanded and with a linear scale of pressure in Fig. 11.10. To understand this topic it is necessary to be clear about the thermodynamics of these phases. For every conceivable structure there is a Gibbs free energy function $G(p, T) = E + pV - TS$, which can be represented by a curved surface over the co-ordinates of pressure p and temperature T. If several such surfaces are constructed, the one with the lowest G at any p and T represents the stable phase, and the line of intersection between the surfaces of two stable phases constitutes the boundary line between these phases on the phase diagram. This is illustrated in Fig. 11.11, which shows schematic plots of G as a function of T for a fixed pressure

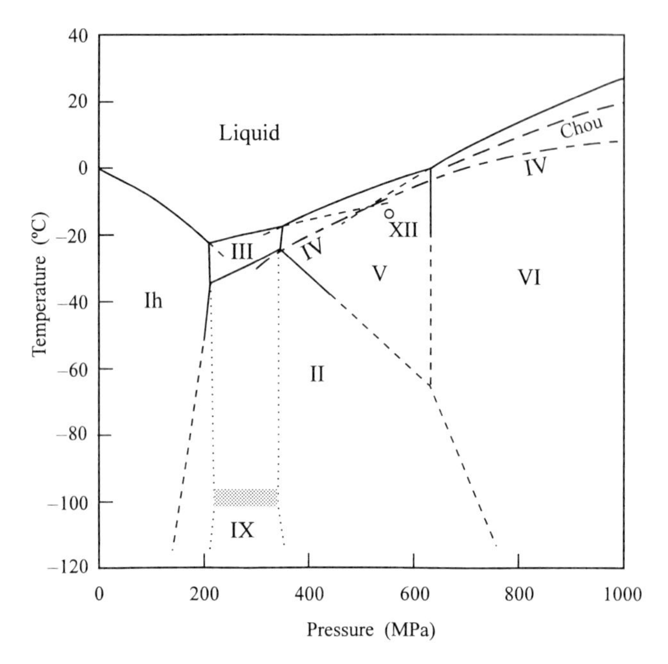

Fig. 11.10 Central region of the phase diagram of the ice–water system, showing regions where metastable phases are observed and including the liquidus lines for ice IV and the new phase observed by Chou *et al.* (1998). In the experiments by Lobban *et al.* (1998) ice XII was formed at the point marked 'XII'.

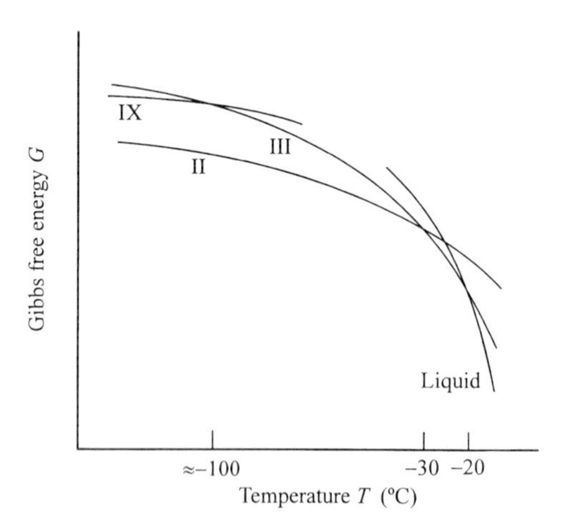

Fig. 11.11 Schematic dependence of the Gibbs free energy on temperature for the phases of the ice–water system at a pressure of ~300 MPa. The stable phase at any temperature is the one with the lowest free energy. The curves for ices III and IX are drawn assuming a sharp phase transition; if this is not so a smooth curve must be drawn just below the two in the region of their intersection.

$p \approx 300$ MPa passing through the range where ice III is stable. Over most of the temperature range either ice II or the liquid has the lowest free energy and so is stable, but there is a tiny range over which ice III has a lower free energy and is therefore the stable phase. However, minimization of the free energy is not the only consideration. There is often some difficulty in nucleating a new phase, so that one phase can continue to exist under conditions where another phase would have a lower free energy. In the example in Fig. 11.11 ice III can be cooled relatively quickly without transforming into the completely different structure of ice II, and it is then said to be metastable. If it is further cooled to around $-100\,°C$ proton ordering can occur even though oxygen reordering does not, and this leads to the formation of ice IX. Ice IX is probably always metastable with respect to ice II, but provided the temperature is kept below $-100\,°C$ this transformation does not occur. Such changes of phase cannot properly be represented on the equilibrium phase diagram, but some indication of the formation of ice IX is given by the dotted lines in Fig. 11.10.

Figure 11.10 also includes some broken lines representing the equilibrium between two phases within the region of stability of another. For example, if ice III has not formed, ice Ih can convert to liquid on the extrapolation of the ice Ih–liquid line. When the liquid is cooled there is often supercooling and the phase eventually formed may depend on the nucleation site rather than which of several possible phases has the lower free energy. Ice IV has no region of stability but under suitable conditions, or when an appropriate nucleating agent is used (L.F. Evans 1967), it can be formed from the liquid over most of the pressure range where ice III, V, or VI would have lower free energy. Ice IV is in equilibrium with the liquid along the liquidus line marked IV in Fig. 11.10 (Engelhardt and Whalley 1972; Chou and Haselton 1998). Lobban *et al.* (1998) have identified and determined the structure of a metastable phase denoted 'ice XII' at the point marked within the zone of stability of ice V. Chou *et al.* (1998) have reported a new phase of unknown structure formed metastably within the region of stability of ice VI, and the liquidus for this phase is marked 'Chou' in Fig. 11.10. This could also be ice XII, but it appears that there may be yet further metastable phases possible close to the liquid phase boundary, and no reliable assignments can be made without a full structure determination.

All phases in this region are made up from hydrogen-bonded water molecules with complicated packings of similar density, and it is not surprising that such phases have closely similar free energies. Where structures are simple as in ice Ih or ice VII there are not so many similar possibilities. We will now consider the nature of these phases, but without giving detailed structural information, for which the original papers should be consulted.

11.6.1 *Ices III and IX*

Ice III is the least dense of the high-pressure phases of ice, but it is more dense than the liquid so that the melting temperature rises with pressure. The crystalline arrangement of oxygen atoms was determined by Kamb and Prakash (1968) on samples recovered by quenching and releasing the pressure. The unit cell is tetrahedral, and the increase in density compared with ice Ih is achieved by a complicated arrangement of five-membered rings of hydrogen-bonded molecules with several different bond lengths and angles. The space group symmetry does not require the hydrogen positions on the bonds to be equivalent so that all possibilities between full order and full disorder are allowed.

In a series of dielectric measurements Whalley *et al.* (1968) showed that ice III has the high permittivity characteristic of hydrogen disorder, and that this permittivity falls gradually between -65 and $-108\,^{\circ}C$ with the formation of the ordered structure which they named 'ice IX'. Such gradual ordering is permitted by the symmetry and we therefore show the transition as a diffuse band in Fig. 11.10. Neutron diffraction experiments on ices III and IX under pressure (Londono *et al.* 1993) have established that the ordering is antiferroelectric, as had been predicted, and have shown that there is appreciable residual order in ice III (Kuhs *et al.* 1998). Thermodynamic evidence concerning the entropy difference between ice III and ice IX is not consistent (Handa *et al.* 1988).

11.6.2 *Ice V*

Ice V has the most complicated structure of all the ice phases, with a monoclinic unit cell of 28 molecules that includes some four-membered rings (Kamb *et al.* 1967). Dielectric measurements indicate that ice V is proton disordered (Wilson *et al.* 1965), but there is evidence from neutron diffraction (Hamilton *et al.* 1969; Kuhs *et al.* 1998), Raman spectroscopy (Minceva-Sukarova *et al.* 1986), and calorimetry of recovered samples (Handa *et al.* 1987) that partial ordering occurs at lower temperatures. As in the case of the ice Ih–ice XI transition the rate of ordering is enhanced by doping with KOH. No ordered phase has been identified, and extrapolation of the phase boundaries in Fig. 11.10 suggests that the requirements of the third law of thermodynamics could be met in principle by the transformation of ice V to ice II.

11.6.3 *Ice IV*

Ice IV exists only as a metastable phase, and it is not easily formed without the aid of a nucleating agent (L.F. Evans 1967; Engelhardt and Whalley 1972). Nevertheless its structure, as determined by Engelhardt and Kamb (1981), is perhaps the most beautiful of all the ice phases. It is shown as a stereo pair in Fig. 11.12, and it is worth making some effort to view this. The structure is rhombohedral with the threefold (hexagonal c-) axis vertical in the figure. Almost-planar six-membered rings of molecules lie perpendicular to this axis, and there is a hydrogen bond between a pair of different molecules passing through the centre of each ring! Yet all the molecules belong to a single hydrogen-bonded network, not to interpenetrating sublattices as in the case of ices VI or VII. The length of the bond threading the ring is 2.92 Å, which is appreciably longer than the other

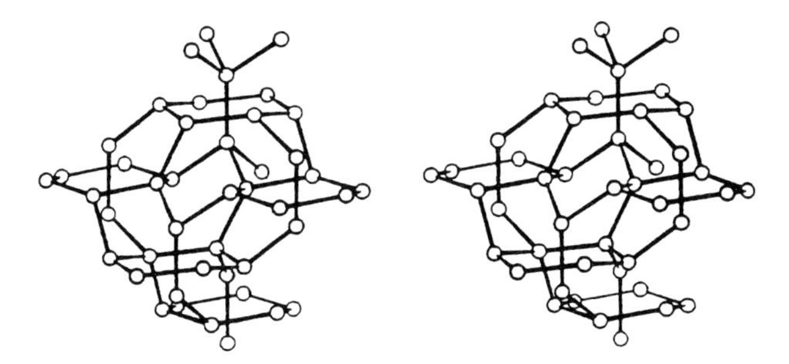

Fig. 11.12 Pair of stereo images of the structure of ice IV. (From Engelhardt and Kamb 1981.) View as described for Fig. 11.9.

bonds in the structure or the bonds in ice Ih; it is about the same as that in ice VII at the lower end of the pressure range. The structure appears to be fully proton disordered.

11.6.4 *Ice XII*

Lobban *et al.* (1998) have identified a metastable phase within the region of stability of ice V and made a full determination of its structure by neutron powder diffraction of D_2O under pressure. The phase has a tetragonal lattice with the parameters in Table 11.2, and it appears to be fully disordered. The structure is simpler than that of ice V, and the slightly higher density is achieved not by interpenetration of the network as in ice IV but by larger departures of the O—O—O angles from tetrahedral. As already stated there may be several metastable phases in this part of the phase diagram, but this and ice IV are the only ones with established structures. We therefore adopt the proposal of Lobban *et al.* (1998) that this phase be called ice XII (see §11.1).

11.7 Ice X and beyond

The structure of ice VII is formed from distinct molecules. In their original paper on its structure Kamb and Davis (1964) recognized that if the molecules were squeezed closer together than about 2.35 Å the ice should transform to a phase with the protons at the mid-points of the bonds, but they could not predict whether the transition would be gradual or accompanied by an abrupt change in volume. Modelling of this problem by Holzapfel (1972) and Stillinger and Schweizer (1983) predicted that the two possible locations for the proton on a hydrogen bond would merge into one for values of r_{OO} less than about 2.4 Å. Because of the ice rules, the minimum energy position for the proton on a given bond is of course determined by the positions of protons on neighbouring bonds, and a bond between two isolated molecules has only one position for the proton. The question about the symmetrization of the hydrogen bond can therefore only be asked for the whole crystal, in which the two hydrogen sites coalesce at the centre of all bonds simultaneously; the ice rule requiring that there be distinct H_2O molecules then becomes inapplicable. The X-ray diffraction experiments of Hemley *et al.* (1987) show that the body-centred cubic oxygen lattice of ice VII is retained up to at least 128 GPa, at which pressure r_{OO} has fallen to 2.24 Å (Fig. 11.6). The predicted transition to symmetric bonds is therefore well within the observed range, but the symmetrization does not alter the space group and would only be detectable by X-rays if there were a change in volume. Neutron diffraction experiments cannot yet be performed at sufficient pressure to observe the transition, but its effect on the lattice vibration spectrum should be appreciable.

Hirsch and Holzapfel (1984) using Raman scattering and Polian and Grimsditch (1984) using Brillouin scattering were the first experimenters to observe features which might indicate the transition to a symmetric phase at around 40–45 GPa, and they called it 'ice X'. However, their identification of the transition was not conclusive. The significant features to study are the modes ν_1, ν_2, and ν_3, because these are essentially properties of the molecules as described for ice Ih in §3.4.5, and they are observable in ice VII. Their pressure dependence has been studied by Goncharov *et al.* (1996), Aoki *et al.* (1996), and Pruzan *et al.* (1997). The infrared bond-stretching mode frequencies, which are shown for the experiment of Aoki *et al.* in Fig. 11.13, fall towards zero with increasing pressure and a new mode ν_T characteristic of the symmetrical structure then appears. The

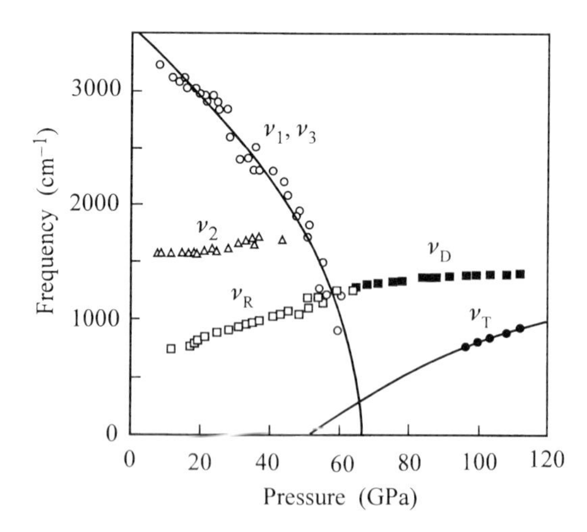

Fig. 11.13 Frequencies of the infrared modes of ice VII (H_2O) as a function of pressure at ambient temperature, indicating the transformation to ice X at ~62 GPa. Reproduced from K. Aoki, H. Yamawaki, M. Sakashita, and H. Fujihisa, *Physical Review*, **B54**, 15673–7, Copyright 1996 by the American Physical Society.

librational mode ν_R in ice VII merges continuously into the mode ν_D of ice X. The spectral range of the instrument used did not extend below 700 cm^{-1}, so that the graphs have to be extrapolated to lower frequencies, but it appears that ice X is being formed at about 62 GPa at ambient temperature. Extrapolation of the Raman data of Pruzan *et al.* (1997) in the ice VII phase predicts that the transformation occurs at about 70 GPa for H_2O and 85 GPa for D_2O.

Pressures of 62–70 GPa are close to that at which ice VIII ceases to exist at 0 K (Fig. 11.2). It is not yet possible to determine experimentally whether ice VIII converts directly to ice X at low temperatures or how ice VII gradually loses its entropy as it is compressed towards 62 GPa. We therefore avoid extrapolating the ice VII–ice X boundary to low temperatures in the phase diagram of Fig. 11.2. Various possibilities were considered in the papers by Stillinger and Schweizer (1983) and Schweizer and Stillinger (1984). It is important to distinguish two forms of symmetric ice, which differ fundamentally in whether the protons on the hydrogen bonds lie in a single or a double potential well. In a shallow double well the protons may be delocalized quantum mechanically or thermally; we continue to call this 'ice VII' and it is also referred to as 'proton-disordered symmetric ice'. Ice X refers to ice in which the protons are in a single potential well at the mid-point of the bond, and it has also been called 'proton-ordered symmetric ice'. As there need be no discontinuity in entropy or volume at the transition between ice VII and ice X, it will be difficult to locate a precise boundary on the phase diagram, and there are indeed doubts as to whether ice X as defined here has yet been conclusively identified experimentally.

Ab initio molecular dynamic simulations of the ordering of ice VII under increasing pressure have been described by Lee *et al.* (1993) and Benoit *et al.* (1998). The latter paper illustrates beautifully the form of transformation expected when the motion of the protons is treated either by quantum or by classical mechanics. In both cases the transformation to ice X proceeds via the proton-disordered symmetric phase, but a higher pressure is required in the classical case where zero-point motion and tunnelling do not contribute to the delocalization.

Ice X is not the densest possible structure for ice, but the phases formed beyond the pressures considered here are matters of speculation or modelling with extrapolated parameters. A favoured candidate for the next phase is the ionic 'anti-fluorite' structure (like CaF_2 with O^{--} on the face-centred cubic Ca^{++} sites and H^+ on the F^- sites), which has fourfold co-ordination (Demontis *et al.* 1988); this phase has the potential to become a superionic conductor at high temperatures. Modelling by Benoit *et al.* (1996) suggests, however, that a different (orthorhombic) structure may be favoured. Other possibilities are considered by Hama and Suito (1992), with the expectation that the material must eventually become metallic. This is an attractive field for theoretical modelling, and as explained in §11.1 we take the view that such predicted structures should not be assigned Roman numerals until their existence has been clearly established by experiment.

11.8 Cubic ice (Ice Ic)

Cubic ice, also designated 'ice Ic', is a metastable variant of ice Ih in which the oxygen atoms are arranged in the cubic structure of diamond rather than on the hexagonal lattice of ice Ih. Every molecule forms four hydrogen bonds to its neighbours as in ice Ih and the densities are virtually identical, but, if the (0001) layers in ice Ih are stacked in the sequence denoted in §7.6.1 by ABABAB..., cubic ice has the stacking sequence ABCABCA.... Figure 11.14 shows the two structures in projection on a $\{11\bar{2}0\}$ plane of the hexagonal lattice as in Fig. 7.17. The relationship between these structures is the same as that between cubic (f.c.c.) and hexagonal close packing in metals, or between the cristobalite and tridymite structures of SiO_2. In Fig. 11.14 the [0001] direction in the hexagonal structure is unique, but in the cubic case there are four equivalent $\langle 111 \rangle$ directions two of which are in the plane of this projection. In cubic ice there are no open channels parallel to this direction as there are in ice Ih, and all hexagonal rings of molecules are 'chair-like' (see §2.2.2).

Cubic ice was first identified in electron diffraction experiments on vapour-deposited layers by König (1943), and many experiments were subsequently performed to investigate the phase formed in this way (see Blackman and Lisgarten 1958). In general ice Ih is formed above 150 K, cubic ice between about 130 and 150 K, and below 130 K the deposit is amorphous. Bertie *et al.* (1963, 1964) found that whenever any of the high-pressure phases (II to IX) were recovered in liquid nitrogen and subsequently warmed they converted to cubic ice at 120–170 K. This provided a way of preparing larger quantities of cubic ice, and they introduced the nomenclature 'ice Ih' and 'ice Ic'. Ice Ic produced in this way or by vapour deposition transforms to hexagonal ice at around 200 K. There is no sharp transition, and ice Ih never transforms to ice Ic on cooling. Handa *et al.* (1988) have found that the energy released when ice Ic transforms to ice Ih varies between 13 and 50 J mol^{-1}. These are very small energies, but the variations are well outside the experimental error and indicate that the quality of ice Ic

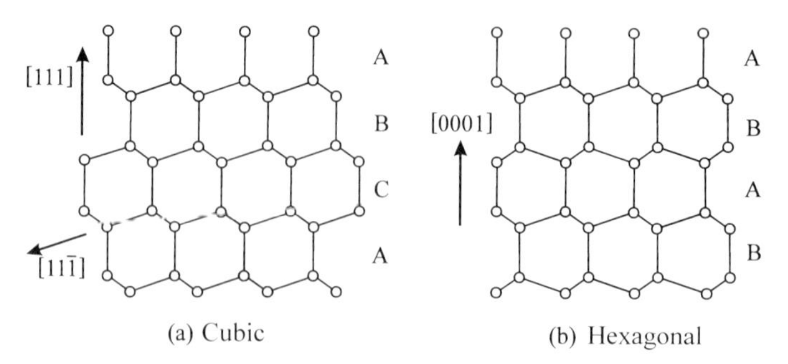

Fig. 11.14 Comparison of the structures of (a) cubic and (b) hexagonal ice I, seen in projection on a $\{11\bar{2}0\}$ plane of the hexagonal lattice.

depends on how it is formed. The evidence is clear that ice Ic is always metastable, and the temperature at which it transforms to ice Ih is determined by the process of molecular rearrangement.

Neutron powder diffraction experiments by Arnold *et al.* (1968) and Kuhs *et al.* (1987) have determined the hydrogen sites in ice Ic and confirmed that the structure is hydrogen disordered exactly as for ice Ih. The single 111 diffraction peak of ice Ic corresponds to the 0002 peak in ice Ih, but in powder diffraction from ice Ih the $10\bar{1}0$ and $10\bar{1}1$ peaks are quite close to it. In powder diffraction from ice Ic the 111 peak is always broadened and there is evidence of an extra peak at the position of the $10\bar{1}0$ reflection from ice Ih, which should not be observed for ice Ic. These facts were interpreted by Kuhs *et al.* (1987) as showing that the ice Ic is far from perfect, containing glide type stacking faults (§7.6.1) and additional disorder as a consequence of the way it has been formed. Single crystals have never been produced and all experiments are on very fine grain polycrystalline material.

Ice Ic is popular for theoretical modelling of ice because its higher symmetry simplifies the calculations. A recent molecular dynamics simulation by Svishchev and Kusalik (1994) of the crystallization of ice from supercooled liquid water always produced ice Ic. This never happens in macroscopic experiments, and this emphasizes that the distinction between the hexagonal and cubic forms of ice I is very subtle.

11.9 Amorphous ices

Amorphous solids resemble liquids in having no long-range order, but they differ from liquids in being frozen into a particular structural arrangement. They are metastable with respect to crystalline phases provided the temperature is sufficiently low, so that molecular rearrangement cannot occur. Amorphous materials produced in different ways may thus be expected to have different properties, in contrast to a liquid which has well-defined properties resulting from a dynamic equilibrium between all possible configurations of the molecules.

The amorphous forms of water have been a topic of increasing interest over the last 15 years. They are referred to as 'amorphous ices'. There are two very distinct forms: low-density amorphs with a density at atmospheric pressure of about $0.94\,\mathrm{Mg\,m^{-3}}$ (less than liquid water but a little larger than ice Ih) and a high-density amorph, which is formed under pressure but on recovery at atmospheric pressure has a density of $1.17\,\mathrm{Mg\,m^{-3}}$. However, recent studies have revealed that the low-density forms produced in different ways are not entirely equivalent. The amorphous ices and the relationships between them and other phases of water are shown diagramatically in Fig. 11.15, in which we use the nomenclature of Johari *et al.* (1996). At the top are the normal solid, liquid, and gaseous phases linked by arrows showing that they can be converted reversibly from one form to another. The amorphous ices are shown in the lower part of the diagram with

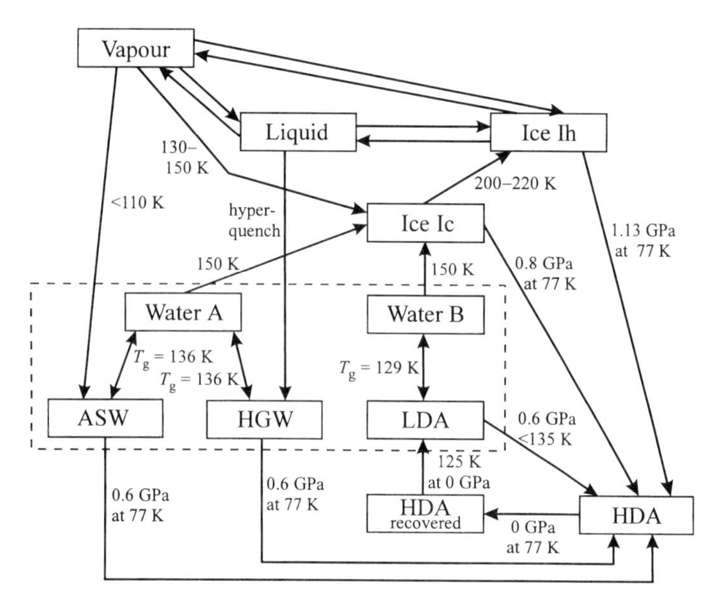

Fig. 11.15 Diagram showing the metastable amorphous forms of ice and their relationship to the stable phases of water and to ice Ic. The conditions stated are only an indication and not a specification of the full range of possibilities. T_g signifies a glass transition. For a simple approach to the subject the contents of the rectangle delineated by the broken lines may be thought of as a single phase and referred to as 'LDA'.

arrows giving an indication of how they are formed or converted to other phases. Ice Ic, which plays an intermediate role in the conversion of amorphous ice back into ice Ih, appears in the middle of the diagram. We will now consider each of the amorphous phases in turn.

For a long time the only known form of amorphous ice was that produced by the deposition of water vapour onto a cold substrate. The first observations were made by Burton and Oliver (1935) who used X-ray diffraction and found that if the substrate was kept below 163 K the deposit formed was amorphous. We stated in the previous section that cubic ice is formed by vapour deposition between 130 and 150 K with amorphous ice below that, so this temperature seems high and their observation is controversial. The significant thing is that an amorphous phase, now to be referred to as 'amorphous solid water' (ASW) can be consistently formed by deposition at liquid nitrogen temperature (77 K). As formed this material is microporous with a high capacity for the adsorption of gases (Mayer and Pletzer 1986). It can be consolidated into a form with lower porosity by annealing (or sintering) at a higher temperature but keeping it below 120 K. Detailed studies, such as those of Jenniskens and Blake (1994) using electron diffraction, reveal a variety of structural changes in ASW during warming. On heating to about 150 K ASW transforms to ice Ic, and this is then converted to ice Ih at 200–220 K.

A liquid which has been supercooled to such an extent that it is effectively a solid is called a glass, but it is not possible to supercool bulk water into a glassy state because crystallization occurs first. However, an amorphous solid known as 'hyperquenched glassy water' (HGW) can be produced by squirting a very fine jet of water into a cryoliquid or by firing droplets of diameter $\leq 3\ \mu m$ at supersonic speed onto a substrate at 77 K (Mayer and Brüggeller 1982; Hallbrucker *et al.* 1989*a*).

For experiments on a given time scale there is a temperature at which molecular rearrangement sets in and a glass regains liquid-like properties. This is known as the glass transition temperature T_g. It is most frequently revealed in calorimetric experiments on material which has been annealed for a long time at a temperature a little below T_g, releasing some energy and acquiring some degree of order. On warming to T_g an endothermic 'melting' then takes place as the material returns to the state characteristic of the liquid. These phenomena were observed by Hallbrucker *et al.* (1989*a*) for HGW, in which case T_g was 136 K. A form of 'water' was produced that could be converted back to HGW by cooling, or transformed to ice Ic by heating to above 150 K, but it is quite unlike what is normally considered to be a liquid. Careful experiments have revealed a similar glass transition in ASW (Hallbrucker *et al.* 1989*b*). The metastable material which is produced in both cases has been named 'Water A'. It is included in Fig. 11.15, but in most experimental situations ASW or HGW will transform direct to ice Ic without this intermediate stage being detected.

We know that at 77 K the high-pressure phases of ice can be recovered at atmospheric pressure, and therefore that rearrangement of molecules cannot occur at this temperature, but this raises the question of what will happen if ice Ih is compressed at 77 K up to the extrapolation of the Ih–liquid boundary line in Fig. 11.10 (i.e. to about 1 GPa). Mishima *et al.* (1984) found that if this is done the low density structure of ice Ih transforms to an amorphous phase of density $\sim 1.31\ Mg\,m^{-3}$. In a suitable pressure cell this transition is very sharp at 1.12 GPa (Whalley *et al.* 1987). On releasing the pressure the density falls to $1.17\ Mg\,m^{-3}$ but the phase remains amorphous. This phase is clearly distinct from the amorphous ices ASW and HGW formed at low pressures, and it is referred to as the 'high-density amorph' (HDA). When HDA is warmed at atmospheric pressure it transforms at about 125 K to a 'low-density amorph' (LDA) with properties superficially very similar to ASW (Handa *et al.* 1986). If this LDA is compressed at 77 K it transforms back to HDA at 0.6 GPa (Mishima *et al.* 1985). Although there is some hysteresis, this is an apparently first-order transition between two amorphous phases. HDA can also be formed by compression of ASW, HGW, or ice Ic (Johari *et al.* 1990).

On warming, LDA undergoes a glass transition at 129 K, which is significantly less than T_g for ASW or HGW. The phase formed above the glass transition of LDA is denoted 'Water B', and Johari *et al.* (1996) show that it has significantly different properties from Water A, although both crystallize to ice Ic at ~150 K. These facts and other evidence show that there are significant differences between LDA and the other low-density amorphous ices. However, some authors see these as minor details and treat all materials within the box shown as a broken line in Fig. 11.15 as minor variants of a single LDA phase.

Under pressure ice Ih normally transforms to ice II, III, or IX, but this can be suppressed by using an emulsion of ice particles of ~5 μm in diameter, and Mishima (1996) has studied the effect of compressing such emulsions over the whole temperature range from the melting point of ice III down to 77 K. Above about 150 K the ice 'melts' to a supercooled liquid, which is recognized because this is an endothermic process. The supercooled liquid can then crystallize or become HDA. At lower temperatures the open structure of ice Ih undergoes mechanical collapse direct to HDA, which is an exothermic process. The nature of the HDA so formed depends on the conditions of its formation, and the range of intermediate processes is quite complicated.

As in the case of liquids, the structures of amorphous phases can only be described in statistical terms. X-ray and neutron diffraction experiments yield a structure factor $S(q)$, where q is the magnitude of the scattering vector. Results from X-ray diffraction (Narten *et al.* 1976) are dominated by the oxygen positions, but (for D_2O) neutrons give data dependent on both the D and the O nuclei. These results can be analysed to give a real-space distribution function $d(r)$, which is defined to emphasize the oscillations and tends to 0 at large r. Plots of this function for ASW (Chowdhury *et al.* 1982) and HGW (Dore 1990) are shown in Fig. 11.16. The curves for the two materials are virtually identical and show more features than measurements on the liquid phase at higher temperatures (Soper 1994). HDA and LDA have been studied both with X-rays (Bizid *et al.* 1987) and with neutrons (Bellissent-Funel *et al.* 1987). The distribution functions for these amorphs differ from one another, and that for LDA seems to be significantly different from ASW.

The structure of ASW or HGW has been modelled very successfully by the continuous random network (CRN) model of Boutron and Alben (1975). A real-space model is constructed from water molecules with fourfold co-ordination and bond lengths close to those in crystalline ice but with no long-range order. The function $d(r)$ calculated from such a structure is included in Fig. 11.16 and has features close to those observed experimentally. HDA cannot be described in this way, but a molecular dynamic (MD) simulation of the hysteretic phase transformation between HDA and LDA has been described by Poole *et al.* (1993). The modelling of the liquid and amorphous solid phases of water is an active topic of research. Particularly for HDA the presence of interstitial molecules within the continuous random network seems to be important (e.g. Svishchev and Kusalik 1995). Devlin and Buch (1995) have modelled the vapour deposition of ASW, and studied its lattice vibrations and the effects of adsorbed gases. Tse (1992) describes a molecular dynamic simulation of the mechanical collapse of ice Ih under pressure to form HDA.

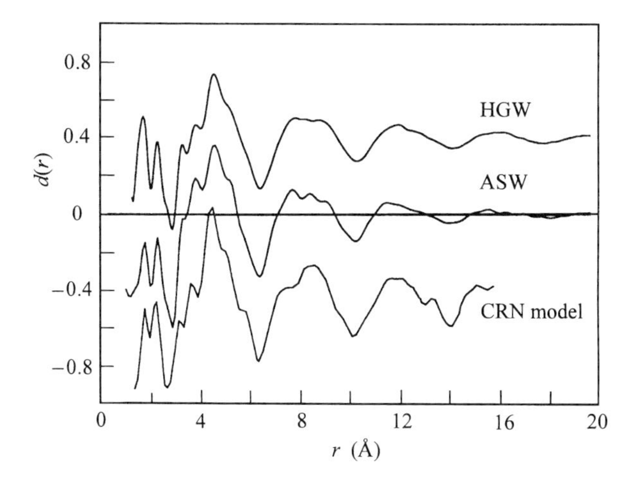

Fig. 11.16 Comparison of the function $d(r)$ determined by neutron scattering for HGW and ASW with that predicted from the continuous random network (CRN) model. The axes apply to ASW, and the other functions are displaced vertically by ± 0.4. Reproduced from *Journal of Molecular Structure*, **237**, J.C. Dore, Hydrogen-bond networks in supercooled liquid water and amorphous/vitreous ices, pp. 221–32, Copyright 1990, with permission from Elsevier Science.

11.10 Clathrate hydrates

The clathrate hydrates are crystalline materials which are so like ice that they are appropriate for inclusion in this chapter. Their structures consist of a three-dimensional framework of hydrogen-bonded water molecules within which are incorporated a smaller number of relatively inert 'guest' molecules. The simplest clathrate hydrates are formed from the rare gases or small molecules like O_2, N_2 or the hydrocarbons. They are mainly stable below about 0 °C and under pressures of up to about 10 MPa, much less than the pressures required to convert ice Ih to ice III. The clathrates of the hydrocarbons are of enormous practical importance, because a large fraction of the Earth's natural gas deposits are in this form, and because they are liable to form blockages in pipelines which contain moisture and natural gas under pressure in cold places. The structures and properties of these materials are reviewed by Davidson (1973), Jeffrey (1984), and Sloan (1990). Calorimetric studies are reviewed by Yamamuro and Suga (1989).

The majority of clathrate hydrates have one of two structures determined by von Stackelberg and Müller (1954) and denoted types I and II. Each water molecule is hydrogen-bonded to four others. The bond lengths and angles are similar to those in ice Ih, though with a spread of values. Both structures are cubic, and they are slightly more open than ices Ih or Ic so that they contain

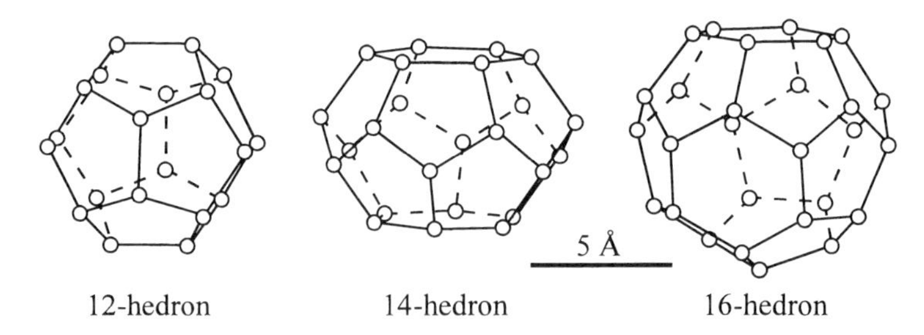

Fig. 11.17 The three polyhedral cages found in the type I and type II clathrate hydrates.

'cages' within which the guest molecules are incorporated. The cages have the form of the 12-, 14-, or 16-faced polyhedra shown in Fig. 11.17. Oxygen atoms at the vertices have three hydrogen bonds along the edges of the cage and one further hydrogen bond which points outwards; this bond forms part of, or links to, another cage. The smallest cage is the pentagonal dodecahedron which is present in both type I and type II structures. The type I clathrates have a structure incorporating both 12- and 14-faced cages, while the type II structure is formed entirely from 12- and 16-faced cages. Details about the structures are given in Table 11.4. The radii of the cages are measured to the oxygen sites, and examples are given of which guest molecules form which structure. It is not necessary that all cages be occupied, with the result that clathrate hydrates have variable composition up to the limiting values stated. These limits depend of course on whether the guest molecules occupy both small and large cages or only the larger ones. The van der Waals binding of the guest molecules in the cages is important in stabilizing the structure.

As in ice Ih, the orientations of the water molecules are disordered subject to the ice rules. This has been confirmed by neutron diffraction (e.g. Hollander and Jeffrey 1977) and gives rise to dielectric relaxation like that observed in ordinary ice (Gough *et al.* 1968; Davidson *et al.* 1977). The relaxation rate is generally faster for more complex guest molecules. On cooling there should be a transition to an ordered phase equivalent to ice XI, and this transition has been observed in clathrates of acetone and the four- and five-membered ring compounds trimethylene oxide (TMO) and tetrahydrofuran (THF) when doped with KOH (Suga *et al.* 1992; Suga 1997). However, for the argon clathrate only a glass transition has been achieved. In cases where the guest molecule is not spherically symmetrical its own orientation must also be considered (Davidson and Ripmeester 1978, 1984). In the case of THF for example, the ordering has been investigated both by calorimetry and by neutron diffraction (Yamamuro *et al.* 1988, 1995).

The molecules He and H_2 do not form type I and type II clathrates but are small enough to be incorporated into ice in different ways. We have already mentioned that He enters the ice II structure, stabilizing it up to the melting line (§11.5).

Table 11.4 Crystal structures of the principal clathrate hydrates

	Type I	Type II
Space group	cubic Pm3n ('body-centred')	cubic Fd3m (diamond)
H_2O molecules per unit cell	46	136
Lattice parameter a (Å)	11.7–12.0	17.0–17.3
Small cages	12-hedra	12-hedra
no. per cell	2	16
average radius (Å)	3.9	3.9
Large cages	14-hedra	16-hedra
no. per cell	6	8
average radius (Å)	4.3	4.7
Limiting composition[†]		
all cages filled	$4M \cdot 23H_2O$	$3M \cdot 17H_2O$
large cages only	$3M \cdot 23H_2O$	$M \cdot 17H_2O$
Examples		
all cages filled	Xe, CH_4 [a]	Ar, Kr [a]
	Cl_2, H_2S, CO_2 [b]	N_2, O_2 [c]
large cages only	C_2H_6 [d]	C_3H_8 [d]
	$(CH_2)_2O$ [e]	$(CH_3)_2CO$, THF [f]

[a]Davidson *et al.* (1984), [b]von Stackelberg and Müller (1954), [c]Davidson *et al.* (1986), [d]Jeffrey (1984), [e]Hollander and Jeffrey (1977), [f]Yamamuro *et al.* (1995).
[†]Assumes single occupancy of cages. Double occupancy of large cages is possible in some cases (Kuhs *et al.* 1997).

A similar hydrate of H_2 is formed at 295 K between 0.75 and 3.1 GPa, but above 2.3 GPa the H_2 can also form a hydrate of 1 : 1 composition within the H_2O structure of ice Ic. The structure is equivalent to that of ice VII with one H_2O sub-lattice replaced by H_2 (Vos *et al.* 1993). Vos *et al.* (1996) have compared the bonds in this hydrate with those in ice VII using Raman spectroscopy and shown that the bonds in the hydrate are equivalent to those in ice VII at half the pressure. A simple interpretation is that the pressure is resisted solely by the H_2O sub-lattice, with the H_2O–H_2 repulsions being comparatively weak. There is speculation that the transition to symmetrical bonds as in ice X occurs in this hydrate at 32 GPa for H_2O or 40 GPa for D_2O.

There are many hydrates of chemical compounds, in some of which the water molecules form hydrogen-bonded networks with residual entropy and other properties characteristic of ice. For an account of such hydrates in an ice context see Parsonage and Staveley (1978).

11.11 Lattice vibrations and the hydrogen bond

In concentrating on structural features most of this chapter has emphasized the differences between the various polymorphs of ice, but except for ice X all phases have the common feature that they are built up from water molecules linked by hydrogen bonds. Experiments which probe the local structure and bonding reveal the similarities between these structures, and these similarities are particularly apparent in the spectra of the lattice vibrations, which were described for ice Ih in §3.4. The molecular modes at around 400 meV (see Fig. 3.5) are present in all phases, and are most highly resolved in Raman spectroscopy where the principal feature is the sharp peak which occurs in ice Ih (at 95 K) at 382.2 meV (3038 cm^{-1}). This in-phase O—H bond-stretching ν_1 mode has been identified in many phases of ice, and the frequencies compiled from a range of experiments are shown as a function of pressure in Fig. 11.18. The results are weakly dependent on temperature and where appropriate they have been standardized for this figure to 246 K, but where this is well outside the range of the original measurements the original values are given at the temperature indicated. The data in the literature are not entirely consistent, but there is a clear trend for the frequency to fall with increasing pressure and thus decreasing bond length r_{OO}. At a phase transition in which the density increases as a result of a denser form of packing there is normally an increase in r_{OO} and a corresponding increase in frequency. The correlation of frequency with bond length is particularly well established for ice VII (Walrafen *et al.* 1982), and similar information can be obtained by studying infrared absorption by the localized mode μ_1 of HOD in D_2O (Klug and Whalley 1984). Klug *et al.* (1987) have used Raman measurements of this localized mode to draw conclusions about the distributions of bond lengths in high- and low-density amorphous phases.

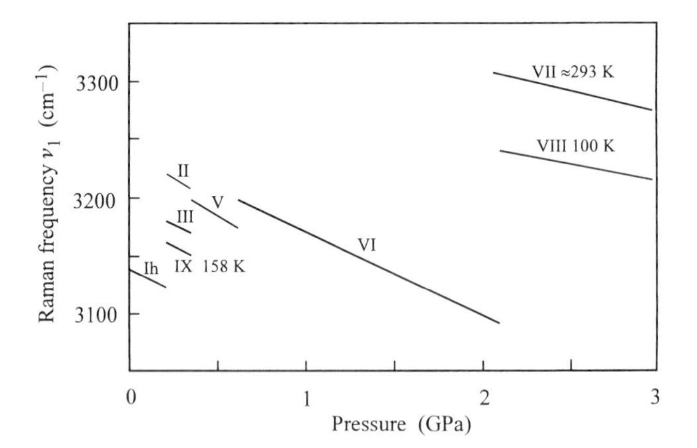

Fig. 11.18 Raman frequencies of the in-phase OH stretching mode ν_1 as a function of pressure in various polymorphs of H_2O ice. Except where indicated, measurements are at or standardized to 246 K. Data for ices Ih, II, III, V, and VI from Minceva-Sukarova *et al.* (1984). Slope for ice VI from Abebe and Walrafen (1979). Ice VII from Walrafen *et al.* (1982). Ice VIII from Hirsch and Holzapfel (1986).

The interpretation of the change of frequency with bond length is that as the oxygen atoms are brought closer together the proton moves very slightly away from the oxygen to which it is covalently bonded, and the restoring force for displacements along the bond is then reduced. The ultimate consequence is the transition to the symmetrical bonding of ice X, as revealed by the continuation of the data of Fig. 11.18 to higher pressures in Fig. 11.13. These effects have been interpreted quantitatively in the case of the localized mode μ_1 by Knuts *et al.* (1993) using *ab initio* modelling of small clusters of molecules.

In the lower frequency part of the spectrum the phonon densities of states of many phases of ice (recovered at zero pressure and low temperature) have been determined by Li *et al.* (1992) and are shown in Fig. 11.19. There are many similarities between these curves but some significant differences, of which the most marked is the more sharply defined structure in the spectra for the ordered phases II, VIII, and to a lesser extent IX. A comparison between ice VII and ice VIII under pressure is reported by Li and Adams (1996). Features corresponding to the two peaks at 28 and 38 meV in ice Ih, which were discussed

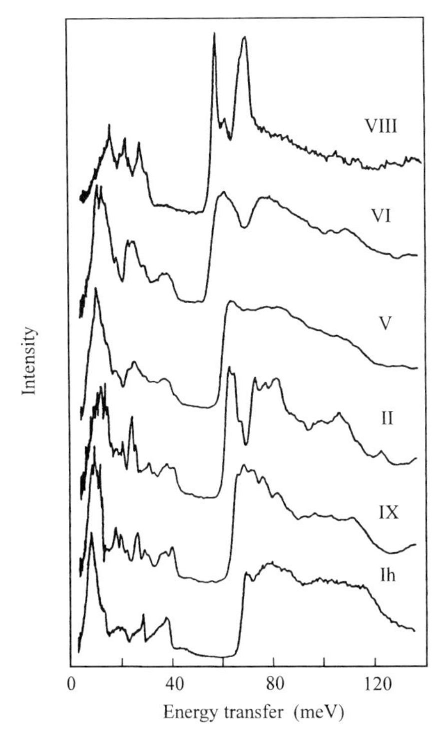

Fig. 11.19 Incoherent inelastic neutron scattering intensity from ice Ih and recovered samples of high pressure polymorphs of H_2O ice measured on the instrument TFXA at 15 K. (From Li *et al.* 1992; see also Li 1996.)

in §3.5.1, can be found in many of the other phases. However, there is clearly only one peak in ice VIII, whereas a second peak at a higher frequency appears in ice VII. It therefore appears that the problem presented by these two peaks is related to the disorder of the protons. These and further results on amorphous ices are discussed by Li (1996).

At the extreme low-frequency limit Brillouin scattering yields the elastic constants, as described for ice Ih in §3.2, and these have been determined under pressure for monocrystalline ice III, V, and VI by Tulk *et al.* (1994, 1996, 1997) and for ices VI and VII by Shimizu *et al.* (1996) and Baer *et al.* (1998).

12
Ice in nature

The aims and style of this chapter are different from those of the other chapters in the book. Its purposes are to survey the occurrence of ice both on Earth and elsewhere in the Solar System, to introduce readers to areas of ice research different from their own, and to show the relevance of the basic physics of ice in these fields. The chapter is intended to be of general interest and to point readers to more authoritative sources of information both at the introductory and the more specialist level.

12.1 Lake and river ice

In winter many lakes and rivers become frozen. The ice forms on the surface, both because water has a maximum density at 4 °C, with the result that water with its free surface at 0 °C is stable against convection, and also because the ice that forms floats on the water. This book is not concerned with such topics as the heat balance in the growth of the ice cover (see for example Lock 1990), but the polycrystalline structure of the ice is relevant. The classification of such ice devised by Michel and Ramseier (1971) (see also Michel 1978) is used elsewhere in ice physics and has been referred to earlier. The formation of ice on lakes and more generally in the natural environment has been reviewed by Shumskii (1964).

We must first distinguish between three layers that are formed on a lake or river as the water freezes. The 'primary' layer consists of that ice which first forms a complete covering on the surface. The 'secondary' layer is that which grows downwards from the primary layer in the temperature gradient, and finally there may be a layer of 'superimposed ice' formed on top of the original surface by flooding from any source and subsequent freezing. The types of primary ice, as classified by Michel and Ramseier (1971), are as follows:

- *P1 ice.* If the primary ice forms slowly on still water, platelets are nucleated which float with the *c*-axes vertical. These platelets grow until the surface is covered by large grains, all with the *c*-axes vertical but with the *a*-axes randomly oriented in the plane.
- *P2 ice.* In a large temperature gradient many more crystals are nucleated and needles are formed as well as platelets. This produces an ice cover with a smaller scale and more complex grain structure.
- *P3 ice.* When the surface is agitated by wind or flow, ice crystals, which may be needles or platelets, are formed and remain in suspension. This is known as

'frazil ice', and can form quite a thick layer of slush. If it freezes solid this becomes a layer of P3 ice. The orientations of the grains are random.
- *P4 ice.* This type of ice is produced when the ice is nucleated by snow falling on the water surface. The grains are of small size with completely random orientation.

As the secondary ice grows downwards from the primary ice it forms long columnar crystals, which get larger in width and fewer in number as the less favourably oriented ones are squeezed out. Figure 12.1 shows the columnar grains in a vertical section through a layer of lake ice, together with horizontal sections at the depths A, B, and C. The primary layer is thin and not seen in this figure; the fine-grained ice in the top of the vertical section is superimposed ice. The grains in columnar ice are not randomly oriented and can usually be classified into one of the following types:

- *S1 ice.* The *c*-axes are all approximately vertical. Such ice forms naturally under a primary layer of type P1.
- *S2 ice.* The *c*-axes all lie in the horizontal plane, but in random directions. S2 ice forms from a primary layer of type P2, P3, or P4. It can also be produced in the laboratory by seeding the surface of a tank of water with powdered ice or snow. The grains with horizontal *c*-axes then grow preferentially and squeeze out other orientations.
- *S3 ice.* This is like S2 but with a preferred direction of the *c*-axes in the horizontal plane.

In natural lakes both S1 and S2 ice are commonly observed, and the determining factors are far from clearly established (Gow 1986). In some cases a single lake may form patches of the two types, which have a different surface appearance (Knight 1962).

The important form of superimposed ice is T1 ice, which is formed when the frozen surface is covered with snow, flooded and then frozen. The grains have random orientations and similar dimensions in all three directions. An example of such ice is seen as the top layer in Fig. 12.1. Its formation is the natural equivalent of the laboratory method for preparing randomly oriented 'granular ice' referred to in §§2.1.1 and 8.3.2. Such ice cannot be obtained simply by freezing water in a container.

The classification of Michel and Ramseier includes a number of further types of ice, but as these apply to ice formed in more specific natural conditions we do not list them here.

Freshwater ice formed on lakes and rivers differs from laboratory ice in containing trace impurities, gas bubbles, and solid particles originally in suspension in the water. Dissolved impurities have a large effect on the steady-state electrical conductivity and inclusions degrade the optical transparency, but mechanical properties are determined primarily by the polycrystalline structure.

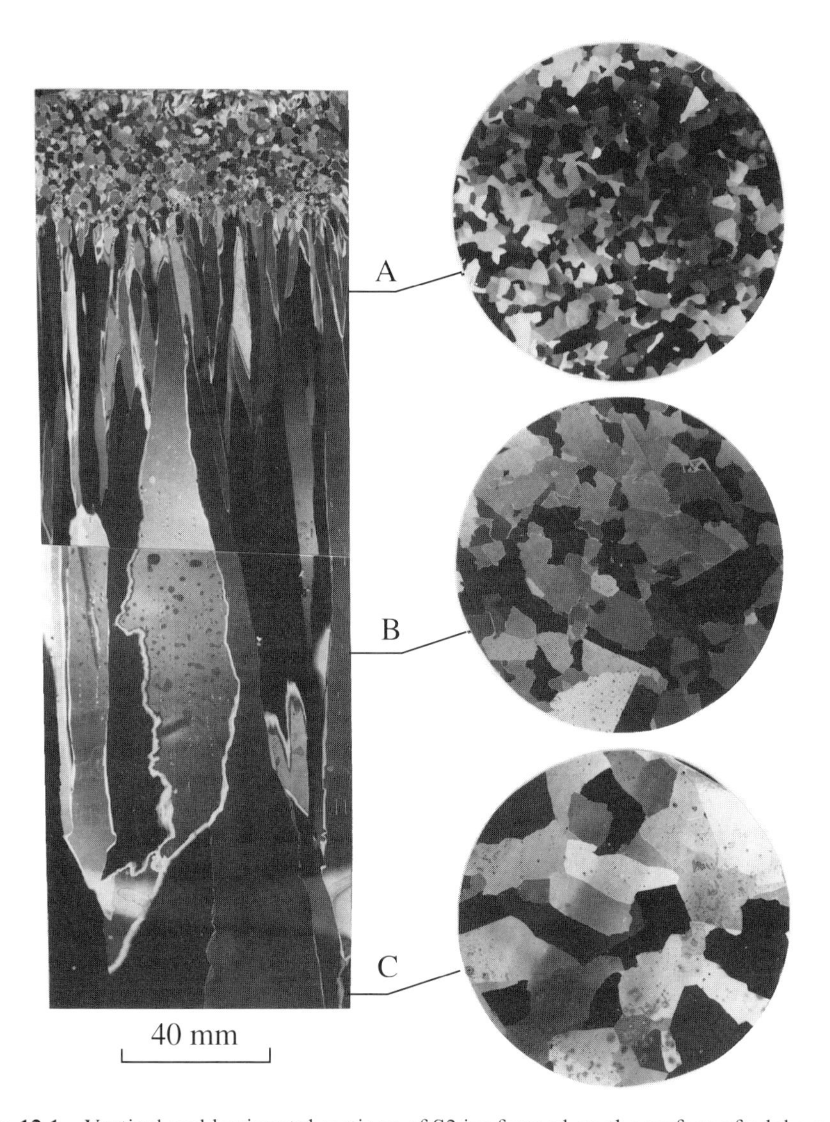

Fig. 12.1 Vertical and horizontal sections of S2 ice formed on the surface of a lake and viewed between crossed polarizers. The primary layer is too thin to be distinguished, and the fine grained ice at the top of the vertical section is superimposed T1 ice. Reproduced from *Journal of Crystal Growth*, **74**, A.J. Gow, Orientation textures in ice sheets of quietly frozen lakes, pp. 247–258, Copyright 1986, with permission from Elsevier Science.

12.2 Sea ice

Sea ice is formed by freezing sea water and differs from freshwater ice because of the large quantities of salts present. Much of the ice cover that forms on the oceans in the polar regions lasts for just one winter before it melts, and this is called 'first-year ice'. In the Arctic, however, there are large areas of ice that do not melt completely in summer and here 'multi-year ice' builds up, with formations and structures that result from its evolution. The structure and properties of sea ice have been the subject of much research, and this work has been reviewed by Weeks and Assur (1967), Schwarz and Weeks (1977), Weeks and Ackley (1982, 1986), and Sanderson (1988).

Sea water contains approximately 35 parts per thousand by weight of ionic salts (principally Na^+, Mg^{2+}, Cl^-, and SO_4^{2-}), and has a freezing point depending on the precise composition of about $-2\,°C$. Unlike pure water, sea water has no maximum density at a temperature above the melting point, so that it cannot be convectionally stable before freezing commences. However, once ice has nucleated at the cold surface it floats and crystals accumulate at the surface. The eventual formation of a solid layer is a complex process very dependent on the ambient conditions such as temperature, wind, and wave action. Once a solid polycrystalline layer is established, the ice that grows from it develops, in a distance of order 0.3 m, into an S2-type columnar structure as described for freshwater ice in §12.1. In one season first-year ice cover can reach a thickness of about 2 m before it begins to melt.

As indicated in §§5.4 and 6.8 salts like NaCl are almost completely insoluble in ice, with the result that the ice which freezes from sea water is almost pure, with the excess salts remaining in the liquid. Under such conditions constitutional supercooling occurs, and the freezing interface becomes sharply corrugated as illustrated in Fig. 12.2 (Harrison and Tiller 1963). The pointed plates of monocrystalline pure ice grow downwards into the water, and the rejected salts increase the concentration of the surrounding solution, thus lowering its lower freezing point. The concentrated brine lying between the plates of ice remains in equilibrium with them, even after the tips of the ice plates have moved well beyond them. In sea ice the plates are formed within the grains of the polycrystalline structure and lie perpendicular to the [0001] crystallographic direction. The spacing is typically 0.5–1.0 mm, and the structures have been illustrated and analysed by, for example, Arcone *et al.* (1986) and Perovich and Gow (1996). Gases are also insoluble in ice, and they too are rejected as the ice freezes, forming bubbles which are trapped together with the brine between the plates of pure ice. If the temperature of the ice is lowered after the plate structure has been formed, more water freezes onto the plates and the layers of trapped brine become thinner and more concentrated. Eventually solid salts crystallize from the brine layers, and Fig. 12.3 illustrates how the proportions of ice, brine, and solid salts vary with temperature in sea ice of a standard composition. The circles in this figure

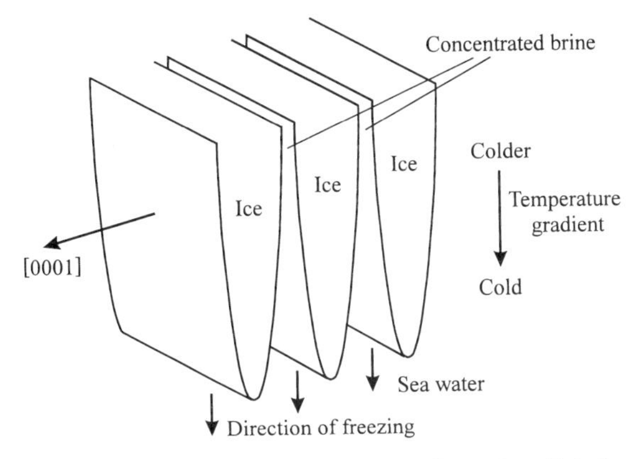

Fig. 12.2 The growth of sea ice in a temperature gradient. Parallel plates of pure ice are formed with concentrated brine in layers between them.

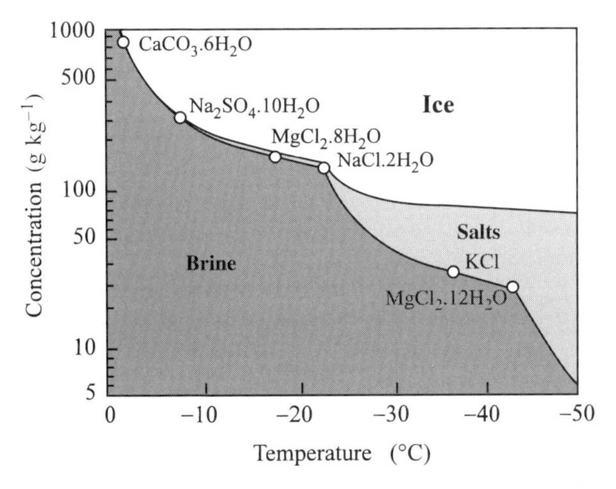

Fig. 12.3 Simplified diagram showing the temperature dependence of the proportions of ice, salts, and brine in standard sea ice (35‰ total salts). The small circles on the brine–salt boundary line indicate the temperatures at which particular salts crystallise from the brine. (Data from Assur 1958.)

indicate the points at which particular hydrated salts crystallize from the brine. The most significant are $Na_2SO_4 \cdot 10H_2O$ and $NaCl \cdot 2H_2O$ which crystallize at $-8.2\,^{\circ}C$ and $-22.9\,^{\circ}C$ respectively. Richardson (1976) used the n.m.r. properties of ice and water referred to in §10.5 to determine the temperature dependence of the liquid fraction within sea ice, confirming the results in Fig. 12.3 and verifying that a small quantity of liquid is still present at temperatures as low as $-50\,^{\circ}C$. The brine can drain out of the ice through drainage channels that develop within the structure, and isolated brine pockets migrate in the temperature gradient, downwards in winter and upwards in summer. Hence the total salt content does not remain constant in time. The salinity profiles of samples of sea ice have been studied by Nakawo and Sinha (1981).

Because of the importance of sea ice in engineering situations, a large amount of research effort has been devoted to studying its mechanical properties, and this is one of the main topics in the reviews cited at the beginning of this section. In many respects the properties of sea ice are determined by the properties of the ice matrix itself, with modifications due to the brine inclusions. The results of laboratory compressive tests on saline ice are not fundamentally different from those on pure ice with similar columnar structure, and in particular the brine pockets do not nucleate fracture (e.g. Gratz and Schulson 1997). A recent analysis which correlated the mechanical properties of sea ice with its structure is given by Cole (1997). An important and controversial question concerns the relation between laboratory-scale and field-scale experiments. This subject has been addressed by theoretical solid mechanics (Bazant and Li 1994), by very large-scale modelling (Zhang and Hibler 1997), and by experiment (Adamson and Dempsey 1998). Schulson and Hibler (1991) have drawn attention to the similarity between wing cracks like those in Fig. 8.18 and the ~100 km long cracks known as 'leads' seen in satellite photographs of ice cover on the Beaufort Sea.

The heat capacity of sea ice differs from that of pure ice because of the latent heat associated with the temperature dependence of the fraction of brine present (Schwerdtfeger 1963). The thermal conductivity can be modelled as an aggregate of the ice, brine, and salt components, and the thermal properties of sea ice are reviewed by Yen (1981) and Yen *et al.* (1991).

From the point of view of its electrical properties sea ice is a very complex material. As explained in §5.4.4 ice grown from a saline solution is acid-doped, and the bulk ice gives a Debye dispersion with the corresponding relaxation time. In addition, isolated brine pockets act as conducting inclusions which produce a Maxwell–Wagner–Sillars relaxation (see von Hippel 1954). Continuous channels of brine between the plates and at grain boundaries contribute an additional d.c. conductivity. These three effects have been observed experimentally by Addison (1969, 1970), who found that the Maxwell–Wagner–Sillars relaxation occurred at frequencies of about 200 MHz. Microwave measurements by Arcone *et al.* (1986) were also interpreted as arising from this high-frequency relaxation.

Optically sea ice looks milky because of the inclusions of brine, salt, and gas bubbles, with the appearance of a particular sample depending on its internal structure. The optical properties of sea ice are described by Perovich (1996).

12.3 Ice in the atmosphere

12.3.1 *Ice crystals in clouds*

The formation of ice plays an essential role in the physics of clouds and hence in the production of our weather, even when the precipitation from a cloud does not reach the ground in frozen form. The physical processes occurring in clouds are described by, for example, Rogers and Yau (1989) and in more detail by Mason (1971) and Pruppacher and Klett (1997). In his book on ice physics Hobbs (1974) devotes particular attention to ice crystals grown from the vapour in the laboratory and formed naturally in clouds.

A cloud is formed when a parcel of moist air rises in the atmosphere and cools adiabatically, with the result that the water vapour becomes supersaturated and finally condenses. The temperature is frequently such that the condensation is in the form of droplets, and these grow by further condensation to sizes of the order of 10 μm. Such tiny droplets are very stable: they do not grow much bigger in the lifetime of the cloud, their aerodynamic properties are such that they do not coalesce, and their terminal velocity is so low that they do not fall out of the cloud. The formation of ice is a necessary precursor to any significant precipitation from the cloud, but these droplets can be supercooled to around −40 °C before they freeze spontaneously. Freezing is more commonly produced on ice-forming nuclei already present in the air, and it may occur either by the direct nucleation of an ice crystal from the vapour or by the nucleation of a pre-existing droplet. Extrapolation of the liquid–vapour equilibrium line below 0 °C on the phase diagram (Fig. 2.1) shows that ice has a lower vapour pressure than water, and, once nucleated, ice crystals will grow by vapour transfer from the droplets. As they grow larger the ice crystals become increasingly able to aggregate with the smaller droplets or other ice crystals, eventually forming entities heavy enough to precipitate from the cloud. In cold conditions they remain frozen, but in warmer air they melt to reach the ground as rain.

At this point we should take note of an important problem of applied ice physics. If the supercooled water droplets in a cloud strike the surface of an aircraft wing or a helicopter blade, they will adhere and freeze immediately. This can cause very dangerous icing, which has to be avoided.

Ice crystals grown from the vapour have widely varying forms, which commonly display the hexagonal symmetry of the crystal structure. Bentley and Humphries (1931) published a beautiful collection of over 2000 photographs of almost perfect ice crystals collected at ground level, and an even more beautiful collection is that of Kobayashi and Furukawa (1991). Another perhaps more representative compilation is that of Nakaya (1954). Good ice crystals, which are typically 1–5 mm across, can be broadly classified into the three types shown in Fig. 12.4. Plate-like crystals are the commonest and are formed when growth is most rapid perpendicular to the [0001] axis, while prisms or needles form when the most rapid growth is parallel to [0001]. Dendritic growth perpendicular to [0001] forms the beautiful six-pointed stars, in which the dendrites are along the $\langle 2\bar{1}\bar{1}0 \rangle$ directions (Glen and Perutz 1954). Bentley and Humphries state 'only one in a multitude of snow crystals is so nearly symmetrical as to show no obvious irregularities', and Nakaya (1954) and LaChapelle (1969) show that while most crystals show hexagonal features they are usually quite irregular. However, observers are always impressed by those few crystals that have highly symmetrical

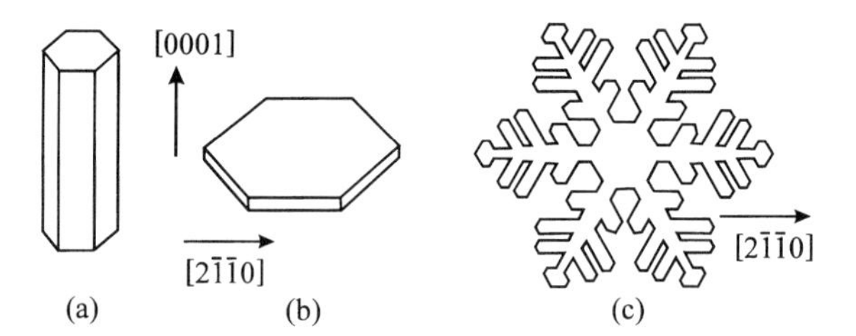

Fig. 12.4 The three basic forms of ice crystal formed by condensation from the vapour. (a) Needles or columns, (b) plates, and (c) dendrites or stars.

forms of great complexity. The explanation must be that the temporal fluctuations in the conditions which generate the pattern are spatially uniform over the small volume of the crystal itself. The larger and often fluffy snowflakes which are observed in some snowstorms are aggregates of many ice (or snow) crystals. Snow crystals and other forms of frozen precipitation are classified by meteorologists into 80 types according to a scheme developed by Magono and Lee (1966) (reproduced in LaChapelle 1969; Hobbs 1974; Pruppacher and Klett 1997).

Artificial snow crystals were first grown by Nakaya in 1936, and since then the dependence of the form of the crystal on the conditions of growth has been extensively studied (Nakaya 1954; Hallett and Mason 1958; Kobayashi 1961). The principal factor determining the form of the crystal is the temperature, with the favoured form changing from plates to prisms and back twice between 0 and −40 °C. The dendritic stars are only produced at high supersaturation between about −13 and −17 °C.

In a cloud ice crystals that are large enough grow further by the accretion of smaller droplets, which freeze on contact forming 'rime' on the surface. As this accretion accumulates, a porous lump of ice is formed, which is referred to as 'graupel' or 'soft hail'. If the accumulated water is absorbed as liquid before it freezes, denser 'hailstones' are produced, and these can become so large (several centimetres in diameter) as to cause damage on reaching the ground. When examined, hailstones have an inhomogeneous highly polycrystalline structure resulting from their complex history of temperature changes in the cloud (Knight and Knight 1968*a*,*b*, 1971).

High altitude clouds are formed at low temperatures by direct condensation of ice crystals from water vapour. The presence of these crystals is apparent from the formation of haloes at specific angles when the Sun or Moon is observed through such a cloud. These haloes are produced by the refraction of light passing through the facets of the ice crystals (Greenler 1980). The commonest is the 22° halo, which is formed by refraction between a pair of prismatic facets at 60° to one another. A 46° halo is produced from the 90° angle between basal and prismatic facets.

An important question concerns the nature of the ice-forming nuclei, and hence the extent to which weather may be affected by atmospheric pollution or deliberately modified by seeding a cloud with nuclei scattered from an aircraft. The effectiveness of particles in the atmosphere as ice nuclei depends very specifically on their crystal structure. The commonest natural ice nuclei are particles of the clay minerals such as kaolinite (Kumai 1961), but there is much debate over the possible role of biogenic nuclei (Szyrmer and Zawadzki 1997). The most effective material which has been used as an artificial nucleating agent is silver iodide (Vonnegut 1947).

12.3.2 *Thunderstorm electricity*

In a thundercloud there is a strong updraught of air which sweeps small droplets and ice crystals upwards while heavier ice particles (graupel pellets) fall downwards. This mechanically driven system acts as an electrostatic generator, charging the upper portion of the cloud positively relative to the bottom to such a point that a lightning discharge may strike between points in the cloud or between one cloud and another or the ground. Innumerable models for the charging process have been proposed and the current state of the subject is reviewed by Saunders (1993). In some circumstances ions from outside the cloud may be swept into it leading to 'convective' charging, but more usually the charge separation occurs within the cloud itself as a result of collisions between the falling graupel and smaller crystals of ice. Two kinds of model are illustrated in Fig. 12.5. In 'inductive' models the large particle is polarized by the electric field in the cloud, or initially by the ambient field in the atmosphere. The smaller particle then picks up positive charge from the lower surface of the larger one and carries this charge upward, thus increasing the dipole moment of the cloud. In 'non-inductive' models the charge separation in the collision is a property of the ice particles themselves, independent of the local electric field. Extensive laboratory experiments, from Reynolds *et al.* (1957) to Keith and Saunders (1990) and Saunders *et al.* (1991), have been reviewed by Saunders (1994). These show that such charge separation does occur, with the graupel pellets becoming negatively charged under conditions to be found in most thunder clouds, but the sign of the charge exchange depends in a complicated way on both temperature and moisture content. There is evidence of a corresponding reverse polarization in the lower and thus warmer parts of some thunderclouds, and the weight of evidence supports the non-inductive model of thundercloud charging.

The key problem from the point of view of the physics of ice is to explain why charge should be transferred in a particular direction between two pieces of ice. It has been shown that the effect is not the result of a temperature difference between the particles but of their differing size and surface properties. The existence of surface charges on ice was recognized in Chapter 10, but current understanding of the surface has not yet been used to explain the charge transfer processes that lead to the phenomenon of lightning.

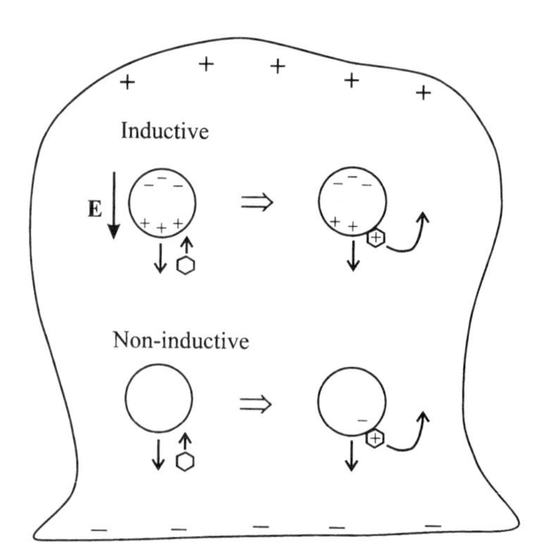

Fig. 12.5 Illustration of two mechanisms for the electrification of a thunder cloud, as explained in the text.

12.3.3 *Polar stratospheric clouds*

In the stratosphere the partial pressure of water vapour is very low, but over the polar regions in winter the temperature can fall below about 190 K, at which point condensation occurs to form polar stratospheric clouds (PSCs) at altitudes of 10–25 km. Clouds designated 'type II' are mainly composed of ice crystals about 10 μm in diameter, while 'type I' PSCs contain nitric acid. The chemical composition of PSCs is now recognized as being quite complex (Peter 1997, 1998). These clouds have recently become a topic of great importance, because they have been identified as catalysts of chemical processes, such as the reaction

$$ClONO_2 + HCl \rightarrow Cl_2 + HNO_3$$

which are involved in the depletion of the ozone layer. It has been observed that the onset of ozone depletion during the Antarctic winter is correlated with the formation of PSCs (Vömel *et al.* 1995). The subject has been reviewed by Solomon (1990) and in the Nobel lecture by Molina (1996). It has stimulated experimental and theoretical studies of the properties of molecules such as HCl and HOCl on the surface of ice (e.g. Thibert and Dominé 1997; Gertner and Hynes 1996; Brown and Doren 1997).

12.4 Snow

In this section we describe the properties of snow that has fallen to the ground. These properties have been the subject of extensive fundamental and applied research for many years, and the history of the subject has been reviewed by Colbeck (1987). Snow is composed of ice crystals, air, and, at temperatures close to the melting point, also of water. The density of freshly fallen snow is typically 50–100 kg m^{-3}. Depending on the ambient conditions and particularly on whether the snow is wet or dry, the crystals become compacted and bonded to one another. Settled snow has a density of 200–300 kg m^{-3}, and its density continues to rise until, if it survives long enough, it is eventually converted to solid ice with pores through it. In this state it is called 'firn', which is the first stage in the formation of glacier ice, to be discussed in the following section. An international classification has been developed for the description of fallen snow, based on such parameters as grain form, grain size, density and water content (Colbeck *et al.* 1990).

The densification of dry snow, sometimes referred to as 'metamorphosis', resembles the sintering of ceramics and involves many processes. In the early stages touching crystals become bonded together by the formation of necks, crystals become more spherical, and larger grains grow at the expense of smaller ones. As pointed out by Colbeck (1998) the interface between adjacent grains is a grain boundary, and as the neck forms the initial sharp angle between the ice–vapour surfaces should increase towards its equilibrium value of 145° (see §7.7.1). An example of such a neck in well-annealed snow is shown in Fig. 12.6. The thermodynamic driving force is the reduction of the combined ice–vapour and grain-boundary surface energy, and mass transfer may occur by diffusion in the surface layer, through the vapour, or along the newly formed boundary. Later stages of densification require an external pressure, which, in the case of glacier and polar snow, results just from the weight of the snow cover itself. These processes have been reviewed by Maeno and Ebinuma (1983), McClung and Schaerer (1993), and Colbeck (1997, 1998).

The mechanical properties of snow are particularly important in such different fields as construction projects and avalanche prediction, but understandably they are strongly dependent on the state of the snow and on the temperature. These properties have been reviewed by Mellor (1975), Salm (1982), and Shapiro *et al.* (1997). Snow is a very unusual material because of its loosely compacted structure. As long as the stress is kept below some limit it can be treated like a normal solid, but if the compressive stress exceeds this limit the snow undergoes a large irreversible change in volume to a more compact state with completely different properties; this compaction can occur more than once. Below the limit a Young's modulus E can be determined, and it increases with the density ρ roughly as $E \propto \exp(a\rho)$, where a is a constant, from $E \approx 10^6$ N m^{-2} at $\rho = 200$ kg m^{-3} to $E \approx 10^9$ N m^{-2} at $\rho = 500$ kg m^{-3}. For comparison, in solid ice $E \approx 10^{10}$ N m^{-2} and $\rho = 917$ kg m^{-3}. The low-strain-rate rheology of snow can be described approximately by viscosities defined for steady-state creep in various modes of deformation. Like Young's modulus, the failure stress under uniaxial load varies with density, from approximately 6×10^3 N m^{-2} at $\rho = 200$ kg m^{-3} to 10^6 N m^{-2} at $\rho = 500$ kg m^{-3} in compression. In tension the failure stress is about a third

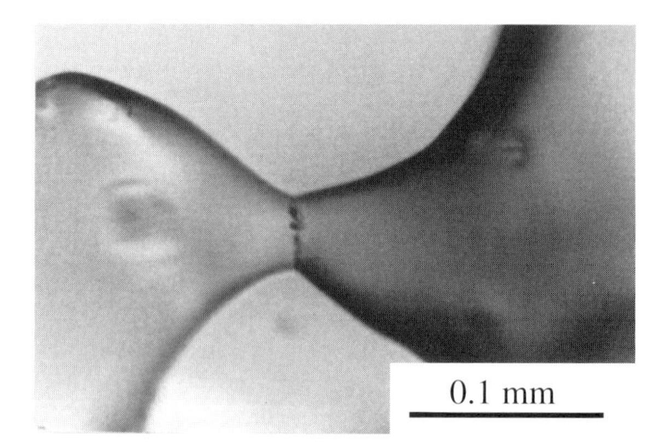

Fig. 12.6 The neck formed between two snow grains after annealing for several years at $-31\,^{\circ}$C. Note that the angle between the ice–air surfaces meeting at the grain boundary is approximately 145°. (From Colbeck 1997; 1998.)

of that in compression. Voitkovsky *et al.* (1975) have shown that the cohesive strength of different samples of snow is very poorly correlated with the density, but that there is a much better correlation with the total surface area of grain contacts per unit volume of the snow.

The thermal conductivity of snow is almost independent of temperature but varies with density from 0.02–0.07 $\mathrm{W\,m^{-1}\,K^{-1}}$ at $\rho = 100\,\mathrm{kg\,m^{-3}}$ to 0.35–0.7 $\mathrm{W\,m^{-1}\,K^{-1}}$ at $\rho = 400\,\mathrm{kg\,m^{-3}}$ (Mellor 1975; Langham 1981). The dielectric properties of snow have been discussed by Glen and Paren (1975), who show that the permittivity ε_∞ measured at microwave frequencies varies with the volume fraction of ice v according to the equation predicted theoretically for a dielectric of permittivity ε_1 containing empty spherical holes:

$$\varepsilon_\infty^{1/3} - 1 = v\left(\varepsilon_1^{1/3} - 1\right). \tag{12.1}$$

This equation was found experimentally to be valid for densities as low as $300\,\mathrm{kg\,m^{-3}}$. At low frequencies snow shows a modified Debye dispersion and steady-state conductivity.

The most important optical property of snow is the albedo, which, within a specified waveband, is the ratio of the total reflected or re-radiated energy flux to that incident on the surface. It is particularly relevant to the energy budget of a snow-covered ice sheet, which receives solar radiation by day and continuously emits thermal radiation characteristic of its own temperature. As always a high albedo or reflectivity implies a low emissivity and vice versa. These optical properties are reviewed by Warren (1982), and depend on the optical properties of ice described in Chapter 9. Over the visible spectrum ice is highly transparent so that light falling on snow is multiply reflected and refracted many times amongst the grains of the snow with very little absorption. If the snow is deep enough the result is that most of the light is diffusely scattered back out of the surface and this yields a high albedo. For wavelengths around 500 nm the albedo of clean Antarctic snow is about 0.98, but absorption begins in the near infrared resulting in a mean albedo of clean snow over the incident solar spectrum of about 0.80–0.85 (Grenfell *et al.* 1994). Laboratory measurements giving somewhat lower values on different types of snow are described and modelled by Sergent *et al.* (1993). Further into the infrared the absorption coefficient of ice becomes very high, and at these wavelengths snow is one of the blackest materials known! The emissivity for thermal radiation from snow is about 0.99 (Salisbury *et al.* 1994).

12.5 Glacier and polar ice

Glacier ice is formed from snow by the long and complex processes of consolidation and sintering under the pressure of accumulated layers. Such ice is formed on mountains, from which it flows down into valleys to melt away at the glacier snout. It also forms the Antarctic and Greenland ice sheets, which in places flow out over the sea and break off to form icebergs; these two ice sheets contain 99% of the ice and 80% of the freshwater on the Earth. Beautifully illustrated accounts of glaciers and ice sheets are given by Sharp (1988) and Hambrey and Alean (1992); more comprehensive treatments are those of Lliboutry (1965) and Paterson (1994).

The behaviour of a glacier depends very much on its temperature. Temperate glaciers are ones in which the ice is in equilibrium with meltwater. This water can flow through channels in and under the ice, and within about 10 m from the surface there may be melting in summer followed by freezing in winter. Such melting and refreezing plays an important role in consolidating the snow into ice. In contrast, the temperature in the higher regions of the Antarctic and Greenland ice sheets is always well below the melting point, and in these regions the formation of solid ice is much slower. The partially consolidated snow is 'firn' as defined in §12.4. There are continuous air channels through it, but eventually these channels are closed off, leaving isolated air bubbles. The transition from firn to ice containing bubbles occurs at a density of about 830 kg m^{-3}, compared with 917 kg m^{-3} (at 0 °C) for ordinary bubble-free ice. At a depth of about 1000 m the bubbles disappear due to the formation of a gas clathrate (§11.10), but when such ice is brought back to atmospheric pressure the dissolved air is released (Gow and Williamson 1975; Shoji and Langway 1982). The time scale for the conversion of firn to ice varies between a few years in temperate glaciers to hundreds of years at places in the Antarctic. In the ice core from the Vostok station in Antarctica, which is an extreme case, the firn–ice transition occurred at a depth of 95 m, where the temperature was −57 °C and the age of the ice is estimated to be 2500 years (Paterson 1994).

The seasonal variations of temperature only penetrate a glacier to a depth of about 10 m. Thereafter there is generally a rise in temperature with depth, which may or may not reach the melting point at the base. The grain size of ice formed from the firn is initially very small, but grain growth and recrystallization occur, gradually resulting in grains several millimetres across. Growth rates depend both on temperature and impurity content (Alley and Woods 1996). In exceptional cases very large crystals are formed, as in the Mendenhall Glacier, which has been used as a source of large ice crystals for many laboratory experiments. Where the ice is flowing, dynamic recrystallization (§8.3.4) results in smaller grains with a preferred orientation. Rigsby (1968) has shown that the grains in actively deforming ice have complicated interlocking shapes, such that a single grain may appear as several separate regions in a cross-section through the ice.

In a steady state there is equilibrium between the annual accumulation of snow on the top of the glacier and the loss of ice by ablation around the perimeter. In reality most glaciers are either advancing or retreating due to an imbalance in these terms, which fluctuate due to changes in global and local climate. It is fundamental to the mass balance of a glacier that there must be flow of ice from the accumulation to the ablation regions. A wide range of velocities is encountered, from almost zero on ice domes to several kilometres per year at the margins. An exceptional case is the surge of a glacier; for example, Variegated Glacier in Alaska advanced by about 1 km during the month of June 1983 (Kamb *et al.* 1985). There are two components to the motion of a glacier: the sliding of ice over the bedrock and the plastic flow of the ice mass itself.

The basic mechanism for the sliding of ice close to its melting point over a rigid but uneven bed was analysed by Weertman (1957). The ice can itself flow plastically around large irregularities, and it can overcome small ones by regelation (i.e. melting under pressure on one side of the obstacle and refreezing on the other side). The greatest resistance to flow arises from intermediate size irregularities, usually taken to be on the scale of 0.01–0.1 m. These processes are modified when there is meltwater present or if the bed is covered with loose deformable rock debris known as 'till' (Alley *et al.* 1986). The plastic flow of the ice itself is governed by the principles described in Chapter 8, but on time scales and for accumulated strains very much larger than those achieved in laboratory experiments (Budd and Jacka 1989). In steady-state tertiary creep under the shear stresses in a glacier, dynamic recrystallization leads to a fabric of grain orientations favourable to the continuation of shear in the same orientation. The modelling of glacier flow is based on the work of Nye (1952, 1965) and is described in detail by Hutter (1983). The polycrystalline fabric is now recognized as being very important for such modelling. The flow law $\dot{\varepsilon} \propto \sigma^n$ (eqn (8.6)) is assumed to apply, usually with $n=3$. With this stress dependence, the flow profile differs from that for the streamline flow of a viscous liquid in that the velocity gradients are larger toward the fixed boundaries. In these regions of high shear stress the ice is in tension at 45° to the direction of flow and crevasses are formed by fracture perpendicular to the tensile stress. Crevasses are also formed across the glacier where the ice flows over a region of increasing gradient. The fracture mechanics of Antarctic ice is described by Rist *et al.* (1996).

12.5.1 *Ice cores*

The Greenland and Antarctic ice sheets reach thicknesses of over 3 km and flow relatively slowly, with the result that the ice in the lower regions can be 100–500 ka (i.e. thousand years) old, extending throughout the last four glaciations of the Pleistocene epoch (Petit *et al.* 1997). This ice has stored within it much information about global conditions at the time it was formed, and great importance has been attached to drilling projects to recover and study continuous cores of ice from these regions. This work is described by Paterson (1994) and Souchez and Lorrain (1991). There are three main forms of information: isotopic composition, gas content, and trace impurities.

When water evaporates the vapour contains about 1% less ^{18}O and 10% less deuterium than the liquid. Further and larger isotopic separation occurs in the formation of clouds and precipitation, so that the snow falling in the polar regions is significantly depleted of these heavier isotopes. It has been shown that the variations in isotope concentrations with depth at a given site reflect variations in the local temperature at the time the snow accumulated (Johnsen *et al.* 1989). In the case of ^{18}O, observations are expressed in terms of the parameter $\delta^{18}O$ which is the deviation of the concentration from that of 'standard mean ocean water' expressed in parts per thousand (‰); it is a negative quantity in polar ice. The variations in $\delta^{18}O$ show both an annual periodicity, which can be used in dating, and long-term changes. Figure 12.7 shows data from a 3028 m deep ice core obtained in the Greenland Ice Core Project (GRIP). This shows a large change in $\delta^{18}O$ as a result of the global warming (possibly up to 20 °C in Greenland) at the end of the last glacial period between 10 and 18 ka ago. Because the oceans show an opposite variation in isotopic composition to the ice caps, similar patterns can be identified in sediments on the ocean floor, though the effects are not closely correlated.

The amount of air in the ice is related to the conditions at the time the air in the firn became trapped as bubbles in the ice. The concentration of CO_2 and CH_4 within this air is measurable and shows, for example, that the CO_2 concentration in the Earth's atmosphere rose from about 200 to 280 p.p.m. by volume at the end of the last glaciation (Raynaud *et al.* 1993). It also shows that the rise in concentration of these gases in the past few decades has taken them significantly higher than at any time in the past 100 000 years.

Trace impurities can be analysed chemically, and Fig. 12.8 shows such measurements on a short section of the core obtained in the Greenland Ice Sheet Project 2 (GISP2). The data show seasonal variations, with a particularly clear annual cycle in nitrate concentration in this section of the core. There is a strong peak in the sulphuric acid content at a depth of 101.3 m, which is correlated with the presence of volcanic ash; such events are very good time markers in the cores.

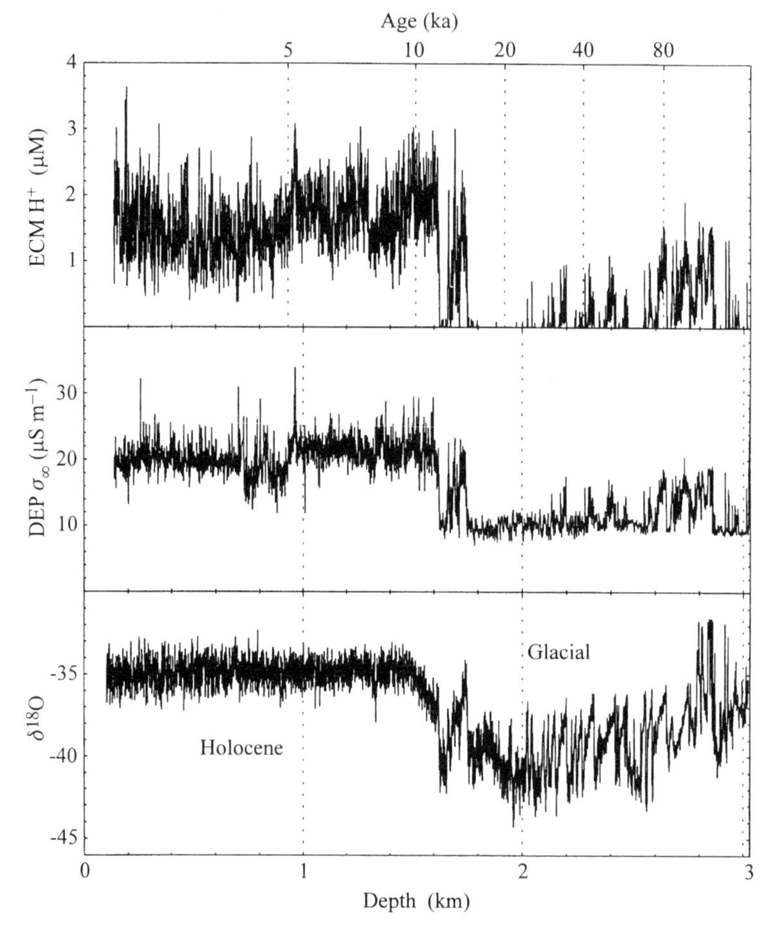

Fig. 12.7 Profiles of the electrical conductivity (ECM, expressed as equivalent acid concentration), DEP (high-frequency conductivity), and deviation of ^{18}O concentration (in ‰) for 3000 m length of the GRIP Greenland ice core. Reproduced from Wolff *et al.* *Journal of Geophysical Research*, **100**, 16249–63, 1995, Copyright by the American Geophysical Union; the ^{18}O data is that of Dansgaard *et al.* (1993).

At 103.9 m there is a peak in NH_4^+ with a corresponding loss of acidity, and this has been attributed to extensive biomass burning in North America at that time. Such features are common in Greenland, but Antarctic ice is always acidic.

From the ice physics point of view one of the most interesting aspects is the effect of these impurities on the electrical properties of the ice. One of the simplest measurements on an ice core is the 'electrical conductivity measurement' (ECM) devised by Hammer (1980) and referred to in §5.2.2. A p.d. of 1.25 kV is applied to a pair of electrodes which are dragged along a flat surface of the ice core.

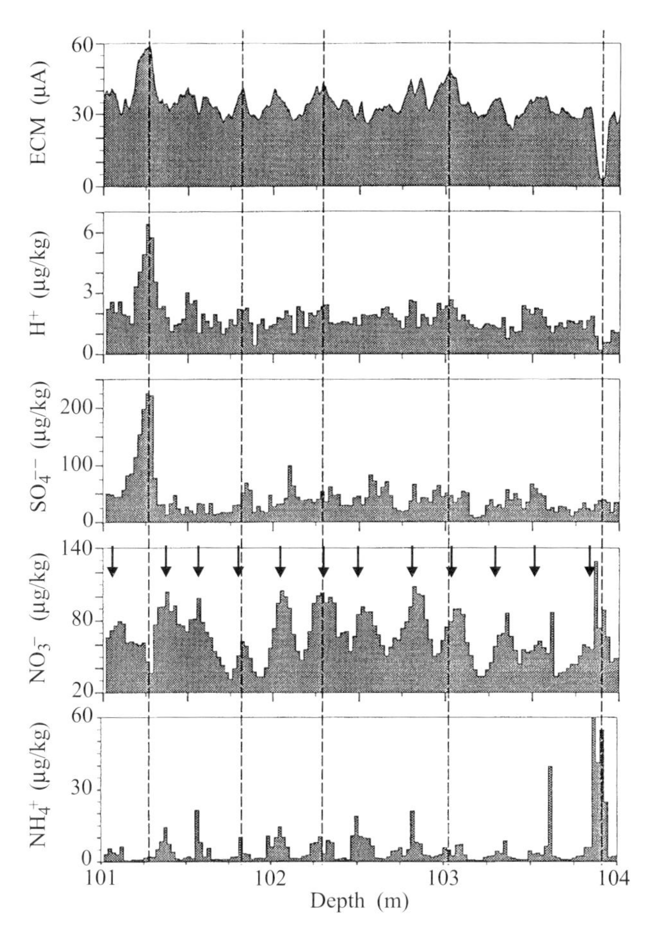

Fig. 12.8 ECM profile (representing steady-state conductivity) together with chemical analysis for acidity (H^+), SO_4^{--}, NO_3^-, and NH_4^+ in a 3 m length of the GISP2 Greenland ice core. The arrows indicate the annual cycle based on this and other evidence; the broken vertical lines are to facilitate comparison (Taylor *et al.* 1992). Reproduced by courtesy of the International Glaciological Society from *Journal of Glaciology*, **38**(130), 1992, 325–32, Fig. 3.

The resulting current yields a profile dependent on the steady-state conductivity σ_s with a resolution of a few millimetres. Figure 12.8 includes an example of this profile, revealing strong correlation with the chemical impurities, and a clear annual periodicity. By far the dominant term in the ECM profile is attributable to the acidity of the ice.

Alternating current measurements on an ice core are made by clamping electrodes around it while it is kept in its protective polythene sleeve as described by

Moore and Paren (1987), Moore *et al.* (1992), and Wilhelms *et al.* (1998). The ice shows the normal Debye dispersion, from which a profile dependent on the high-frequency conductivity σ_∞ can be derived. This method is known as 'dielectric profiling' (DEP), and data for the GRIP core are shown together with the ECM and $\delta^{18}O$ data in Fig. 12.7. Correlation with observations of chemical composition shows that there are contributions from hydrogen, ammonium, and chloride ions that can be represented at $-15\,^\circ C$ by the empirical equation

$$\sigma_\infty = 9 + 4[H^+] + 1[NH_4^+] + 0.55[Cl^-], \tag{12.2}$$

where σ_∞ is in $\mu S\ m^{-1}$ and the concentrations [...] are in μM (Wolff *et al.* 1995); different coefficients apply to different ice cores measured under different conditions (e.g. Moore *et al.* 1992). The interpretation of these measurements in terms of the electrical properties of ice is reviewed by Wolff *et al.* (1997) and considered in §5.4.4.

The data in Fig. 12.7 span a very long period of time. The data points are averaged over several years and short-term fluctuations look like noise, but the longer-term changes in global conditions are very clear, and plotting the data at higher resolution shows strong correlations between features seen by different techniques. During periods of extensive glaciation not only was the temperature lower but in Greenland there was much deposition of dust which neutralized the atmospheric acidity. Provided that the core is taken from a region which has not been disrupted by plastic flow, the time resolution of the historical record stored in it is ultimately limited by the diffusion rates of impurities in the ice, and the existence of the record is a demonstration of how slow such diffusion processes are. A simple calculation illustrates the point using the molecular self-diffusion measurements on single crystals described in §6.4.1. At $-20\,^\circ C$ $D_s \approx 7 \times 10^{-16}\,m^2\,s^{-1}$, and in a time period $t = 1000$ years the diffusion length $\sqrt{(4D_s t)} \approx 10$ mm.

12.5.2 *Radio echo sounding*

It was shown in §9.2 that for frequencies well above the Debye dispersion the attenuation distance α^{-1} for electromagnetic waves in ice is independent of frequency up to the onset of the infrared absorption band at about 10 GHz. This window in the radio/microwave region makes it possible to use this waveband for radio echo sounding of ice sheets, and extensive surveys have been carried out since the 1950s using both ground and airborne instruments producing pulsed signals at carrier frequencies of 30–500 MHz. These surveys can determine not only the thickness of the ice and snow cover, but also the presence of reflecting layers of volcanic dust, acidity, or anything which changes ε' or ε''. The resolution is typically tens of metres, but higher resolution can be obtained with smaller penetration using radar waves up to 30 GHz. This work is reviewed by Bogorodsky *et al.* (1985) and there have been many subsequent technical developments.

The penetration distance α^{-1} over which the intensity falls by a factor of e depends on the high-frequency conductivity σ_∞ according to equation (9.15).

At a temperature of $-10\,^{\circ}\mathrm{C}$ laboratory measurements on pure ice give $\sigma_{\infty} = 16\,\mu\mathrm{S\,m^{-1}}$ (Fig. 5.5), from which $\alpha^{-1} = 300\,\mathrm{m}$ corresponding to an attenuation of $0.014\,\mathrm{dB\,m^{-1}}$. Actual attenuations measured on Antarctic ice at this temperature are somewhat larger at about $0.042\,\mathrm{dB\,m^{-1}}$, but it has still been possible to make soundings to the base of the Antarctic ice sheet at a depth of over 4 km.

12.6 Frozen ground

Very large quantities of terrestrial ice are hidden in the form of frozen ground. Ground which remains frozen from one year to the next is referred to as 'permafrost', and it underlies some quarter of the Earth's land area extending to depths of hundreds of metres; in Siberia depths of over 1400 m have been recorded. Figure 12.9 shows the distribution of various kinds of permafrost in the northern hemisphere. Most of this permafrost has been in place for thousands or tens of thousands of years, and the time scale for any change involving the transport of latent heat to such depths is very long. In those places on Earth where the surface temperature undergoes an annual cycle passing both above and below $0\,^{\circ}\mathrm{C}$, the temperature oscillation can only penetrate to a depth of 2–3 m, and this is the maximum depth of the 'active layer' which freezes and thaws annually. In such regions, if the annual mean temperature at the surface is below $0\,^{\circ}\mathrm{C}$, the active layer will extend over a region of permafrost, but if the mean temperature is above $0\,^{\circ}\mathrm{C}$, the ground below the active layer will never become frozen.

Most of the early research on permafrost was carried out in Russia, and this work is described by Muller (1947). A general survey of the subject has been written by Mackay (1972), who includes the classification of frozen ground originally devised in Russia by Shumskii. The most useful modern account is the book by Williams and Smith (1989), and the physical and thermal properties are also described by D.M. Anderson *et al.* (1978).

The ground of interest to us is porous, with water in the interstices between the solid mineral particles, and the properties of permafrost depend on the nature of these particles. However, frozen ground is not produced simply by the solidification of the water already present, with the consequent increase of a few per cent in volume. In the process of freezing, additional water is often drawn into the ground, and it freezes both in the interstices between the soil particles and as horizontal layers of solid ice. These layers are referred to as 'ice lenses'. Their formation is the principal cause of frost heaving, in which the volume of the frozen ground can be as much as doubled, and serious structural damage can be caused to roadways and buildings. Diverse other phenomena are observed, of which some of the most remarkable are 'pingos' or hillocks some tens of metres high that consist of solid ice formed just below the surface. When frozen ground melts and the water drains away, the ground level sinks and the soil becomes very soft. A variety of strange surface features can result from this melting. When

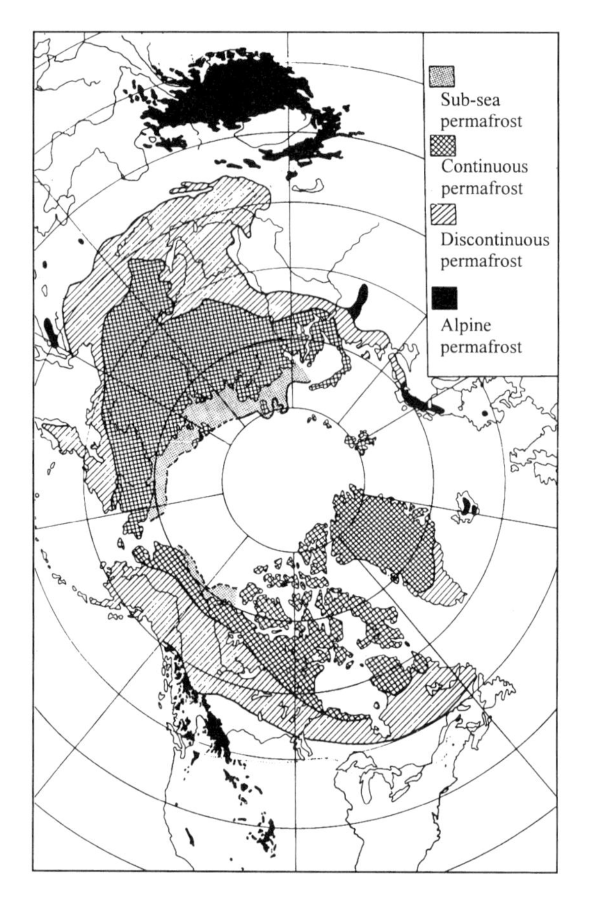

Fig. 12.9 Distribution of various kinds of permafrost in the Northern hemisphere. (From Péwé 1983.)

erecting a building above frozen ground, it is important to ensure that the presence of the building does not lead to thawing of the permafrost underneath it.

The freezing point of water in a soil is depressed below 0 °C, by a small amount due to dissolved salts, but more by the confinement of the water in narrow channels with a very high surface area. The effect falls into the category of the surface effects discussed in Chapter 10, and means that soils freeze over a range of a few degrees Celsius, with the liquid fraction decreasing with falling temperature. Nersesova (1950) for example found that in montmorillonite clay 16% of the water remained unfrozen at −10 °C, but in quartz sand there was no water below −4 °C. Where both ice and water are present in equilibrium the material has a very large apparent heat capacity that includes the latent heat of melting. This greatly decreases the thermal diffusivity and hence the penetration depth of seasonal variations in temperature. The physics of the melting and freezing processes are outside the scope of this book, but are the dominant considerations in understanding the characteristic properties of frozen ground. The growth of ice lenses resulting in frost heaving has been interpreted in thermodynamic terms, including the effect of the quasi-liquid layer at the interface between the ice and the soil particles (Gilpin 1980; Dash *et al.* 1995). However, other features which involve large volumes of solid ice, such as the pingo, are thought to arise from the gradual intrusion of bulk water from elsewhere which then freezes.

The mechanical properties of frozen ground have been reviewed by Sayles (1988) and Andersland (1989). Under rapid loading, deformation leading to fracture is determined largely by the properties of the ice that binds the material together, but under long-term loading the creep that would occur in pure ice is greatly impeded by the rigidity introduced by the skeleton of soil particles.

12.7 Ice in the Solar System

Astronomers use the word 'ice' to refer to the frozen forms not just of water but of volatile compounds such as ammonia, carbon monoxide, carbon dioxide, methane, and nitrogen. There are vast quantities of these ices in the low-temperature regions of the Solar System, and of these water ice is the most abundant. We focus attention here on the satellites of the outer planets and on the comets, although ice is also believed to be an important constituent of interstellar dust. The occurrence and properties of the Solar System ices are the subject of an important series of reviews edited by Schmitt *et al.* (1998*a*). An older but more general overview is that of Klinger (1983), and the topic is described in a wider astronomical context by Lewis (1995).

12.7.1 *Satellites of the outer planets*

The largest quantities of ice in the Solar System are to be found in the satellites, or moons, of the planets from Jupiter outwards. An excellent introduction to the observation and geophysical properties of these satellites is to be found in a book by Rothery (1992). The principal satellites are listed in Table 12.1, together with similar data on the Earth and the Moon for comparison. Approximate masses and radii have been known for some time from Earth-based astronomical observations, but much better values of these quantities and detailed observations of the surface terrain, composition, and temperature were obtained between 1979 and 1989 by the Voyager 1 and Voyager 2 spacecraft. Observations of the moons of Jupiter have been made with much higher resolution from the Galileo spacecraft which went into orbit around that planet in 1995. The most significant parameter for our purposes is the mean density, derived simply from the ratio of mass to volume. The mean density of the Moon is 3.34 Mg m^{-3}, and the innermost of Jupiter's moons, Io, has a density similar this. The densities of all the other moons in Table 12.1 (including the planet Pluto itself) are appreciably

Table 12.1 The principal moons of the outer planets

Planet and moon	Orbital radius (10^3 km)	Radius (km)	Mean density (Mg m^{-3})	Surface	Character
Jupiter					
Io	422	1815	3.57	Rock and sulphur	Volcanic. No ice
Europa	671	1569	2.97	Smooth ice	Probably active
Ganymede	1070	2631	1.94	Ice and dirt	Recently active
Callisto	1883	2400	1.86	Ice and dirt	Dead
Saturn					
Mimas	186	197	1.17	Ice	Dead
Enceladus	238	251	1.24	Ice	Probably active
Tethys	295	524	1.26	Ice	Recently active
Dione	377	559	1.44	Ice	Recently active
Rhea	527	764	1.33	Ice	Dead
Titan	1222	2575	1.88	Opaque atmosphere	Unknown
Iapetus	3561	718	1.21	Ice and dirt	Dead
Uranus					
Miranda	130	236	1.3	Ice	Recently active
Ariel	191	579	1.7	Ice	Recently active
Umbriel	266	586	1.5	Ice	Dead
Titania	436	790	1.7	Ice	Recently active
Oberon	583	762	1.6	Ice	Dead
Neptune					
Triton	355	1350	2.07	N_2 and CO_2	Active (Retrograde orbit)
Pluto		1145–1200	1.92–2.06	Methane	
Charon	20	600–650	1.51–1.81	Ice	Unknown
Earth		6371	5.52	Rock, water, ice	Active
Moon	384	1738	3.34	Rock	Dead

Numerical data provided in advance from second edition (2000) of Rothery (1992).

smaller, and these bodies are believed to contain a large proportion of ice together with some silicate rock. The density of water ice varies from 0.9 Mg m^{-3} at zero pressure to 1.6 Mg m^{-3} for ice VII at 2.4 GPa, which is typical of the highest pressures attainable in the largest moons. Except for Titan, the moons of Saturn have mean densities from 1.17 to 1.44 Mg m^{-3} and clearly contain large proportions of ice. The other moons listed in Table 12.1 also contain ice, but not in such large proportions.

The evidence that the surfaces of the icy satellites of Jupiter and Saturn consist of water ice comes from measurements of the infrared reflectance spectra of these bodies (Cruikshank *et al.* 1998). Such measurements also showed that the rings of Saturn were made of ice, or at least covered in water frost (Pilcher *et al.* 1970). The optical properties of water and other ices at temperatures relevant to such studies have been reviewed by Schmitt *et al.* (1998*b*). Non-water ices are present in differing proportions in the more distant parts of the Solar System.

It is generally assumed that the icy satellites were formed by condensation and aggregation of material in the nebula that initially surrounded each planet. The more volatile materials would only condense in the cooler regions further from the Sun, so that no bodies permanently within the orbit of Jupiter are made of ice, and materials with a higher vapour pressure than water only appear in the more distant systems. The moons so formed would initially consist of homogeneous mixtures of ice and rock, and some of the smaller ones such as Mimas have probably remained so ever since. However, there is a tendency for the rock to sink within the ice matrix, and if the temperature becomes high enough for this to occur, the rearrangement of material forms a rocky core surrounded by ice. The necessary increase in temperature can be produced by tidal friction as the rotation of the satellite is brought into synchronism with its orbital motion (i.e. until it always presents the same face towards the planet). In some cases tidal heating continues beyond this stage merely as a consequence of the perturbations of the orbit of one satellite by the others. Radioactive heating may also contribute but to a lesser extent. For some of the moons the temperature in the interior will, at some point during the evolution, have been high enough for ice to melt, in which case the rock will easily sink to the centre, releasing yet more heat. A moon is said to be 'differentiated' if it has a rocky core surrounded by ice, and partial differentiation is also possible with, for example, a rocky core surrounded by a mantle of ice and an outer crust of an unseparated mixture of ice and rock. The temperature at the surface will be maintained at a value typical of the distance from the Sun, being around 100 K at Jupiter and 75 K at Saturn.

Following the work of Consolmagno and Lewis (1976) on the icy moons of Jupiter there has been intense interest in modelling the evolution of the internal structure and temperature distribution within the moons, all of which have their own particular features. Fresh observational evidence is continually emerging and has to be incorporated in the models. The modelling is fundamentally dependent on the properties of ice, or in some cases ice mixtures or clathrate hydrates, at the extreme temperatures and pressures present in these bodies.

The first consideration is that of pressure. For a spherical body of radius R and uniform density ρ in internal hydrostatic equilibrium, the pressure at radius r is

$$p(r) = \frac{2\pi G\rho^2}{3}(R^2 - r^2), \tag{12.3}$$

where G is the gravitational constant. For an undifferentiated satellite with the average properties of Callisto this gives $p(0) = 2.8$ GPa, which is well into the range of stability of ice VII or VIII. The icy portions of the larger satellites will therefore consist of shells of ice in different phases, and the transition from ice Ih to ice II occurs at depths of only $\sim$100 km. Assuming they are undifferentiated, Iapetus and Rhea with $R \approx 700$–750 km are just about large enough to form ice II at the centre. The existence of more than one phase can have significant geophysical effects. For example, if ice Ih, ice V, or ice VI are cooled they convert to ice II, with an associated change in volume that could fracture the crust of the moon leading to features apparent on the surface.

In the process of condensation from the planetary nebula the temperature would have been so low that water molecules would form amorphous ice. After aggregation of the amorphous particles and some heating this amorphous ice would convert to ice Ic with the release of considerable heat. This and further phase changes will have had a significant effect on the heat balance of the evolving moon.

The mechanical properties of ice are very important for modelling the differentiation, the possibility of internal convection, and the extrusion of ice from the interior to the surface in volcanic-type processes. Laboratory studies of the rheology of ice under pressure have been reviewed in this context by Durham *et al.* (1997, 1998). The steady-state creep of all phases for strains of more than a few percent is fitted to a flow law of the form of equation (8.7) with $n \approx 4$ or 5, and different values of the other parameters for each phase. Ice III flows most easily and ice II most slowly. For astronomical modelling ice is given a Newtonian viscosity η defined as $\sigma/3\dot{\varepsilon}$, where $\dot{\varepsilon}$ is the strain rate under a uniaxial stress σ. This is not really applicable for a material with $n \neq 1$, and extrapolations of laboratory results using equation (8.7) down to geophysical strain rates of the order of $10^{-11}\,\text{s}^{-1}$ are subject to serious uncertainty.

Observations of the individual moons show that each has its own particular characteristics, and these must result from very different patterns of evolution. The diversity is well illustrated by the four principal (Galilean) satellites of Jupiter on which we now have remarkably detailed information from the Galileo spacecraft. The innermost, Io, is similar in size and density to our Moon, but differs from it in being actively volcanic. It is the only one without ice.

The next is Europa, which appears particularly fascinating to anyone interested in ice. The image in Fig. 12.10 shows a smooth icy surface criss–crossed by dark bands, which represent ridges up to about 200 m high that were probably formed by fracture of the crust allowing the extrusion of warmer ice or even

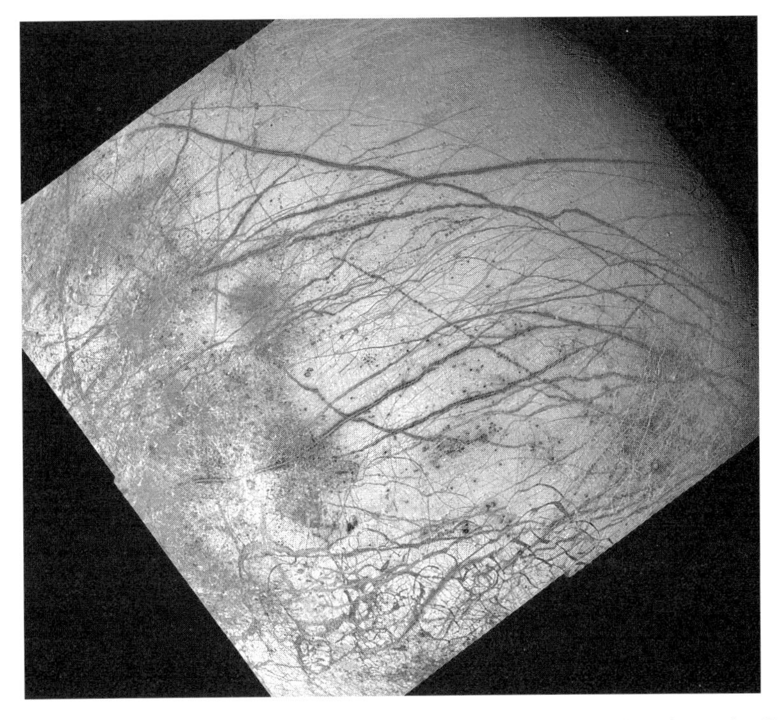

Fig. 12.10 Image of part of the Northern hemisphere of Europa taken from the Galileo spacecraft at a distance of 156 000 km. The Sun illuminates the surface from the left and the sunset terminator can be seen on the right. (From NASA National Space Science Data Center.)

water. Particularly near these bands there is spectroscopic evidence of mineral salts, perhaps from a brine-like interior. Figure 12.11 shows on a much larger scale an untypical but very interesting region in which the old surface, identified by its criss-cross markings, has been broken into pieces that are then moved about and the spaces between them filled in with a newer smoother material (Carr *et al.* 1998). The older pieces resemble icebergs standing 100–200 m high and the region superficially resembles the break-up of the Arctic ice cover in Spring. At the time this surface structure was formed, which could be 1–100 million years ago, it is argued that the icy crust of Europa was only 1–2 km thick and floating on an ocean of water. It may still be so today, but evidence of current geophysical activity on Europa has not yet been seen. The interior of Europa is strongly differentiated, and the dense core may even have a central region that is metallic, which could account for this moon's observed magnetic moment (J.D. Anderson *et al.* 1997*a*).

The third of Jupiter's moons, Ganymede, is also differentiated and it has a magnetic core (J.D. Anderson *et al.* 1996). The formation of metallic cores implies

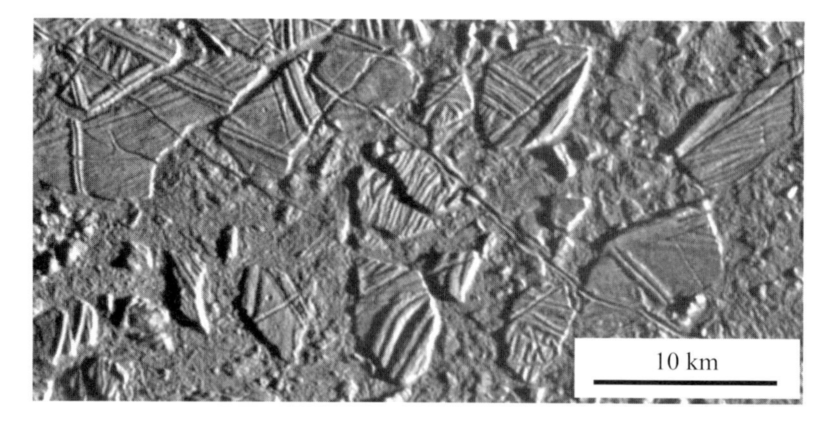

Fig. 12.11 High-resolution image of a portion of the surface of Europa taken from the Galileo spacecraft at a distance of 5340 km. The terrain consists of high-standing blocks on which the pre-existing surface texture is preserved with low-lying smoother material between them. (From NASA National Space Science Data Center; also published in Carr *et al.* 1998.)

that at some stage in the history of the moon the temperature at the centre was high enough to melt iron. The surface of Ganymede is grooved and cratered, showing evidence of past geological activity. Finally, Callisto has a heavily cratered surface of great age. It is a dead world with at most only some partial differentiation of the original mixture of ice and rock (J.D. Anderson *et al.* 1997*b*, 1998). These and the moons of the more distant planets are classified in Table 12.1 as 'dead', 'recently active', or 'active' worlds according to the scheme used by Rothery (1992), where activity refers to tectonic or volcanic action the results of which can be seen on the surface. For interpreting such activity the relevant properties of ice are more like those of a mineral (as studied by a geologist) than those of the terrestrial ice studied by glaciologists.

In addition to the moons listed in Table 12.1 many smaller icy bodies have been observed in orbit around the planets from Jupiter outwards. There are also thought to be a large number of others, with sizes varying from a few 100 km downwards, orbiting the Sun in what is known as the Kuiper belt beyond the orbit of Neptune.

12.7.2 *Comets*

Comets are icy bodies typically 1–50 km across with highly eccentric orbits in which they spend most of their time in very cold regions far from the Sun. They become visible during the comparatively short periods when they approach within a few astronomical units (i.e. Earth orbit radii) of the Sun. During this time they are heated by solar radiation, and the rise in temperature causes them to emit vapours and dust. These form the bright coma around the nucleus and are swept away by the solar wind to form the tail. Good introductory accounts of comets are those of Whipple (1985) and Hartmann (1993). The icy composition of the cometary nucleus has been reviewed by Delsemme (1983) and in more depth by Rickman (1998).

The model of a comet as an aggregate of frozen gases and stony meteoric material or dust (sometimes referred to as a 'dirty snowball') was proposed by Whipple (1950), who realized that the recoil arising from the emission of gases from the warmer faces of rotating cometary nuclei would be sufficient to account for observed changes in their orbits. Ground-based spectroscopic observations have revealed a wide range of chemical species in the coma, but close observations were first made in the encounters of the Vega and Giotto spacecraft with Comet Halley in 1986 (Sagdeev *et al.* 1986; Reinhard 1986 and companion papers). These encounters revealed an irregular shaped nucleus approximately $14 \times 7.5 \times 7.5\,km^3$ in size. The surface was dark in appearance and gas from the heated icy interior emerged as jets at an estimated rate of $40\,Mg\,s^{-1}$. H_2O was the dominant molecule in the coma, and is believed to be the predominant kind of ice in the nucleus. The other molecules observed in the coma or tail are thought to be present in the nucleus as hydrates, not as pure ices. The observations confirm that the nucleus is a rigid lump of material and not a loosely held aggregate that could easily fall apart.

At the temperature of its formation the ice in a comet must have been amorphous, and it is assumed to have had a composition typical of the dust in outer space. The nucleus is too small to become differentiated like the larger planetary moons, but the extent to which ice close to the surface or in the interior becomes consolidated or is transformed to ice Ic, ice Ih, or clathrate hydrates are questions yet to be resolved.

13
Adhesion and friction

Readers familiar with life in cold climates will know well that when water freezes on solid bodies, like car windscreens, power cables, aircraft, or the superstructure of ships, it is very difficult to remove except by melting. This strong adhesion of ice to other materials is a property of the ice–solid interface. There have been attempts to reduce it by the use of special coatings (e.g. Croutch and Hartley 1992), but the fundamental physics of the adhesion of ice is not yet well understood. In contrast, skating and skiing depend on the very small degree of sticking (i.e. low coefficient of friction) between ice and a solid body to which it has not been frozen. These two topics form the subject of this brief final chapter.

13.1 Experiments on adhesion

Some of the earliest studies of the properties of ice concerned the adhesion of ice to itself. Faraday (1850, 1860) observed that if two pieces of ice were brought into contact at 0 °C with negligible force they adhered to one another with the formation of a neck. This was attributed by him to the presence of a liquid-like layer on the surface, such as has been discussed in Chapter 10. More recent experiments over the temperature range from 0 to −80 °C are described by Hosler *et al.* (1957). The lowest temperature at which adhesion was measurable in a water-saturated atmosphere was −25 °C, whereas in a dry atmosphere the adhesion of ice to ice practically ceased at −3 °C. This difference could be accounted for by the rapid evaporation of the quasi-liquid layer, which therefore ceased to have its equilibrium thickness. Hobbs and Mason (1964) have, however, interpreted their quantitative measurements on the adhesion of very small particles of ice as supporting a model in which the neck grows by diffusion within the vapour phase. Adhesion is the first stage in the metamorphosis of snow, for which further mechanisms are mentioned in §12.4.

Experiments to measure the adhesion of ice to other materials can be of various kinds. A simple tensile test with the stress normal to the interface frequently results in fracture within the ice, known as a 'cohesive break'. An 'adhesive break' on the interface itself is often observed when a shear stress is applied in the plane of the interface, and this geometry is therefore more commonly used for studying the adhesion of ice.

The adhesion of ice to various materials was studied in a series of experiments by Jellinek (1959, 1962). In tensile tests he always observed cohesive breaks at ice–metal interfaces, but adhesive breaks were frequently observed for interfaces with

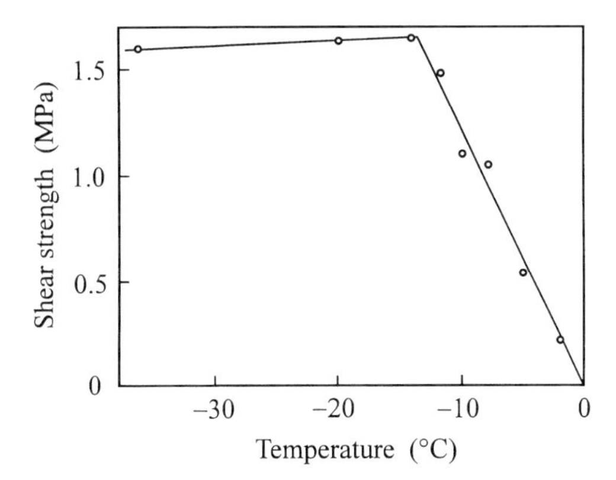

Fig. 13.1 Temperature dependence of the shear strength of the adhesion of stainless steel to snow-ice. Breaks were adhesive above −13 °C and cohesive below −13 °C. Reproduced from *Journal of Colloid Science*, **14**, (3), H.H.G. Jellinek, Adhesive properties of ice, pp. 268–80, Copyright 1959, by permission of the publisher Academic Press.

polystyrene or polymethylmethacrylate at stresses of 0.2–1.0 MPa. Results of shear tests between stainless steel and ice in the form of frozen snow are shown in Fig. 13.1. Above −13 °C adhesive breaks were observed, but the stress required became larger with falling temperature until at −13 °C it reached the cohesive strength of the ice, and below this temperature failure occurred within the ice. In a range of experiments he found that the adhesive strength depended on the rate of loading, the surface finish, and the type of material. Essentially similar observations were made by Raraty and Tabor (1958), who also observed that adhesion to PTFE was weaker than that to harder plastics. Jellinek (1967) interpreted his observations of the shear strength of ice–solid interfaces as due to sliding on a quasi-liquid layer in the interface, and Churaev *et al.* (1993) have described observations of such sliding at ice–quartz interfaces.

The strength of the adhesion between two materials is defined in terms of the 'work of adhesion' W_A, which is the free energy required to separate a boundary of unit area between the media. If γ_1, γ_2, and γ_{12} are the surface free energies per unit area of medium 1, medium 2, and the interface,

$$W_A = \gamma_1 + \gamma_2 - \gamma_{12}. \tag{13.1}$$

In the case of the liquid–solid interface shown in Fig. 10.11 we may write $\gamma_1 = \gamma_{sv}$, $\gamma_2 = \gamma_{lv}$, and $\gamma_{12} = \gamma_{sl}$, and the angle of contact θ is given by equation (10.9). The work of adhesion between the liquid and the solid is then

$$W_A = \gamma_{lv}(1 + \cos\theta), \tag{13.2}$$

and for water on different solids the values of W_A can be deduced from the angles of contact. Because the binding energies of H_2O molecules to different solids are

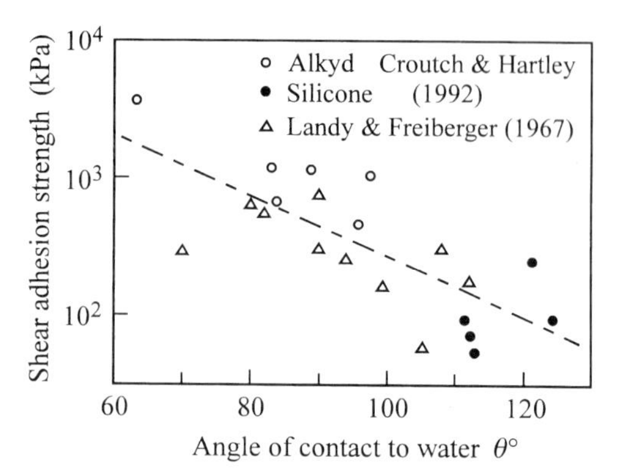

Fig. 13.2 Correlation between the shear strength of adhesion of ice to various plastics (logarithmic scale) and the angle of contact of the plastic with liquid water. The straight line is a best fit to the points.

expected to be similar in water and ice, it may be supposed that the values of W_A on different ice–solid interfaces will be correlated with the angles of contact of water on these solids. A hydrophilic solid is one which is easily wetted and has θ near to zero, while one with $\theta > \sim 90°$ is hydrophobic. Landy and Freiberger (1967), Bascom *et al.* (1969), and Croutch and Hartley (1992) made measurements of the shear strength of adhesion of ice on various plastics and coatings, and some of their results are collected in Fig. 13.2. There is a correlation with the angle of contact, with the adhesion to hydrophobic surfaces being weaker, but the correlation is not good. This is not surprising, because surfaces cannot be equivalently smooth on an atomic scale, the values of W_A may not be the same for water and ice, and the assumption that the force required for adhesive failure depends simply on W_A takes no account of the mechanism by which failure occurs. In a comparable situation the fracture strength of ice discussed in §8.4 depends on more than just the cohesive energy, represented in that case by K_{Ic}.

Wei *et al.* (1996) developed a technique for observing the delamination of an ice coating on a metal plate. The ice was deposited on the metal over a thin narrow plastic strip, a cut was made in the ice down to the strip, and the strip was pulled out to leave a pre-crack on the ice–metal interface. The plate was then deformed in bending so as to open the cut, causing the ice to peel off from the metal. Observation of the fracture surfaces showed an initial adhesive break which led subsequently to brittle fracture within the ice itself. From such experiments they obtained values of W_A for ice on steel, varying from 1 J m^{-2} for icing by water droplets freezing on the surface to 19 J m^{-2} for a single crystal frozen to the surface. These values of W_A are many times larger than the free energy of the two surfaces so formed, and this indicates that there must have been appreciable plastic deformation in the ice adjacent to the interface.

Adhesion on a quite different scale has been observed in the force required to withdraw an SFM tip (§10.6) after it had made contact with and become bonded to the ice surface. For an Si_3N_4 tip Petrenko (1997) deduced that $W_A = 80 \pm 12\,\text{mJ}\,\text{m}^{-2}$ at $-10\,°\text{C}$, and related experiments have been described by Slaughterbeck *et al.* (1996) and Pittenger *et al.* (1998).

It is difficult to deduce the work of adhesion for the interface between two solids, particularly because of the plastic deformation that is associated with fracture on the interface, but in the case of a solid and a liquid the work of adhesion can be determined reliably by conventional surface tension techniques. The interface between ice and mercury is therefore of special interest as an example of the adhesion between ice and a metal, and, by measuring contact angles or the rise of mercury in an ice capillary, Petrenko (1998) found $W_A = 400$, 360, and $290\,\text{mJ}\,\text{m}^{-2}$ at $-10\,°\text{C}$ for ice doped with NaCl, HF, and KOH respectively. The ice was doped to make it electrically conducting, and Petrenko then showed that the application of a p.d. of a few volts between the ice and the mercury changed the value of W_A, causing it to pass through a sharp minimum for a bias of $\sim$+1 V applied to the ice.

In a possibly related experiment referred to in §7.5 Petrenko and Schulson (1993) found that potential differences of $\sim$50 V applied across thin (10–200 μm) monocrystalline samples of ice frozen between stainless steel plates can completely halt the creep occurring in shear between the plates. Although interpreted at the time in terms of the plastic flow of the ice, this may be due to an effect of the p.d. on sliding at the interface.

Finally, from a practical point of view we should mention the importance of differential thermal expansion. The coefficient of expansion of ice (Fig. 3.2) is greater than that of metals and less than that of plastics. Ice which is frozen onto such materials will almost always break off if cooled so quickly or so far that the differential stress cannot be accommodated by plastic flow in the ice.

13.2 Physical mechanisms of adhesion

The physical processes of adhesion can be classified into three categories: covalent or chemical bonding, dispersion or Lifshitz–van der Waals forces, and direct electrostatic interactions (Israelachvili 1991).

Chemical bonding involves a chemical reaction between the molecules on opposite sides of the interface, and will be specific to the nature of the materials concerned. Chemisorption is usually defined as requiring bonding energies in excess of 0.5 eV, which for perfect contact will yield a work of adhesion greater than $\sim 0.5\,\text{J}\,\text{m}^{-2}$. These forces act only over distances of the order of 0.1–0.2 nm. For some materials water molecules are strongly adsorbed on the surface, but for others there is no affinity between water molecules and the surface.

In contrast, the Lifshitz–van der Waals forces are of longer range and act between all kinds of materials. The generalized theory of these forces has been formulated by Dzyaloshinskii *et al.* (1961). It has been applied to interfaces between ice and several metals and insulators by Wilen *et al.* (1995), who conclude that it is not the dominant source of adhesion (see also §10.9.4).

In addition to these forces, solids which contain non-compensated spatial distributions of charge will exert electrostatic forces on one another. The possible importance of these forces was recognized by Stoneham and Tasker (1985) and has been reviewed by Hays (1991). The presence of such space charge layers at the surface of ice is described in §§10.4.2–3, and Ryzhkin and Petrenko (1997) have shown that, depending on the conditions at the interface, electrostatic forces could contribute up to $500\,\mathrm{mJ\,m^{-2}}$ to the work of adhesion. The observation described in the previous section that the application of a small p.d. makes a large change in the work of adhesion of ice to mercury demonstrates the likely importance of the electrostatic contribution to adhesion processes in ice. The electrostatic component of W_A will probably be different for water and ice, and this may contribute to the poor correlation in Fig. 13.2.

While these concepts are relevant to the energetics of the adhesion process, the actual process of breaking ice off another material is much more complicated. It is important to realize that macroscopic experiments are never performed with atomically clean or flat interfaces, and that even what appears to be a clean adhesive break may well be leaving many water molecules on the non-ice side of the fracture.

13.3 Friction

In elementary mechanics the coefficient of friction μ between two bodies is defined as the ratio of the tangential friction force to the normal reaction across the interface, but in practice it is not a constant. For ice it depends on the normal reaction and the speed of sliding, as well as on the temperature. The property of most interest, because of its relevance in sports like skating, is the remarkably low value of the kinetic or sliding friction at speeds higher than $\sim 0.1\,\mathrm{m\,s^{-1}}$. This was studied in experiments designed to simulate skating by D.C.B. Evans *et al.* (1976), and their results for steel, copper, and perspex sliding on ice at a speed v are shown in Fig. 13.3. This figure shows a linear dependence of μ on $v^{-1/2}$, and similar observations have been reported for ice on ice (Oksanen and Keinonen 1982) and a range of other materials (Akkok *et al.* 1987). The temperature dependence of μ at a constant speed is shown in Fig. 13.4.

In these experiments the sliding interface is lubricated by a thin layer of water, and it was argued by Bowden and Hughes (1939) that this liquid was produced as a result of heating due to the work dissipated by the friction in the interface (see also Colbeck 1995). This heat raises the temperature at the interface, and is conducted away from the interface through the ice and through the slider. Only a small fraction of the energy is used to melt the ice, and the water formed refreezes as it is left behind the slider. The liquid film reaches an equilibrium thickness,

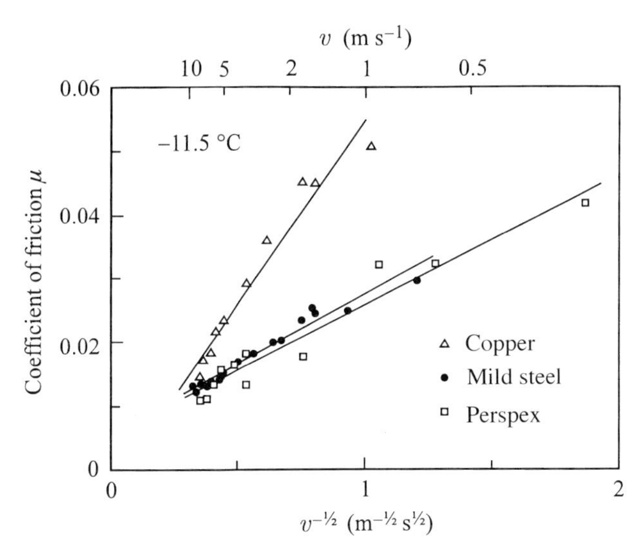

Fig. 13.3 The coefficient of friction of various materials on ice as a function of the speed v, plotted against $v^{-1/2}$ in accordance with a model of frictional melting. Figures 13.3 and 13.4 are reproduced from D.C.B. Evans, J.F. Nye, and K.J. Cheeseman, The kinetic friction of ice, *Proceedings of the Royal Society of London*, **A347**, 493–512, Figs 3 and 4 (1976).

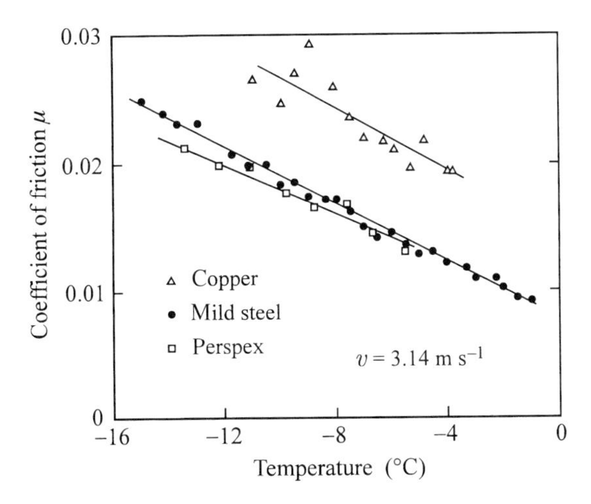

Fig. 13.4 Temperature dependence of the coefficient of friction μ of various sliders on ice at a fixed speed. (From Evans *et al.* 1976 as Fig. 13.3.)

because if it becomes too thin the friction and so the heat generated is increased, and likewise if it becomes too thick the friction is reduced. Experiments demonstrating that the temperature rises under a skate blade have been described by Colbeck *et al.* (1997). An alternative mechanism that has to be considered is that the melting point of the ice is depressed under the pressure at the interface. For this to be effective the melting point would have to be reduced below the ambient temperature, and the temperature at the interface would decrease to this lower value by the absorption of latent heat. This mechanism can be rejected, both because experiments show that the temperature rises, and because the depression of the melting point under pressure (Table 2.2) is far too small. The pressure required to melt ice at $-10\,°C$ is more than 100 MPa, which is many times the yield point of ice. Such pressures could not be sustained at points of local contact, and are well above the mean pressure possible under the blade of an ice skate.

Evans *et al.* (1976) and Oksanen and Keinonen (1982) developed semi-quantitative models for the frictional melting process which predict the dependence as $v^{-1/2}$ shown in Fig. 13.3. In the case of the copper slider shown in that figure heat is lost through the copper, causing an increase in friction as compared with steel and perspex for which heat is conducted away almost entirely through the ice. For a given speed μ rises with falling temperature (Fig. 13.4), because more work has to be done to raise the interface temperature to the melting point. The value of μ decreases with increasing load L roughly as $L^{-1/3}$. In these models the mere production of a lubricant does not explain the friction. The dissipation of energy arises from viscosity in the lubricant combined with the usual processes of shearing at points of adhesion and the ploughing of asperities into the softer material (Tabor 1987; Hutchings 1992). These processes result in abrasion of the slider, which is particularly marked on a soft material like perspex.

It is clear from the nature of the model that no melting will occur if the sliding speed is too small, and the value of μ should then be much larger. In the low-speed limit sliding involves breaking of whatever adhesion has been established between the slider and the ice. P. Barnes *et al.* (1971) performed experiments for speeds between 10^{-1} and $10^{-8}\,\mathrm{m\,s^{-1}}$. For sliding of ice on glass and granite μ increased with falling speed to high values (~ 0.3 and ~ 0.9 respectively) and finally returned to smaller values at the lowest speeds. At these very low speeds the 'sliding' was shown to be due to plastic creep in the ice. High values of μ were not observed by Barnes *et al.* for the sliding of ice on steel, but D.E. Jones *et al.* (1991) observed a large increase in μ for the sliding of ice on ice as the speed fell below about $10^{-4}\,\mathrm{m\,s^{-1}}$.

Petrenko and Colbeck (1995) demonstrated the separation of electrostatic charge in the frictional rubbing of materials on ice, and Petrenko (1994*b*) showed that the coefficient of friction at $-30\,°C$ and $2\,\mathrm{m\,s^{-1}}$ is increased by $\sim 50\%$ on applying a potential difference of 3 kV between the ice and the slider. This effect may be a consequence of the additional electrostatic pressure between the materials, but it could equally well be related to the effect of the potential difference on adhesion as described in §13.1.

The processes involved in friction on snow are the same as those on ice, with the added complications that fresh snow is compacted under the slider as it moves, and that snow only makes contact at more localized points. The grains that run against the slider become polished and show evidence of surface melting and subsequent refreezing. This topic and its application to the design of skis has been reviewed by Colbeck (1992, 1994).

The low friction of materials on ice is not always a benefit. In the design of automobile tyres the opposite is required (Ahagon *et al.* 1988). Other fields where the friction of ice is important include the sliding of ice floes over or against one another, the design of ice breakers, the brittle shear fracture of ice under compression (Rist 1997), and the sliding of a mode II crack to form wing cracks (§8.4.2).

Bibliography

1. General books and reviews

The following references, listed in order of publication, provide general information about the physics of ice. They are not specifically referred to in the text of the book.

Dorsey, N.E. (1940). *Properties of ordinary water-substance*, Part IIC. *Ice*, pp. 395–510. Reinhold Publishing Corporation, New York. A comprehensive compilation of the data available at that time.

Shumskii, P.A. (1964). *Principles of structural glaciology* (trans. D. Kraus), 497 pages. Dover, New York. A broad survey of the properties of ice in a glaciological context.

Pounder, E.R. (1965). *The physics of ice*, 151 pages. Pergamon Press, Oxford. Short survey with particular emphasis on floating ice.

Fletcher, N.H. (1970). *The chemical physics of ice*, 271 pages. Cambridge University Press, Cambridge. A clear exposition of selected fundamental topics.

Hobbs, P.V. (1974). *Ice physics*, 837 pages. Clarendon Press, Oxford. The most comprehensive work available, very useful as a source of references up to 1973. Includes chapters on ice formation and ice in the atmosphere, which are not treated in detail in the present book.

Glen, J.W. (1974). *The physics of ice*, 79 pages. US Army Corps of Engineers Cold Regions Research and Engineering Laboratory Monograph II–C2a. Together with the following report this is a broad survey of basic physical properties.

Glen, J.W. (1975). *The mechanics of ice*, 43 pages. US Army Corps of Engineers Cold Regions Research and Engineering Laboratory Monograph II–C2b.

Maeno, N. (1980). *Kori no kagaku* (*Physics of ice*), 222 pages (in Japanese). Hokkaido University Press, Sapporo. Translation into Russian by A.I. Leonov, edited by V.F. Petrenko (1988), Nauka, Moscow.

Lock, G.S.H. (1990). *The growth and decay of ice*, 434 pages. Cambridge University Press, Cambridge. A serious treatment of many aspects complementary to the present book.

2. The International Symposia on the Physics and Chemistry of Ice

For fuller information see Glen (1997). Many papers from these proceedings are included in the list of references.

1962 Erlenbach, Switzerland. No published proceedings.
1966 Sapporo, Japan. (Not always included as one of the series.)
Physics of snow and ice, Vol. 1, Parts 1 and 2. (ed. H. Ôura) 1967. The Institute of Low Temperature Science, Hokkaido University, Sapporo.
1968 Munich, Germany.
Physics of ice (ed. N. Riehl, B. Bullemer, and H. Engelhardt) 1969. Plenum Press, New York.
1972 Ottawa, Canada.
Physics and chemistry of ice (ed. E. Whalley, S.J. Jones, and L.W. Gold) 1973. Royal Society of Canada, Ottawa.
1977 Cambridge, England.
Journal of Glaciology, **21**(85), 1–714 (1978).
1982 Rolla, Missouri, USA.
Journal of Physical Chemistry, **87**(21), 4015–340 (1983).
1986 Grenoble, France.
Journal de Physique, **48**, Colloque C1, supplement to no. 3 (1987).
1991 Sapporo, Japan.
Physics and chemistry of ice (ed. N. Maeno and T. Hondoh) 1992. Hokkaido University Press, Sapporo.
1996 Hanover, New Hampshire, USA.
Journal of Physical Chemistry, **101**(32), 6079–312 (1997).
2001 Projected next symposium in Canterbury, UK.

3. The CRREL Special Reports

The following series of 'Special Reports' by V.F. Petrenko were published by the US Army Corps of Engineers Cold Regions Research and Engineering Laboratory, Hanover, NH, USA between 1993 and 1996. They constitute the embryo from which this book was born (see Preface).

93-25 *Structure of ordinary ice Ih*, Part I: *Ideal structure of ice*.
94-4 *Structure of ordinary ice Ih*, Part II: *Defects in ice*, Vol. 1: *Point defects*. (Co-author R.W. Whitworth.)
94-12 *Structure of ordinary ice Ih*, Part II: *Defects in ice*, Vol. 2: *Dislocations and plane defects*. (Co-author R.W. Whitworth.)
93-20 *Electrical properties of ice*.
94-22 *The surface of ice*.
96-2 *Electromechanical phenomena in ice*.

These reports are obtainable from the National Technical Information Service, 5285 Port Royal Road, Springfield, VA 22161, USA.

References

The page number(s) on which each reference is cited are given in square brackets at the end of each reference.

Abbati, I., Braicovich, L., and De Michelis, B. (1979). Investigating ultraviolet photo-electron spectroscopy of ice. *Solid State Communications*, **29**, 511–14. [223]

Abe, H. and Kawada, S. (1991). The dependency of dielectric relaxation time on alkali metal ion (Li^+, Na^+, K^+, Rb^+) in alkali-hydroxide-doped ice. *Journal of Physics and Chemistry of Solids*, **52**, 617–21. [108]

Abebe, M. and Walrafen, G.E. (1979). Raman studies of ice VI using a diamond anvil cell. *Journal of Chemical Physics*, **71**, 4167–9. [284]

Abragam, A. (1961). *The principles of nuclear magnetism*. Clarendon Press, Oxford. [147]

Adams, D.J. (1984). Monte Carlo calculations for the ice-rules model, with and without Bjerrum defects. *Journal of Physics C: Solid State Physics*, **17**, 4063–70. [71]

Adamson, R.M. and Dempsey, J.P. (1998). Field-scale in-situ compliance of Arctic first-year sea ice. *Journal of Cold Regions Engineering*, **12**, 52–63. [292]

Addison, J.R. (1969). Electrical properties of saline ice. *Journal of Applied Physics*, **40**, 3105–14. [292]

Addison, J.R. (1970). Electrical relaxation in saline ice. *Journal of Applied Physics*, **41**, 54–63. [292]

Ahagon, A., Kobayashi, T., and Misawa, M. (1988). Friction on ice. *Rubber Chemistry and Technology*, **61**, 14–35. [321]

Ahmad, S. and Whitworth, R.W. (1988). Dislocation motion in ice: a study by synchrotron X-ray topography. *Philosophical Magazine*, **A57**, 749–66. [165,168,171]

Ahmad, S., Ohtomo, M., and Whitworth, R.W. (1986). Observation of a dislocation source in ice by synchrotron radiation topography. *Nature*, **319**, 659–60. [168]

Ahmad, S., Shearwood, C., and Whitworth, R.W. (1992). Dislocation multiplication mechanisms in ice. In *Physics and chemistry of ice* (ed. N. Maeno and T. Hondoh), pp. 492–6. Hokkaido University Press, Sapporo. [168–9]

Akkok, M., Ettles, C.M.McC., and Crabtree, S.J. (1987). Parameters affecting the kinetic friction of ice. *Transactions of the American Society of Mechanical Engineers: Journal of Tribology*, **109**, 552–61. [318]

Allen, L.C. (1975). A simple model of hydrogen bonding. *Journal of the American Chemical Society*, **97**, 6921–40. [7]

Alley, B.H., Blankenship, D.D., Bentley, C.R., and Rooney, S.T. (1986). Deformation of till beneath ice stream B, West Antarctica. *Nature*, **322**, 57–9. [300]

Alley, R.B. and Woods, G.A. (1996). Impurity influence on normal grain growth in the GISP2 ice core, Greenland. *Journal of Glaciology*, **42**, 255–60. [300]

Andersland, O.B. (1989). General report on mechanical properties of frozen soil. In *Ground freezing 88. Proceedings of the Fifth International Symposium on Ground Freezing, Nottingham, 26–28 July 1988*, Vol. 2 (ed. R.H. Jones and J.T. Holden), pp. 433–41. A.A. Balkema, Rotterdam. [307]

Anderson, D.M., Pusch, R., and Penner, E. (1978). Physical and thermal properties of frozen ground. In *Geotechnical engineering for cold regions* (ed. O.B. Andersland and D.M. Anderson), pp. 37–102. McGraw-Hill, New York. [305]

Anderson, J.D., Lau, E.L., Sjogren, W.L., Schubert, G., and Moore, W.B. (1996). Gravitational constraints on the internal structure of Ganymede. *Nature*, **384**, 541–3. [311]

Anderson, J.D., Lau, E.L., Sjogren, W.L., Schubert, G., and Moore, W.B. (1997*a*). Europa's differentiated internal structure: Inferences from two Galileo encounters. *Science*, **276**, 1236–9. [311]

Anderson, J.D., Lau, E.L., Sjogren, W.L., Schubert, G., and Moore, W.B. (1997*b*). Gravitational evidence for an undifferentiated Callisto. *Nature*, **387**, 264–6. [312]

Anderson, J.D., Schubert, G., Jacobson, R.A., Lau, E.L., Moore, W.B., and Sjogren, W.L. (1998). Distribution of rock, metals, and ices in Callisto. *Science*, **280**, 1573–6. [312]

Anderson, T.L. (1995). *Fracture mechanisms. Fundamentals and applications* (2nd edn), p. 629. CRC Press, Boca Raton. [207]

Andersson, O. and Suga, H. (1994). Thermal conductivity of the Ih and XI phases of ice. *Physical Review*, **B50**, 6583–8. [45, 262]

Andersson, P., Ross, R.G., and Bäckström, G. (1980). Thermal resistivity of ice Ih near the melting point. *Journal of Physics C: Solid State Physics*, **13**, L73–76. [43]

Antonchenko, V.Y., Davydov, A.S., and Zolotariuk, A.S. (1983). Solitons and proton motion in ice-like structures. *Physica Status Solidi* (*b*), **115**, 631–40. [145]

Aoki, K., Yamawaki, H., Sakashita, M., and Fujihisa, H. (1996). Infrared aborption study of the hydrogen-bond symmetrization in ice to 110 GPa. *Physical Review*, **B54**, 15673–7. [274]

Apekis, L. and Pissis, P. (1987). Study of the multiplicity of dielectric relaxation times in ice at low temperatures. *Journal de Physique*, **48**, Colloque C1, 127–33. [123]

Arcone, S.A., Gow, A.J., and McGrew, S. (1986). Structure and dielectric properties at 4.8 and 9.5 GHz of saline ice. *Journal of Geophysical Research* **91**(C12), 14281–303. [290, 292]

Arias, D., Levi, L., and Lubart, L. (1966). Electrical properties of ice doped with NH_3. *Transactions of the Faraday Society*, **62**, 1955–62. [108]

Arnold, G.P., Finch, E.D., Rabideau, S.W., and Wenzel, R.G. (1968). Neutron-diffraction study of ice polymorphs. III. Ice Ic. *Journal of Chemical Physics*, **49**, 4365–9. [277]

Ashby, M.F. and Hallam, S.D. (1986). The failure of brittle solids containing small cracks under compressive stress states. *Acta Metallurgica*, **34**, 497–510. [208–9]

Ashcroft, N.W. and Mermin, N.D. (1976). *Solid State Physics*. Holt, Rinehart and Winston, New York. [41]

Askebjer, P., Barwick, S.W., Bergström, L., Bouchta, A., Carius, S., Coulthard, A. *et al.* (1995). Optical properties of the South Pole ice at depths between 0.8 and 1 kilometer. *Science*, **267**, 1147–50. [220]

Askebjer, P., Barwick, S.W., Bergström, L., Bouchta, A., Carius, S., Dalberg, E. *et al.* (1997). Optical properties of deep ice at the South Pole: absorption. *Applied Optics*, **36**, 4168–80. [220–1]

Assur, A. (1958). Composition of sea ice and its tensile strength. In *Arctic sea ice*, pp. 106–38. US National Academy of Sciences—National Research Council. Publication no. 598. [291]

Ayrton, W.E. and Perry, J. (1877). Ice as an electrolyte. *Proceedings of the Physical Society*, **2**, 171–82 and 199–201. [91]

Baer, B.J., Brown, J.M., Zaug, J.M., Schiferl, D., and Chronister, E.L. (1998). Impulsive stimulated scattering in ice VI and VII. *Journal of Chemical Physics*, **108**, 4540–4. [286]

Baianu, C., Bowden, N., Lightowlers, D., and Mortimer, M. (1978). A new approach to the structure of concentrated aqueous electrolyte solutions using pulsed NMR methods. *Chemical Physics Letters*, **54**, 169–75. [32]

Baker, R.W. (1978). The influence of ice-crystal size on creep. *Journal of Glaciology*, **21**, 485–500. [201]

Barer, S.S., Kvlividze, V.I., Kurzaev, A.B., Sobolev, V.D., and Churaev, N.V. (1977). Thickness and viscosity of thin unfrozen layer between ice and quartz surfaces. *Doklady Akademii Nauk SSSR*, **235**, 601–3. [*Akademiia Nauk SSSR, Proceedings Physical Chemistry*, **232–236**, 701–4.] [240]

Barkema, G.T. and de Boer, J. (1993). Properties of a statistical model of ice at low temperatures. *Journal of Chemical Physics*, **99**, 2059–67. [261]

Barnaal, D.E. and Lowe, I.J. (1967). Experimental free-induction-decay shapes and theoretical second moments for hydrogen in hexagonal ice. *Journal of Chemical Physics*, **46**, 4800–9. [32]

Barnaal, D. and Slotfeldt-Ellingsen, D. (1983). Pulsed nuclear magnetic resonance studies of doped ice Ih. *Journal of Physical Chemistry*, **87**, 4321–5. [147]

Barnes, P., Tabor, D., and Walker, J.C.F. (1971). The friction and creep of polycrystalline ice. *Proceedings of the Royal Society of London*, **A324**, 127–55. [190, 198, 201, 205, 320]

Barnes, W.H. (1929). The crystal structure of ice between 0 °C and −183 °C. *Proceedings of the Royal Society of London*, **A125**, 670–93. [17]

Bascom, W.D., Cottington, R.L., and Singleterry, C.R. (1969). Ice adhesion to hydrophilic and hydrophobic surfaces. *Journal of Adhesion*, **1**, 246–63. [316]

Bass, R. (1958). Zur Theorie der mechanischen Relaxation des Eises. *Zeitschrift für Physik*, **153**, 16–37. [138]

Bazant, Z.P. and Li, Y.N. (1994). Penetration fracture of sea ice plate: analysis and size effect. *Journal of Engineering Mechanics*, **120**, 1304–21. [292]

Beaglehole, D. and Nason, D. (1980). Transition layer on the surface of ice. *Surface Science*, **96**, 357–63. [232]

Beaglehole, D. and Wilson, P. (1993). Thickness and anisotropy of the ice–water interface. *Journal of Physical Chemistry*, **97**, 11053–5. [232]

Beaglehole, D. and Wilson, P. (1994). Extrinsic premelting at the ice glass interface. *Journal of Physical Chemistry*, **98**, 8096–100. [233]

Bellissent-Funel, M.-C., Teixeira, J., and Bosio, L. (1987). Structure of high-density amorphous water. II. Neutron scattering study. *Journal of Chemical Physics*, **87**, 2231–5. [280]

Benedict, W.S., Gailar, N., and Plyler, E.K. (1956). Rotation–vibration spectra of deuterated water vapor. *Journal of Chemical Physics*, **24**, 1139–65. [5, 51]

Benoit, M., Bernasconi, M., Focher, P., and Parrinello, M. (1996). New high-pressure phase of ice. *Physical Review Letters*, **76**, 2934–6. [275]

Benoit, M., Marx, D., and Parrinello, M. (1998). Tunnelling and zero-point motion in high-pressure ice. *Nature*, **392**, 258–61. [275]

Bentley, W.A. and Humphreys, W.J. (1931). *Snow crystals*. McGraw-Hill, New York. [Republished Dover, New York 1962.] [293]

Bernal, J.D. and Fowler, R.H. (1933). A theory of water and ionic solutions, with particular reference to hydrogen and hydroxyl ions. *Journal of Chemical Physics*, **1**, 515–48. [8, 9, 17, 29, 50, 258]

Bertie, J.E. and Whalley, E. (1964). Infrared spectra of ices II, III, and V in the range 4000 to 350 cm^{-1}. *Journal of Chemical Physics*, **40**, 1646–59. [269]

Bertie, J.E. and Whalley, E. (1967). Optical spectra of orientationally disordered crystals. II. Infrared spectrum of ice Ih and ice Ic from 360 to 50 cm^{-1}. *Journal of Chemical Physics*, **46**, 1271–84. [57]

Bertie, J.E., Calvert, L.D., and Whalley, E. (1963). Transformations of ice II, ice III, and ice V at atmospheric pressure. *Journal of Chemical Physics*, **38**, 840–6. [276]

Bertie, J.E., Calvert, L.D., and Whalley, E. (1964). Transformations of ice VI and ice VII at atmospheric pressure. *Canadian Journal of Chemistry*, **42**, 1373–8. [262, 276]

Bertie, J.E., Labbé, H.J., and Whalley, E. (1969). Absorptivity of ice I in the range 4000–30 cm^{-1}. *Journal of Chemical Physics*, **50**, 4501–20. [46–7]

Besson, J.M., Pruzan, Ph., Klotz, S., Hamel, G., Silvi, B., Nelmes, R.J. *et al.* (1994). Variation of interatomic distances in ice VIII to 10 GPa. *Physical Review*, **B49**, 12540–50. [265]

Besson, J.M., Klotz, S., Hamel, G., Marshall, W.G., Nelmes, R.J., and Loveday, J.S. (1997). Structural instability in ice VIII under pressure. *Physical Review Letters*, **78**, 3141–4. [252]

Beverley, M.N. and Nield, V.M. (1997). Extensive tests on the application of reverse Monte Carlo modelling to single-crystal neutron diffuse scattering from ice Ih. *Journal of Physics: Condensed Matter*, **9**, 5145–56. [34]

Bilgram, J., Wenzl, H., and Mair, G. (1973). Perfection of zone refined ice single crystals. *Journal of Crystal Growth*, **20**, 319–21. [13]

Bilgram, J.H., Roos, J., and Gränicher, H. (1976). Spin–lattice relaxation in HF and NH_3 doped ice and the out diffusion of impurities. *Zeitschrift für Physik*, **B23**, 1–9. [147]

Bishop, P.G. and Glen, J.W. (1969). Electrical polarization effects in pure and doped ice at low temperatures. In *Physics of ice* (ed. N. Riehl, B. Bullemer, and H. Engelhardt), pp. 492–501. Plenum Press, New York. [123]

Bizid, A., Bosio, L., Defrain, A., and Oumezzine, M. (1987). Structure of high-density amorphous water. I. X-ray diffraction study. *Journal of Chemical Physics*, **87**, 2225–30. [280]

Bjerrum, N. (1951). Structure and properties of ice. *Kongelige Danske Videnskabernes Selskab Matematisk-fysiske Meddelelser*, **27**, 1–56. [26, 29, 73]

Bjerrum, N. (1952). Structure and properties of ice. *Science*, **115**, 385–90. [173]

Blackman, M. and Lisgarten, N.D. (1958). Electron diffraction investigation into the cubic and other forms of ice. *Advances in Physics*, **7**, 189–98. [276]

Blicks, H., Dengel, O., and Riehl, N. (1966). Diffusion von Protonen (Tritonen) in reinen und dotierten Eis-Einkristallen. *Physik der Kondensierten Materie*, **4**, 375–81. [133]

Bogorodsky, V.V., Bentley, C.R., and Gudmandsen, P.E. (1985). *Radioglaciology*. Reidel, Dordrecht. [304]

Bohren, C.F. (1983). Colors of snow, frozen waterfalls, and icebergs. *Journal of the Optical Society of America*, **73**, 1646–52. [220]

Boresi, A.P., Schmidt, R.J., and Sidebottom, O.M. (1993). *Advanced mechanics of materials* (5th edn). Wiley, New York. [197, 199]

Boutron, P. and Alben, R. (1975). Structural model for amorphous solid water. *Journal of Chemical Physics*, **62**, 4848–53. [280]

Bowden, F.P. and Hughes, T.P. (1939). The mechanism of sliding on ice and snow. *Proceedings of the Royal Society of London*, **A172**, 280–98. [318]

Bragg, W.H. (1922). The crystal structure of ice. *Proceedings of the Physical Society*, **34**, 98–103. [17]

Braun, C.L. and Smirnov, S.N. (1993). Why is water blue? *Journal of Chemical Education*, **70**, 612–4. [220]

Braun, J., Glebov, A., Graham, A.P., Menzel, A., and Toennies, J.P. (1998). Structure and phonons of the ice surface. *Physical Review Letters*, **80**, 2638–41. [228]

Brewster, D. (1814). On the affections of light transmitted through crystallized bodies. *Philosophical Transactions of the Royal Society of London*, **104**, 187–218. [219]

Brewster, D. (1818). On the laws of polarisation and double refraction in regularly crystallized bodies. *Philosophical Transactions of the Royal Society of London*, **108**, 199–273. [219]

Bridgman, P.W. (1912). Water, in the liquid and five solid forms, under pressure. *Proceedings of the American Academy of Arts and Sciences*, **47**, 441–558. [252, 257]

Bridgman, P.W. (1935). The pressure–volume–temperature relations of the liquid, and the phase diagram of heavy water. *Journal of Chemical Physics*, **3**, 597–605. [254]

Bridgman, P.W. (1937). The phase diagram of water to 45 000 kg/cm^2. *Journal of Chemical Physics*, **5**, 964–6. [262]

Brown, A.R. and Doren, D.J. (1997). HOCl adsorption on ice surface. *Journal of Physical Chemistry*, **B101**, 6308–12. [296]

Brown, D.E. and George, S.M. (1996). Surface and bulk diffusion of $H_2{}^{18}O$ on single-crystal $H_2{}^{16}O$ ice multilayers. *Journal of Physical Chemistry*, **100**, 15460–9. [133, 228]

Buch, V., Delzeit, L., Blackledge, C., and Devlin, J.P. (1996). Structure of the ice nanocrystal surface from simulated versus experimental spectra of adsorbed CF_4. *Journal of Physical Chemistry*, **100**, 3732–44. [229, 249]

Buch, V., Sandler, P., and Sadlej, J. (1998). Simulations of H_2O solid, liquid, and clusters, with an emphasis on ferroelectric ordering transition in hexagonal ice. *Journal of Physical Chemistry*, **B102**, 8641–53. [9, 261]

Budd, W.F. and Jacka, T.H. (1989). A review of ice rheology for ice sheet modelling. *Cold Regions Science and Technology*, **16**, 107–44. [184, 193, 199, 201–3, 205, 300]

Bullemer, B., Engelhardt, H., and Riehl, N. (1969). Protonic conduction of ice. Part I: High temperature region. In *Physics of ice* (ed. N. Riehl, B. Bullemer, and H. Engelhardt), pp. 416–29. Plenum Press, New York. [97]

Burnham, C.J., Li, J.-C., and Leslie, M. (1997). Molecular dynamics calculations for ice Ih. *Journal of Physical Chemistry*, **B101**, 6192–5. [57]

Burton, E.F. and Oliver, W.F. (1935). The crystal structure of ice at low temperatures. *Proceedings of the Royal Society of London*, **A153**, 166–72. [278]

Buser, O. and Jaccard, C. (1978). Charge separation by collision of ice particles on metals: electronic surface states. *Journal of Glaciology*, **21**, 547–57. [240]

Camp, P.R., Kiszenick, W., and Arnold, D. (1969). Electrical conduction in ice. In *Physics of ice* (ed. N. Riehl, B. Bullemer, and H. Engelhardt), pp. 450–70. Plenum Press, New York. [97, 234]

Camplin, G.C. and Glen, J.W. (1973). The dielectric properties of HF-doped single crystals of ice. In *Physics and chemistry of ice* (ed. E. Whalley, S.J. Jones, and L.W. Gold), pp. 256–61. Royal Society of Canada, Ottawa. [105]

Camplin, G.C., Glen, J.W., and Paren, J.G. (1978). Theoretical models for interpreting the dielectric behaviour of HF-doped ice. *Journal of Glaciology*, **21**, 123–41. [99, 100, 105–6]

Cannon, N.P., Schulson, E.M., Smith, T.R., and Frost, H.J. (1990). Wing cracks and brittle compressive fracture. *Acta Metallurgica et Materialia*, **38**, 1955–62. [208]

Caranti, J.M. and Illingworth, A.J. (1983*a*). Frequency dependence of the surface conductivity of ice. *Journal of Physical Chemistry*, **87**, 4078–83. [234, 236]

Caranti, J.M. and Illingworth, A.J. (1983*b*). Transient Workman–Reynolds freezing potential. *Journal of Geophysical Research*, **88**(C13), 8483–9. [240]

Caranti, J.M. and Lamfri, M.A. (1987). Hall effect on the surface of ice. *Physics Letters*, **A126**, 47–51. [235]

Carr, M.H., Belton, M.J.S., Chapman, C.R., Davies, M.E., Geissler, P., Greenberg, R. *et al.* (1998). Evidence for a subsurface ocean on Europa. *Nature*, **391**, 363–5. [311–12]

Casassa, S., Ugliengo, P., and Pisani, C. (1997). Proton-ordered models of ordinary ice for quantum-mechanical studies. *Journal of Chemical Physics*, **106**, 8030–40. [30, 261]

Castelnau, O., Duval, P., Lebensohn, R.A., and Canova, G.R. (1996). Viscoplastic modeling of texture development in polycrystalline ice with a self-consistent approach: Comparison with bound estimates. *Journal of Geophysical Research*, **101**(B6), 13851–68. [205]

Chan, R.K., Davidson, D.W., and Whalley, E. (1965). Effect of pressure on the dielectric properties of ice I. *Journal of Chemical Physics*, **43**, 2376–83. [96, 139]

Chen, M.S., Onsager, L., Bonner, J., and Nagle, J. (1974). Hopping of ions in ice. *Journal of Chemical Physics*, **60**, 405–19. [145]

Chou, I.-M. and Haselton, H.T. (1998). Visual observations of crystal morphologies and melting of ice IV at pressures up to 1 GPa. *The Review of High Pressure Science and Technology*, **7**, 1132–4. [252–3, 271]

Chou, I.-M., Blank, J.G., Goncharov, A.F., Mao, H., and Hemley, R.J. (1998). In situ observations of a high-pressure phase of H_2O ice. *Science*, **281**, 809–12. [270–1]

Chowdhury, M.R., Dore, J.C., and Wenzel, J.T. (1982). The structural characteristics of amorphous D_2O ice by neutron diffraction. *Journal of Non-Crystalline Solids*, **53**, 247–65. [280]

Chrzanowski, J. (1988). Thickness-dependent d.c. electrical conduction in thin polycrystalline ice films deposited on to a dielectric substrate. *Journal of Materials Science Letters*, **7**, 1058–60. [236]

Churaev, N.V., Bardasov, S.A., and Sobolev, V.D. (1993). On the non-freezing water interlayers between ice and a silica surface. *Colloids and Surfaces*, **A79**, 11–24. [315]

Claydon, C.R., Segal, G.A., and Taylor, H.S. (1971). Theoretical interpretation of the optical and electron scattering spectra of H_2O. *Journal of Chemical Physics*, **54**, 3799–816. [5]

Clementi, E., Corongiu, G., and Sciortino, F. (1993). Liquid and solid phases of water: an extensive molecular dynamics simulation with an *ab initio* polarizable potential. *Journal of Molecular Structure*, **296**, 205–13. [58]

Clifford, J. (1967). Proton magnetic resonance data on ice. *Chemical Communications*, **17**, 880–1. [240]

Clough, S.A., Beers, Y., Klein, G.P., and Rothman, L.S. (1973). Dipole moment of water from Stark measurements of H_2O, HDO, and D_2O. *Journal of Chemical Physics*, **59**, 2254–9. [5]

Cohen-Adad, R. and Michaud, M. (1956). Les équilibres liquide–solide du système binaire eau-potasse. *Comptes Rendus Hebdomadaires des Séances de l'Académie des Sciences*, **242**, 2569–71. [109]

Colbeck, S.C. (1987). History of snow-cover research. *Journal of Glaciology*, Special issue, 60–5. [297]

Colbeck, S.C. (1992). *A review of the processes that control snow friction*, US Army Cold Regions Research and Engineering Laboratory Monograph 92–2. [321]

Colbeck, S.C. (1994). A review of the friction of snow skis. *Journal of the Sports Sciences*, **12**, 285–95. [321]

Colbeck, S.C. (1995). Pressure melting and ice skating. *American Journal of Physics*, **63**, 888–90. [318]

Colbeck, S.C. (1997). *A review of sintering in seasonal snow*, US Army Corps of Engineers Cold Regions Research and Engineering Laboratory Special Report 97-10. [297–8]

Colbeck, S.C. (1998). Sintering in a dry snow cover. *Journal of Applied Physics*, **84**, 4585–9. [297–8]

Colbeck, S., Akitaya, E., Armstrong, R., Gubler, H., Lafeuille, J., Lied, K. *et al.* (1990). International classification for seasonal snow on the ground. International Commission on Snow and Ice of the International Association of Scientific Hydrology, World Data Center A for Glaciology, Boulder, Colorado. [297]

Colbeck, S.C., Najarian, L., and Smith, H.B. (1997). Sliding temperature of ice skates. *American Journal of Physics*, **65**, 488–92. [320]

Cole, D.M. (1979). Preparation of polycrystalline ice specimens for laboratory experiments. *Cold Regions Science and Technology*, **1**, 153–9. [12, 186, 198]

Cole, D.M. (1987). Strain-rate and grain-size effects in ice. *Journal of Glaciology*, **33**, 274–80. [198–9, 201]

Cole, D.M. (1996). Observations of pressure effects on the creep of ice single crystals. *Journal of Glaciology*, **42**, 169–75. [191]

Cole, D.M. (1997). On the relationship between the physical and mechanical properties of sea ice. In *13th IAHR International Symposium on Ice, Beijing, China, Aug. 27–31 1996. Post-symposium Proceedings, Vol. 3*, pp. 913–30. [292]

Cole, K.S. and Cole, R.H. (1941). Dispersion and absorption in dielectrics. I—Alternating current characteristics. *Journal of Chemical Physics*, **9**, 341–51. [67]

Collier, W.B., Ritzhaupt, G., and Devlin, J.P. (1984). Spectroscopically evaluated rates and energies for proton transfer and Bjerrum defect migration in cubic ice. *Journal of Physical Chemistry*, **88**, 363–8. [141]

Collins, J.G. and White, G.K. (1964). Thermal expansion of solids. *Progress in Low Temperature Physics*, **4**, 450–79. [41]

Compaan, K. and Haven, Y. (1956). Correlation factors for diffusion in solids. *Transactions of the Faraday Society*, **52**, 786–801. [131]

Consolmagno, G.J. and Lewis, J.S. (1976). Structural and thermal models of icy Galilean satellites. In *Jupiter—studies of the interior, atmosphere, magnetosphere and satellites* (ed. T. Gehrels), pp. 1035–51. University of Arizona Press, Tucson. [309]

Cotterill, R.M.J., Martin, J.W., Nielsen, O.V., and Pedersen, O.B. (1973). Computer studies of perfect-crystal properties and defect structures in ice I. In *Physics and chemistry of ice* (ed. E. Whalley, S.J. Jones, and L.W. Gold), pp. 23–7. Royal Society of Canada, Ottawa. [136]

Coulson, C.A. and Eisenberg, D. (1966). Interactions of H_2O molecules in ice—I. The dipole moment of an H_2O molecule in ice. *Proceedings of the Royal Society of London*, **A291**, 445–53. [30, 102]

Cox, S.F.J. (1992). μSR studies of ice, illustrating the positive muon as a magnetic resonance probe of structure and dynamics. *Physica Scripta*, **T45**, 292–6. [149]

Cox, S.F.J., Smith, J.A.S., and Symons, M.C.R. (1990). Measurement of ^{17}O quadrupole interactions and identification of the diamagnetic fraction in ice by muon level crossing resonance. *Hyperfine Interactions*, **65**, 993–1004. [149]

Cox, S.F.J., Ayres de Campos, N., Mendes, P.J., Gil, J.M., Kratzer, A., Nield, V.M. *et al.* (1994). Level crossing resonances in ice: identification of two diamagnetic muon states. *Hyperfine Interactions*, **86**, 747–52. [149]

Croutch, V.K. and Hartley, R.A. (1992). Adhesion of ice to coatings and the performance of ice release coatings. *Journal of Coatings Technology*, **64**, 41–53. [314, 316]

Cruikshank, D.P., Brown, R.H., Calvin, W.M., Roush, T.L., and Bartholomew, M.J. (1998). Ices on the satellites of Jupiter, Saturn, and Uranus. In *Solar system ices* (ed. B. Schmitt, C. de Bergh, and M. Festou), pp. 579–606. Kluwer Academic Publishers, Dordrecht. [309]

Curtiss, L.A., Frurip, D.J., and Blander, M. (1979). Studies of molecular association in H_2O and D_2O vapors by measurement of thermal conductivity. *Journal of Chemical Physics*, **71**, 2703–11. [8]

Dansgaard, W., Johnsen, S.J., Clausen, H.B., Dahl-Jensen, D., Gundestrup, N.S., Hammer, C.U. *et al.* (1993). Evidence for general instability of past climate from a 250-kyr ice-core record. *Nature*, **364**, 218–20. [302]

Dantl, G. (1968). Die elastischen Moduln von Eis-Einkristallen. *Physik der Kondensierten Materie*, **7**, 390–7. [38–9]

Dantl, G. (1969). Elastic moduli of ice. In *Physics of ice* (ed. N. Riehl, B. Bullemer, and H. Engelhardt), pp. 223–9. Plenum Press, New York. [38–9]

Dash, J.G., Fu, H., and Wettlaufer, J.S. (1995). The premelting of ice and its environmental consequences. *Reports on Progress in Physics*, **58**, 115–67. [227, 247, 307]

Davidson, D.W. (1973). Clathrate hydrates. In *Water—A comprehensive treatise*, Vol. 2 (ed. F. Franks), pp. 115–234. Plenum Press, New York. [281]

Davidson, D.W. and Ripmeester, J.A. (1978). Clathrate ices—recent results. *Journal of Glaciology*, **21**, 33–49. [282]

Davidson, D.W. and Ripmeester, J.A. (1984). NMR, NQR and dielectric properties of clathrates. In *Inclusion compounds*, Vol. 3 (ed. J.L. Atwood, J.E.D. Davies, and D.D. MacNicol), pp. 69–128. Academic Press, London. [282]

Davidson, D.W., Garg, S.K., Gough, S.R., Hawkins, R.E., and Ripmeester, J.A. (1977). Characterization of natural gas hydrates by nuclear magnetic resonance and dielectric relaxation. *Canadian Journal of Chemistry*, **55**, 3641–50. [282]

Davidson, D.W., Handa, Y.P., Ratcliffe, C.I., Tse, J.S., and Powell, B.M. (1984). The ability of small molecules to form clathrate hydrates of structure II. *Nature*, **311**, 142–3. [283]

Davidson, D.W., Handa, Y.P., Ratcliffe, C.I., Ripmeester, J.A., Tse, J.S., Dahn, J.R. *et al.* (1986). Crystallographic studies of clathrate hydrates. Part I. *Molecular Crystals and Liquid Crystals*, **141**, 141–9. [283]

Davydov, A.S. (1991). *Solitons in molecular systems* (2nd edn) (Mathematics and its applications No. 61). Kluwer Academic Publishers, Dordrecht. [145]

Debye, P. (1929). *Polar molecules*. The Chemical Catalog Company, New York. [64, 91]

Decroly, J.C., Gränicher, H., and Jaccard, C. (1957). Caractère de la conductivité électrique de la glace. *Helvetica Physica Acta*, **30**, 465–7. [62]

de Groot, S.R. (1951). *Thermodynamics of irreversible processes*. North-Holland, Amsterdam. [83]

de Haas, M.P., Kunst, M., and Warman, J.M. (1983). Nanosecond time-resolved conductivity studies of pulse-ionized ice. 1. The mobility and trapping of conduction-band electrons in H_2O and D_2O ice. *Journal of Physical Chemistry*, **87**, 4089–92. [151]

Delibaltas, P., Dengel, O., Helmreich, D., Riehl, N., and Simon, H. (1966). Diffusion von ^{18}O in Eis-Einkristallen. *Physik der Kondensierten Materie*, **5**, 166–70. [132]

Delsemme, A.H. (1983). Ice in comets. *Journal of Physical Chemistry*, **87**, 4214–18. [313]

Delzeit, L., Devlin, M.S., Rowland, B., Devlin, J.P., and Buch, V. (1996). Adsorbate-induced partial ordering of the irregular surface and subsurface of crystalline ice. *Journal of Physical Chemistry*, **100**, 10076–82. [229]

Demontis, P., LeSar, R.L., and Klein, M.L. (1988). New high-pressure phases of ice. *Physical Review Letters*, **60**, 2284–7. [275]

Dengel, O. and Riehl, N. (1963). Diffusion von Protonen (Tritonen) in Eiskristallen. *Physik der Kondensierten Materie*, **1**, 191–6. [132]

Dengel, O., Jacobs, E., and Riehl, N. (1966). Diffusion von Tritonen in NH_4F-dotierten Eis-Einkristallen. *Physik der Kondensierten Materie*, **5**, 58–9. [133]

Dennison, D.M. (1921). The crystal structure of ice. *Physical Review*, **17**, 20–2. [17]

Devlin, J.P. (1990). Vibrational spectra and point defect activities of icy solids and gas phase clusters. *International Reviews in Physical Chemistry*, **9**, 29–65. [50, 139]

Devlin, J.P. (1992). Defect activity in icy solids from isotopic exchange rates: Implications for conductance and phase transitions. In *Proton transfer in hydrogen-bonded systems*, NATO Advanced Science Institutes Series B, Vol. 291 (ed. T. Bountis), pp. 249–60. Plenum Press, New York. [141–2, 154]

Devlin, J.P. and Buch, V. (1995). Surface of ice as viewed from combined spectroscopic and computer modelling studies. *Journal of Physical Chemistry*, **99**, 16534–48. [229, 249, 280]

Devlin, J.P. and Buch, V. (1997). Vibrational spectroscopy and modelling of the surface and subsurface of ice and of ice–adsorbate interactions. *Journal of Physical Chemistry*, **B101**, 6095–8. [229]

Devlin, J.P. and Richardson, H.H. (1984). FT-IR investigation of proton transfer in irradiated ice at 90 K in the absence of mobile Bjerrum defects. *Journal of Chemical Physics*, **81**, 3250–5. [141]

Devlin, J.P., Wooldridge, P.J., and Ritzhaupt, G. (1986). Decoupled isotopomer vibrational frequencies in cubic ice: A simple unified view of the Fermi diads of decoupled H_2O, HOD, and D_2O. *Journal of Chemical Physics*, **84**, 6095–100. [50]

Dewar, J. and Fleming, J.A. (1897). Note on the dielectric constant of ice and alcohol at very low temperatures. *Proceedings of the Royal Society of London*, **61**, 2–18. [91]

Dillard, D.S. and Timmerhaus, K.D. (1969). Low temperature thermal conductivity of selected dielectric crystalline solids. In *Thermal conductivity—Proceedings of the eighth conference* (ed. C.Y. Ho and R.E. Taylor), pp. 949–67. Plenum Press, New York. [43]

Ditchburn, R.W. (1963). *Light* (2nd edn), p. 553. Blackie, London. [216]

Dolling, G. (1976). The calculation of phonon frequencies. In *Methods in computational physics*, Vol. 15—Vibrational properties of solids (ed. G. Gilat), pp. 1–40. Academic Press, New York. [57]

Dore, J.C. (1990). Hydrogen-bond networks in supercooled liquid water and amorphous/vitreous ices. *Journal of Molecular Structure*, **237**, 221–32. [280–1]

Dosch, H., Lied, A., and Bilgram, J.H. (1995). Glancing-angle X-ray scattering studies of the premelting of ice surfaces. *Surface Science*, **327**, 145–64. [229–31]

Dosch, H., Lied, A., and Bilgram, J.H. (1996). Disruption of the hydrogen-bonding network at the surface of Ih ice near surface premelting. *Surface Science*, **366**, 43–50. [230]

Dunitz, J.D. (1963). Nature of orientational defects in ice. *Nature*, **197**, 860–2. [142]

Durham, W.B., Kirby, S.H., Heard, H.C., and Stern, L.A. (1987). Inelastic properties of several high pressure crystalline phases of H_2O: ices II, III and V. *Journal de Physique*, **48**, Colloque C1, 221–6. [203]

Durham, W.B., Kirby, S.H., and Stern, L.A. (1997). Creep of water ices at planetary conditions: A compilation. *Journal of Geophysical Research*, **102**(E7), 16293–302. Erratum p. 28725. [203, 310]

Durham, W.B., Kirby, S.H., and Stern, L.A. (1998). Rheology of planetary ices. In *Solar system ices* (ed. B. Schmitt, C. de Bergh, and M. Festou), pp. 63–78. Kluwer Academic Publishers, Dordrecht. [310]

Duval, P. (1978). Anelastic behaviour of polycrystalline ice. *Journal of Glaciology*, **21**, 621–8. [205]

Duval, P. (1981). Creep and fabrics of polycrystalline ice under shear and compression. *Journal of Glaciology*, **27**, 129–40. [204]

Duval, P., Ashby, M.F., and Anderman, I. (1983). Rate-controlling processes in the creep of polycrystalline ice. *Journal of Physical Chemistry*, **87**, 4066–74. [190, 193, 205]

Dyke, T.R., Mack, K.M., and Muenter, J.S. (1977). The structure of water dimer from molecular beam electric resonance spectroscopy. *Journal of Chemical Physics*, **66**, 498–510. [7]

Dzyaloshinskii, I.E., Lifshitz, E.M., and Pitaevskii, L.P. (1961). The general theory of Van der Waals forces. *Advances in Physics*, **10**, 165–209. [250, 318]

Eckener, U., Helmreich, D., and Engelhardt, H. (1973). Transit time measurements of protons in ice. In *Physics and chemistry of ice* (ed. E. Whalley, S.J. Jones, and L.W. Gold), pp. 242–5. Royal Society of Canada, Ottawa. [119]

Ehringhaus, A. (1917). Beiträge zur Kenntnis der Dispersion der Doppelbrechung einiger Kristalle. *Neues Jahrbuch für Mineralogie, Geologie und Paläontologie*, **41**, 342–419. [219–20]

Elbaum, M., Lipson, S.G., and Dash, J.G. (1993). Optical study of surface melting on ice. *Journal of Crystal Growth*, **129**, 491–505. [232, 243, 246]

Eldrup, M., Mogensen, O.E., and Bilgram, J.H. (1978). Vacancies in HF-doped and in irradiated ice by positron annihilation techniques. *Journal of Glaciology*, **21**, 101–13. [136]

Elvin, A.A. and Sunder, S.S. (1996). Microcracking due to grain boundary sliding in polycrystalline ice under uniaxial compression. *Acta Materialia*, **44**, 43–56. [211]

Engelhardt, H. and Kamb, B. (1981). Structure of ice IV, a metastable high-pressure phase. *Journal of Chemical Physics*, **75**, 5887–99. [256, 272–3]

Engelhardt, H. and Riehl, N. (1965). Space-charge limited proton currents in ice. *Physics Letters*, **14**, 20–1. [117]

Engelhardt, H. and Riehl, N. (1966). Zur protonischen Leitfähigkeit von Eis-Einkristallen bei tiefen Temperaturen und hohen Feldstärken. *Physik der Kondensierten Materie*, **5**, 73–82. [117]

Engelhardt, H. and Whalley, E. (1972). Ice IV. *Journal of Chemical Physics*, **56**, 2678–84. [254, 271–2]

Errera, M.J. (1924). La dispersion des ondes Herziennes dans les solides au voisinage du point de fusion. *Journal de Physique et le Radium*, Series 6, **5**, 304–11. [91]

Evans, D.C.B., Nye, J.F., and Cheeseman, K.J. (1976). The kinetic friction of ice. *Proceedings of the Royal Society of London*, **A347**, 493–512. [318–20]

Evans, L.F. (1967). Selective nucleation of the high-pressure ices. *Journal of Applied Physics*, **38**, 4930–2. [271–2]

Evtushenko, A.A. and Petrenko, V.F. (1991). Investigation of the pseudopiezoelectric effect and the stress-potential constants of charge carriers in ice. *Fizika Tverdogo Tela*, **33**, 1509–17. [*Soviet Physics—Solid State*, **33**, 850–55.] [139–40]

Evtushenko, A.A., Martirosyan, M.B., and Petrenko, V.F. (1988). Experimental investigations of electrical properties of ice grown in a static electric field. *Fizika Tverdogo Tela*, **30**, 2133–8. [*Soviet Physics—Solid State*, **30**, 1229–32.] [113]

Falls, A.H., Wellinghoff, S.T., Talmon, Y., and Thomas, E.L. (1983). A transmission electron microscopy study of hexagonal ice. *Journal of Materials Science*, **18**, 2752–64. [162]

Faraday, M. (1850). Report of lecture "On certain conditions of freezing water". *The Athenaeum*, No. 1181, 640–1. [227, 314]

Faraday, M. (1860). Note on regelation. *Proceedings of the Royal Society of London*, **10**, 440–50. [227, 244, 314]

Faure, P. (1969). Étude d'un modèle dynamique du réseau cristallin de la glace. *Journal de Physique*, **30**, 214–20. [54–5]

Faure, P. and Chosson, A. (1978). The translational lattice-vibration Raman spectrum of single-crystal ice Ih. *Journal of Glaciology*, **21**, 65–72. [47]

Fifolt, D.A., Petrenko, V.F., and Schulson, E.M. (1993). Preliminary study of electromagnetic emissions from cracks in ice. *Philosophical Magazine*, **B67**, 289–99. [212]

Finney, J.L., Quinn, J.E., and Baum, J.O. (1985). The water dimer potential surface. *Water Science Reviews*, **1**, 93–170. [7, 9]

Fleming, J.A. and Dewar, J. (1897). On the dielectric constants of pure ice, glycerine, nitrobenzol, and ethylene dibromide at and above the temperature of liquid air. *Proceedings of the Royal Society of London*, **61**, 316–30. [91]

Fletcher, N.H. (1962). Surface structure of water and ice. *Philosophical Magazine*, **7**, 255–69. [250]

Fletcher, N.H. (1968). Surface structure of water and ice II. A revised model. *Philosophical Magazine*, **18**, 1287–300. [250]

Fletcher, N.H. (1973). The surface of ice. In *Physics and chemistry of ice* (ed. E. Whalley, S.J. Jones, and L.W. Gold), pp. 132–6. Royal Society of Canada, Ottawa. [250]

Fletcher, N.H. (1992). Reconstruction of ice crystal surfaces at low temperatures. *Philosophical Magazine*, **B66**, 109–15. [228]

Floriano, M.A., Klug, D.D., Whalley, E., Svensson, E.C., Sears, V.F., and Hallman, E.D. (1987). Direct determination of the intramolecular O–D distance in ice Ih by neutron diffraction. *Nature*, **329**, 821–3. [32]

Flubacher, P., Leadbetter, A.J., and Morrison, J.A. (1960). Heat capacities of ice at low temperatures. *Journal of Chemical Physics*, **33**, 1751–5. [28, 40–1]

Frank, F.C. and Nicholas, J.F. (1953). Stable dislocations in the common crystal lattices. *Philosophical Magazine*, **44**, 1213–35. [176]

Fröhlich, H. (1949). *Theory of dielectrics*. Oxford University Press, Oxford. [71]

Frost, H.J., Goodman, D.J., and Ashby, M.F. (1976). Kink velocities on dislocations in ice. A comment on the Whitworth, Paren and Glen model. *Philosophical Magazine*, **33**, 951–61. [174]

Fujara, F., Wefing, S., and Kuhs, W.F. (1988). Direct observation of tetrahedral hydrogen jumps in ice Ih. *Journal of Chemical Physics*, **88**, 6801–9. [32]

Fukazawa, H., Ikeda, S., and Mae, S. (1998). Incoherent inelastic neutron scattering measurements on ice XI; the proton-ordered phase of ice Ih doped with KOH. *Chemical Physics Letters*, **282**, 215–8. [58]

Fukuda, A. and Higashi, A. (1969). X-ray diffraction topographic studies of dislocations in natural large ice single crystals. *Japanese Journal of Applied Physics*, **8**, 993–9. [163]

Fukuda, A. and Higashi, A. (1973). Dynamical behavior of dislocations in ice crystals. *Crystal Lattice Defects*, **4**, 203–10. [164]

Fukuda, A. and Higashi, A. (1988). Generation, multiplication and motion of dislocations in ice crystals. In *Lattice defects in ice crystals* (ed. A. Higashi), pp. 69–96. Hokkaido University Press, Sapporo. [164, 166]

Fukuda, A. and Shoji, H. (1988). Dislocations in natural ice crystals. In *Lattice defects in ice crystals* (ed. A. Higashi), pp. 13–25. Hokkaido University Press, Sapporo. [163]

Fukuda, A., Hondoh, T., and Higashi, A. (1987). Dislocation mechanisms of plastic deformation of ice. *Journal de Physique*, **48**, Colloque C1, 163–73. [158, 164, 166, 171, 179]

Fukuta, N. (1987). An origin of the equilibrium liquid-like layer on ice. *Journal de Physique*, **48**, Colloque C1, 503–9. [248]

Furukawa, Y. and Nada, H. (1997). Anisotropic surface melting of an ice crystal and its relationship to growth forms. *Journal of Physical Chemistry*, **B101**, 6167–70. [232, 248–9]

Furukawa, Y., Yamamoto, M. and Kuroda, T. (1987). Ellipsometric study of the transition layer on the surface of an ice crystal. *Journal of Crystal Growth*, **82**, 665–77. [232]

Gagnon, R.E., Kiefte, H., Clouter, M.J., and Whalley, E. (1988). Pressure dependence of the elastic constants of ice Ih to 2.8 kbar by Brillouin spectroscopy. *Journal of Chemical Physics*, **89**, 4522–8. [39, 40]

Gammon, P.H., Kiefte, H., and Clouter, M.J. (1983*a*). Elastic constants of ice samples by Brillouin spectroscopy. *Journal of Physical Chemistry*, **87**, 4025–9. [38]

Gammon, P.H., Kiefte, H., Clouter, M.J., and Denner, W.W. (1983*b*). Elastic constants of artificial and natural ice samples by Brillouin spectroscopy. *Journal of Glaciology*, **29**, 433–60. [38–40]

Gao, X.Q. and Jacka, T.H. (1987). The approach to similar tertiary creep rates for Antarctic core ice and laboratory prepared ice. *Journal de Physique*, **48**, Colloque C1, 289–96. [204]

Gemmell, D.S. (1974). Channeling and related effects in the motion of charged particles through crystals. *Reviews of Modern Physics*, **46**, 129–227. [231]

George, A. and Rabier, J. (1987). Dislocations and plasticity in semiconductors. I. Dislocation structures and dynamics. *Revue de Physique Appliquée*, **22**, 941–66. [158, 172]

Gertner, B.J. and Hynes, J.T. (1996). Molecular dynamics simulation of hydrochloric acid ionization at the surface of stratospheric ice. *Science*, **271**, 1563–6. [296]

Giauque, W.F. and Ashley, M.F. (1933). Molecular rotation in ice at 10 K. Free energy of formation and entropy of water. *Physical Review*, **43**, 81–2. [17]

Giauque, W.F. and Stout, J.W. (1936). The entropy of water and the third law of thermodynamics. The heat capacity of ice from 15 to 273 K. *Journal of the American Chemical Society*, **58**, 1144–50. [28, 40]

Gilpin, R.R. (1980). A model for the prediction of ice lensing and frost heave in soils. *Water Resources Research*, **16**, 918–30. [307]

Ginnings, D.C. and Corruccini, R.J. (1947). An improved ice calorimeter—the determination of its calibration factor and the density of ice at 0 °C. *Journal of Research of the National Bureau of Standards*, **38**, 583–91. [12]

Girifalco, L.A. (1973). *Statistical physics of materials*, Chap. 8. Wiley, New York. [131]

Glasel, J.A. (1972). Nuclear magnetic resonance studies on water and ice. In *Water—A comprehensive treatise*, Vol. 1 (ed. F. Franks), pp. 215–53. Plenum Press, New York. [147]

Glen, J.W. (1955). The creep of polycrystalline ice. *Proceedings of the Royal Society of London*, **A228**, 519–38. [193, 198, 201–2, 205]

Glen, J.W. (1958) The flow law of ice. *International Association of Scientific Hydrology*, Publication no. 47, 171–183. [199]

Glen, J.W. (1968). The effect of hydrogen disorder on dislocation movement and plastic deformation of ice. *Physik der Kondensierten Materie*, **7**, 43–51. [173–5]

Glen, J.W. (1997). 34 years of ice physics symposia. *Journal of Physical Chemistry*, **B101**, 6079–81. [4, 227, 322]

Glen, J.W. and Paren, J.G. (1975). The electrical properties of snow and ice. *Journal of Glaciology*, **15**, 15–38. [298]

Glen, J.W. and Perutz, M.F. (1954). The growth and deformation of ice crystals. *Journal of Glaciology*, **2**, 397–403. [157, 186, 293]

Godzik, A. (1990). An estimation of the energy parameters for the soliton movement in hydrogen-bonded chains. *Chemical Physics Letters*, **171**, 217–21. [145]

Gold, L.W. (1958). Some observations of the dependence of strain on stress for ice. *Canadian Journal of Physics*, **36**, 1265–75. [38]

Gold, L.W. (1963). Deformation mechanisms in ice. In *Ice and snow—properties, processes, and applications* (ed. W.D. Kingery), pp. 8–27. MIT Press, Cambridge, Massachusetts. [196]

Gold, L.W. (1966). Dependence of crack formation on crystallographic orientation for ice. *Canadian Journal of Physics*, **44**, 2757–64. [196–7]

Gold, L.W. (1972). The process of failure in columnar-grained ice. *Philosophical Magazine*, **26**, 311–28. [208]

Goldblatt, M. (1964). The density of liquid T_2O. *Journal of Physical Chemistry*, **68**, 147–51. [15]

Goldsby, D.L. and Kohlstedt, D.L. (1997). Grain boundary sliding in fine-grained ice I. *Scripta Materialia*, **37**, 1399–406. [203]

Golecki, I. and Jaccard, C. (1978). Intrinsic surface disorder in ice near the melting point. *Journal of Physics C: Solid State Physics*, **11**, 4229–37. [231]

Goncharov, A.F., Struzhkin, V.V., Somayazulu, M.S., Hemley, R.J., and Mao, H.K. (1996). Compression of ice to 210 Gigapascals: infrared evidence for a symmetric hydrogen-bonded phase. *Science*, **273**, 218–20. [274]

Good, R.J. and Chaudhury, M.K. (1991). Theory of adhesive forces across interfaces: 1. The Lifshitz–van der Waals component of interaction and adhesion. In *Fundamentals of adhesion* (ed. L.-H. Lee), pp. 137–51. Plenum Press, New York. [244]

Good, R.J., Chaudhury, M.K., and van Oss, C.J. (1991). Theory of adhesive forces across interfaces: 2. Interfacial hydrogen bonds as acid–base phenomena and as factors enhancing adhesion. In *Fundamentals of adhesion* (ed. L.-H. Lee), pp. 153–72. Plenum Press, New York. [244]

Goodman, D.J., Frost, H.J., and Ashby, M.F. (1981). The plasticity of polycrystalline ice. *Philosophical Magazine*, **A43**, 665–95. [184]

Gosar, P. (1974). Note on the Hall effect in ice. *Physics of Condensed Matter*, **17**, 183–7. [98]

Gosse, S., Labrie, D., and Chylek, P. (1995). Refractive index of ice in the 1.4–7.8 μm spectral range. *Applied Optics*, **34**, 6582–6. [219]

Goto, K., Hondoh, T., and Higashi, A. (1982). Experimental determinations of the concentration and mobility of interstitials in pure ice crystals. In *Point defects and defect interactions in metals* (ed. J. Takamura, M. Doyama, and M. Kiritani), pp. 174–6. University of Tokyo Press. [134]

Goto, K., Hondoh, T., and Higashi, A. (1986). Determination of diffusion coefficients of self-interstitials in ice with a new method of observing climb of dislocations by X-ray topography. *Japanese Journal of Applied Physics*, **25**, 351–7. [126, 132, 134–5]

Goto, A., Hondoh, T., and Mae, S. (1990). The electron density distribution in ice Ih determined by single-crystal X-ray diffractometry. *Journal of Chemical Physics*, **93**, 1412–17. [18]

Gough, S.R. (1972). A low temperature dielectric cell and the permittivity of hexagonal ice to 2 K. *Canadian Journal of Chemistry*, **50**, 3046–51. [95]

Gough, S.R., Whalley, E., and Davidson, D.W. (1968). Dielectric properties of the hydrates of argon and nitrogen. *Canadian Journal of Chemistry*, **46**, 1673–81. [282]

Gow, A.J. (1986). Orientation textures in ice sheets of quietly frozen lakes. *Journal of Crystal Growth*, **74**, 247–58. [288–9]

Gow, A.J. and Williamson, T.C. (1972). Linear compressibility of ice. *Journal of Geophysical Research*, **77**, 6348–52. [39]

Gow, A.J. and Williamson, T. (1975). Gas inclusions in the Antarctic ice sheet and their glaciological significance. *Journal of Geophysical Research*, **80**, 5101–8. [299]

Gränicher, H. (1958). Gitterfehlordnung und physikalische Eigenschaften hexagonaler und kubischer Eiskristalle. *Zeitschrift für Kristallographie*, **110**, 432–71. [73, 79]

Gränicher, H. (1963). Properties and lattice imperfections of ice crystals and the behaviour of H_2O–HF solid solutions. *Physik der Kondensierten Materie*, **1**, 1–12. [99]

Gränicher, H. (1969). Evaluation of dielectric dispersion data. In *Physics of ice* (ed. N. Riehl, B. Bullemer, and H. Engelhardt), pp. 527–33. Plenum Press, New York. [66]

Gränicher, H., Jaccard, C., Scherrer, P., and Steinemann, A. (1957). Dielectric relaxation and the electrical conductivity of ice crystals. *Discussions of the Faraday Society*, **23**, 50–62. [79]

Gratz, E.T. and Schulson, E.M. (1997). Brittle failure of columnar saline ice under triaxial compression. *Journal of Geophysical Research*, **102**(B3), 5091–107. [212, 292]

Greenler, R. (1980). *Rainbows, halos, and glories*. Cambridge University Press, Cambridge. [294]

Grenfell, T.C. and Perovich, D.K. (1981). Radiation absorption coefficients of polycrystalline ice from 400–1400 nm. *Journal of Geophysical Research*, **86**(C8), 7447–50. [220–1]

Grenfell, T.C., Warren, S.G., and Mullen, P.C. (1994). Reflection of solar radiation by the Antarctic snow surface at ultraviolet, visible, and near-infrared wavelengths. *Journal of Geophysical Research*, **99**(D9), 18669–84. [299]

Gross, G.W. and Johnson, J. (1983). The layered capacitor method for dielectric bridge measurements. Data analysis and interpretation of fluoride doped ice. *IEEE Transactions on Electrical Insulation*, **EI-18**, 485–97. [93]

Gross, G.W. and Svec, R.K. (1997). Effect of ammonium on anion uptake and dielectric relaxation in laboratory-grown ice columns. *Journal of Physical Chemistry*, **B101**, 6282–4. [111]

Gross, G.W., Wong, P.M., and Humes, K. (1977). Concentration dependent solute redistribution at the ice–water phase boundary. III. Spontaneous convection. Chloride solutions. *Journal of Chemical Physics*, **67**, 5264–74. [111]

Gross, G.W., Hayslip, I.C., and Hoy, R.N. (1978). Electrical conductivity and relaxation in ice crystals with known impurity content. *Journal of Glaciology*, **21**, 143–60. [106, 111]

Groves, G.W. and Kelly, A. (1963). Independent slip systems in crystals. *Philosophical Magazine*, **8**, 877–87. [193]

Grundy, W.M. and Schmitt, B. (1998). The temperature-dependent near-infrared absorption spectrum of hexagonal H_2O ice. *Journal of Geophysical Research*, **103**(E11), 25809–22. [219]

Haas, C. (1962). On diffusion, relaxation and defects in ice. *Physics Letters*, **3**, 126–8. [142]

Haida, O., Matsuo, T., Suga, H., and Seki, S. (1974). Calorimetric study of the glassy state X. Enthalpy relaxation at the glass-transition temperature of hexagonal ice. *Journal of Chemical Thermodynamics*, **6**, 815–25. [28, 40–1, 96, 261]

Hallbrucker, A., Mayer, E., and Johari, G.P. (1989*a*). The heat capacity and glass transition of hyperquenched glassy water. *Philosophical Magazine*, **B60**, 179–87. [279]

Hallbrucker, A., Mayer, E., and Johari, G.P. (1989*b*). Glass–liquid transition and the enthalpy of devitrification of annealed vapor-deposited amorphous solid water. A comparison with hyperquenched glassy water. *Journal of Physical Chemistry*, **93**, 4986–90. [279]

Hallett, J. and Mason, B.J. (1958). The influence of temperature and supersaturation on the habit of ice crystals grown from the vapour. *Proceedings of the Royal Society of London*, **A247**, 440–53. [294]

Haltenorth, H. and Klinger, J. (1969). Diffusion of hydrogen fluoride in ice. In *Physics of ice* (ed. N. Riehl, B. Bullemer, and H. Engelhardt), pp. 579–84. Plenum Press, New York. [150]

Hama, J. and Suito, K. (1992). On the metallization of ice under ultra-high pressures. In *Physics and chemistry of ice* (ed. N. Maeno and T. Hondoh), pp. 75–82. Hokkaido University Press, Sapporo. [275]

Hambrey, M. and Alean, J. (1992). *Glaciers*. Cambridge University Press. [299]

Hamilton, W.C., Kamb, B., LaPlaca, S.J., and Prakash, A. (1969). Deuteron arrangements in high-pressure forms of ice. In *Physics of ice* (ed. N. Riehl, B. Bullemer, and H. Engelhardt), pp. 44–58. Plenum Press, New York. [272]

Hammer, C.U. (1980). Acidity of polar ice cores in relation to absolute dating, past volcanism, and radio-echoes. *Journal of Glaciology*, **25**, 359–72. [94, 302]

Handa, Y.P., Mishima, O., and Whalley, E. (1986). High-density amorphous ice. III. Thermal properties. *Journal of Chemical Physics*, **84**, 2766–70. [279]

Handa, Y.P., Klug, D.D., and Whalley, E. (1987). Phase transitions of ice V and ice VI. *Journal de Physique*, **48**, Colloque C1, 435–40. [272]

Handa, Y.P., Klug, D.D., and Whalley, E. (1988). Energies of the phases of ice at low temperature and pressure relative to ice Ih. *Canadian Journal of Chemistry*, **66**, 919–24. [30, 269, 272, 276]

Haneman, D. (1987). Surfaces of silicon. *Reports on Progress in Physics*, **50**, 1045–86. [228]

Hardy, S.C. (1977). A grain boundary groove measurement of the surface tension between ice and water. *Philosophical Magazine*, **35**, 471–84. [244]

Harrison, J.D. and Tiller, W.A. (1963). Controlled freezing of water. In *Ice and snow. Properties, processes, and applications* (ed. W.D. Kingery), pp. 215–31. MIT Press, Cambridge, Massachusetts. [290]

Hartmann, W.K. (1993). *Moons and planets* (3rd edn), Chap. 8. Wadsworth Publishing Company, Belmont, California. [313]

Hartshorne, N.H. and Stuart, A. (1970). *Crystals and the polarising microscope* (4th edn). Edward Arnold, London. [221]

Hassan, R. and Campbell, E.S. (1992). The energy and structure of Bjerrum defects in ice Ih determined with an additive and a nonadditive potential. *Journal of Chemical Physics*, **97**, 4326–35. [143]

Hays, D.A. (1991). Role of electrostatics in adhesion. In *Fundamentals of adhesion* (ed. L.-H. Lee), pp. 249–78. Plenum Press, New York. [318]

Heggie, M.I., Maynard, S.C.P., and Jones, R. (1992). Computer modelling of dislocation glide in ice Ih. In *Physics and chemistry of ice* (ed. N. Maeno and T. Hondoh), pp. 497–501. Hokkaido University Press, Sapporo. [161]

Hemley, R.J., Jephcoat, A.P., Mao, H.K., Zha, C.S., Finger, L.W., and Cox, D.E. (1987). Static compression of H_2O-ice to 128 GPa (1.28 Mbar). *Nature*, **330**, 737–40. [256, 263, 274]

Higashi, A. (1974). Growth and perfection of ice crystals. *Journal of Crystal Growth*, **24/25**, 102–7. [163]

Higashi, A. (1978). Structure and behaviour of grain boundaries in polycrystalline ice. *Journal of Glaciology*, **21**, 589–605. [181]

Higashi, A. (ed.) (1988). *Lattice defects in ice crystals*. Hokkaido University Press, Sapporo. [157, 163]

Higashi, A. and Sakai, N. (1961). Movement of small angle boundary of ice crystal. *Journal of the Physical Society of Japan*, **16**, 2359–60. [181]

Higashi, A., Koinuma, S., and Mae, S. (1964). Plastic yielding in ice crystals. *Japanese Journal of Applied Physics*, **3**, 610–16. [188, 190]

Higashi, A., Koinuma, S., and Mae, S. (1965). Bending creep of ice single crystals. *Japanese Journal of Applied Physics*, **4**, 575–82. [189–90]

Higashi, A., Oguro, M., and Fukuda, A. (1968). Growth of ice single crystals from the melt, with special reference to dislocation structure. *Journal of Crystal Growth*, **3/4**, 728–32. [163]

Higuchi, K. (1958). The etching of ice crystals. *Acta Metallurgica*, **6**, 636–42. [14, 162]

Hirsch, K.R. and Holzapfel, W.B. (1984). Symmetric hydrogen bonds in ice X. *Physics Letters*, **A101**, 142–4. [274]

Hirsch, K.R. and Holzapfel, W.B. (1986). Effect of high pressure on the Raman spectra of ice VIII and evidence for ice X. *Journal of Chemical Physics*, **84**, 2771–5. [284]

Hirth, J.P. and Lothe, J. (1982). *Theory of dislocations* (2nd edn). Wiley, New York. [156, 172, 176, 181]

Hobbs, P.V. (1974). *Ice Physics*. Clarendon Press, Oxford. [257, 293–4]

Hobbs, P.V. and Mason, B.J. (1964). The sintering and adhesion of ice. *Philosophical Magazine*, **9**, 181–97. [314]

Hollander, F. and Jeffrey, G.A. (1977). Neutron diffraction study of the crystal structure of ethylene oxide deuterohydrate at 80 K. *Journal of Chemical Physics*, **66**, 4699–705. [282–3]

Hollins, G.T. (1964). Configurational statistics and the dielectric constant of ice. *Proceedings of the Physical Society*, **84**, 1001–16. [27, 68]

Holzapfel, W.B. (1972). On the symmetry of the hydrogen bonds in ice VII. *Journal of Chemical Physics*, **56**, 712–15. [273]

Homer, D.R. and Glen, J.W. (1978). The creep activation energies of ice. *Journal of Glaciology*, **21**, 429–44. [189–90]

Hondoh, T. (1988). Observations of large-angle grain boundaries in ice crystals. In *Lattice defects in ice crystals* (ed. A. Higashi), pp. 129–46. Hokkaido University Press, Sapporo. [181–2]

Hondoh, T. (1992). Glide and climb processes of dislocations in ice. In *Physics and chemistry of ice* (ed. N. Maeno and T. Hondoh), pp. 481–7. Hokkaido University Press, Sapporo. [135, 173]

Hondoh, T. and Higashi, A. (1978). X-ray diffraction topographic observations of the large-angle grain boundary in ice under deformation. *Journal of Glaciology*, **21**, 629–38. [181]

Hondoh, T. and Higashi, A. (1979). Anisotropy of migration and faceting of large-angle grain boundaries in ice bicrystals. *Philosophical Magazine*, **A39**, 137–49. [181]

Hondoh, T. and Higashi, A. (1983). Generation and absorption of dislocations at large-angle grain boundaries in deformed ice crystals. *Journal of Physical Chemistry*, **87**, 4044–50. [168]

Hondoh, T., Itoh, T., and Higashi, A. (1981). Formation of stacking faults in pure ice single crystals by cooling. *Japanese Journal of Applied Physics*, **20**, L737–740. [183]

Hondoh, T., Itoh, T., Amakai, S., Goto, K., and Higashi, A. (1983). Formation and annihilation of stacking faults in pure ice. *Journal of Physical Chemistry*, **87**, 4040–4. [179]

Hondoh, T., Iwamatsu, H., and Mae, S. (1990). Dislocation mobility for non-basal glide in ice measured by *in situ* X-ray topography. *Philosophical Magazine*, **A62**, 89–102. [171]

Hondoh, T., Hoshi, R., Goto, A., and Yamagami, H. (1991). A new method using synchrotron-radiation topography for determining point-defect diffusivity under hydrostatic pressure. *Philosophical Magazine Letters*, **63**, 1–5. [136]

Hooke, R. LeB., Mellor, M., Budd, W.F., Glen, J.W., Higashi, A., Jacka, T.H. *et al.* (1980). Mechanical properties of polycrystalline ice: an assessment of current knowledge and priorities for research. *Cold Regions Science and Technology*, **3**, 263–75. [193]

Hopkins, W. (1862). On the theory of the motion of glaciers. *Philosophical Transactions of the Royal Society of London*, **152**, 677–745. [184]

Hore, P.J. (1995). *Nuclear magnetic resonance*. Oxford University Press, Oxford. [32]

Hosler, C.L., Jensen, D.C., and Goldshlak, L. (1957). On the aggregation of ice crystals to form snow. *Journal of Meteorology*, **14**, 415–20. [314]

Howe, R. (1987). The possible ordered structures of ice Ih. *Journal de Physique*, **48**, Colloque C1, 599–604. [258]

Howe, R. and Whitworth, R.W. (1987). The configurational entropy of partially ordered ice. *Journal of Chemical Physics*, **86**, 6443–5. [260]

Howe, R. and Whitworth, R.W. (1989*a*). A determination of the crystal structure of ice XI. *Journal of Chemical Physics*, **90**, 4450–3. [259]

Howe, R. and Whitworth, R.W. (1989*b*). The electrical conductivity of KOH-doped ice from 70 to 250 K. *Journal of Physics and Chemistry of Solids*, **50**, 963–5. [110]

Hubmann, M. (1978). Effect of pressure on the dielectric properties of ice Ih single crystals doped with NH_3 and HF. *Journal of Glaciology*, **21**, 161–72. [102, 108, 139]

Hubmann, M. (1979). Polarization processes in the ice lattice I. Approach by thermodynamics of irreversible processes. New experimental verification by means of a universal relation. *Zeitschrift für Physik*, **B32**, 127–39. [79, 83–4, 101–3]

Huckaby, D.A., Pitis, R., Kincaid, R.H., and Hamilton, C. (1993). Inclusion–exclusion calculation of the dipole–dipole energy of hexagonal ice and cubic ice. *Journal of Chemical Physics*, **98**, 8105–9. [30]

Humbel, F., Jona, F., and Scherrer, P. (1953). Anisotropie der Dielektrizitätskonstante des Eises. *Helvetica Physica Acta*, **26**, 17–32. [96]

Hutchings, I.M. (1992). *Tribology: Friction and wear of engineering materials*, Chap. 3. Edward Arnold, London. [320]

Hutchinson, J.W. (1977). Creep and plasticity of hexagonal polycrystals as related to single crystal slip. *Metallurgical Transactions*, **A8**, 1465–9. [193]

Hutter, K. (1983). *Theoretical glaciology*. Reidel, Dordrecht. [300]

Ignat, M. and Frost, H.J. (1987). Grain boundary sliding in ice. *Journal de Physique*, **48**, Colloque C1, 189–95. [196]

Israelachvili, J.N. (1991). *Intermolecular and surface forces* (2nd edn), Chap. 15. Academic Press, London. [243, 317]

Itagaki, K. (1967). Self diffusion in single crystal ice. *Journal of the Physical Society of Japan*, **22**, 427–31. [132]

Itagaki, K. (1970). X-ray topographic study of vibrating dislocations in ice under an AC electric field. *Advances in X-ray Analysis*, **13**, 526–38. [176]

Itoh, H., Kawamura, K., Hondoh, T., and Mae, S. (1996*a*). Molecular dynamics studies of self-interstitials in ice Ih. *Journal of Chemical Physics*, **105**, 2408–13. [126, 136]

Itoh, H., Kawamura, K., Hondoh, T., and Mae, S. (1996*b*). Molecular dynamics studies of proton ordering effects on lattice vibrations in ice Ih. *Physica*, **B219 & 220**, 469–72.[58]
Jaccard, C. (1959). Étude théorique et expérimentale des propriétés de la glace. *Helvetica Physica Acta*, **32**, 89–128. [78, 99, 105–6]
Jaccard, C. (1964). Thermodynamics of irreversible processes applied to ice. *Physik der Kondensierten Materie*, **3**, 99–118. [78, 83]
Jacka, T.H. (1984*a*). The time and strain required for development of minimum strain rates in ice. *Cold Regions Science and Technology*, **8**, 261–8. [190, 200–1]
Jacka, T.H. (1984*b*). Laboratory studies on relationships between ice crystal size and flow rate. *Cold Regions Science and Technology*, **10**, 31–42. [201]
Jacka, T.H. and Li, J. (1994). The steady-state crystal size of deforming ice. *Annals of Glaciology*, **20**, 13–18. [202, 204]
Jacka, T.H. and Lile, R.C. (1984). Sample preparation techniques and compression apparatus for ice flow studies. *Cold Regions Science and Technology*, **8**, 235–40. [198]
Jacka, T.H. and Maccagnan, M. (1984). Ice crystallographic and strain rate changes with strain in compression and extension. *Cold Regions Science and Technology*, **8**, 269–86. [204]
Jackson, S.M. and Whitworth, R.W. (1995). Evidence for ferroelectric ordering of ice Ih. *Journal of Chemical Physics*, **103**, 7647–8. [259]
Jackson, S.M., Nield, V.M., Whitworth, R.W., Oguro, M., and Wilson, C.C. (1997). Single-crystal neutron diffraction studies of the structure of ice XI. *Journal of Physical Chemistry*, **B101**, 6142–5. [258]
Jaeger, J.C. (1956). *Elasticity, fracture and flow*, p. 93. Methuen, London. [199]
Jeffrey, G.A. (1984). Hydrate inclusion compounds. In *Inclusion compounds*, Vol. 1 (ed. J.L. Atwood, J.E.D. Davies, and D.D. MacNicol), pp. 135–90. Academic Press, London. [281, 283]
Jellinek, H.H.G. (1959). Adhesive properties of ice. *Journal of Colloid Science*, **14**, 268–80. [314–15]
Jellinek, H.H.G. (1962). Ice adhesion. *Canadian Journal of Physics*, **40**, 1294–309. [314]
Jellinek, H.H.G. (1967). Liquid-like (transition) layer on ice. *Journal of Colloid and Interface Science*, **25**, 192–205. [315]
Jellinek, H.H.G. and Brill, R. (1956). Viscoelastic properties of ice. *Journal of Applied Physics*, **27**, 1198–209. [205]
Jenniskens, P. and Blake, D.F. (1994). Structural transitions in amorphous water ice and astrophysical implications. *Science*, **265**, 753–6. [278]
Johari, G.P. (1976). The dielectric properties of H_2O and D_2O ice Ih at MHz frequencies. *Journal of Chemical Physics*, **64**, 3998–4005. [95]
Johari, G.P. (1981). The spectrum of ice. *Contemporary Physics*, **22**, 613–42. [215–16]
Johari, G.P. and Chew, H.A.M. (1984). Pressure and temperature dependence of the O—H and O—D stretching vibrations in the Raman spectrum of ice. *Philosophical Magazine*, **B49**, 647–60. [51–2]
Johari, G.P. and Jones, S.J. (1975). Study of the low-temperature "transition" in ice Ih by thermally stimulated depolarization measurements. *Journal of Chemical Physics*, **62**, 4213–23. [123–4]
Johari, G.P. and Jones, S.J. (1976). Dielectric properties of polycrystalline D_2O ice Ih (hexagonal). *Proceedings of the Royal Society of London*, **A349**, 467–95. [95–6]
Johari, G.P. and Jones, S.J. (1978). The orientation polarization in hexagonal ice parallel and perpendicular to the *c*-axis. *Journal of Glaciology*, **21**, 259–76. [95–6]

Johari, G.P. and Whalley, E. (1976). Dielectric properties of ice VI at low temperatures. *Journal of Chemical Physics*, **64**, 4484–9. [267]

Johari, G.P. and Whalley, E. (1979). Evidence for a very slow transformation in ice VI at low temperatures. *Journal of Chemical Physics*, **70**, 2094–7. [267]

Johari, G.P. and Whalley, E. (1981). The dielectric properties of ice Ih in the range 272–133 K. *Journal of Chemical Physics*, **75**, 1333–40. [63, 96]

Johari, G.P., Lavergne, A., and Whalley, E. (1974). Dielectric properties of ice VII and VIII and the phase boundary between ice VI and VII. *Journal of Chemical Physics*, **61**, 4292–300. [254, 264]

Johari, G.P., Hallbrucker, A., and Mayer, E. (1990). Calorimetric study of pressure-amorphized cubic ice. *Journal of Physical Chemistry*, **94**, 1212–14. [279]

Johari, G.P., Hallbrucker, A., and Mayer, E. (1996). Two calorimetrically distinct states of liquid water below 150 Kelvin. *Science*, **273**, 90–2. [277, 279]

Johnsen, S.J., Dansgaard, W., and White, J.W.C. (1989). The origin of Arctic precipitation under present and glacial conditions. *Tellus*, **B41**, 452–68. [301]

Jones, D.E., Kennedy, F.E., and Schulson, E.M. (1991). The kinetic friction of saline ice against itself at low sliding velocities. *Annals of Glaciology*, **15**, 242–6. [210, 320]

Jones, S.J. (1967). Softening of ice crystals by dissolved fluoride ions. *Physics Letters*, **25A**, 366–7. [175, 191–2]

Jones, S.J. and Chew, H.A.M. (1983). Creep of ice as a function of hydrostatic pressure. *Journal of Physical Chemistry*, **87**, 4064–6. [203]

Jones, S.J. and Gilra, N.K. (1972). Increase of dislocation density in ice by dissolved hydrogen fluoride. *Applied Physics Letters*, **20**, 319–20. [191]

Jones, S.J. and Gilra, N.K. (1973). X-ray topographical study of dislocations in pure and HF-doped ice. *Philosophical Magazine*, **27**, 457–72. [164]

Jones, S.J. and Glen, J.W. (1969*a*). The mechanical properties of single crystals of pure ice. *Journal of Glaciology*, **8**, 463–73. [189–90]

Jones, S.J. and Glen, J.W. (1969*b*). The effect of dissolved impurities on the mechanical properties of ice crystals. *Philosophical Magazine*, **19**, 13–24. [175, 191]

Jones, W.M. (1952). The triple point temperature of tritium oxide. *Journal of the American Chemical Society*, **74**, 6065–6. [15]

Jordaan, I.J. and Timco, G.W. (1988). Dynamics of the ice-crushing process. *Journal of Glaciology*, **34**, 318–26. [212]

Jorgensen, W.L., Chandrasekhar, J., Madura, J.D., Impey, R.W., and Klein, M.L. (1983). Comparison of simple potential functions for simulating liquid water. *Journal of Chemical Physics*, **79**, 926–35. [8]

Kamb, W.B. (1961). The glide direction in ice. *Journal of Glaciology*, **3**, 1097–106. [157, 186]

Kamb, B. (1964). Ice II: A proton-ordered form of ice. *Acta Crystallographica*, **17**, 1437–49. [256, 268]

Kamb, B. (1965). The structure of ice VI. *Science*, **150**, 205–9. [266]

Kamb, B. (1973). Crystallography of ice. In *Physics and chemistry of ice* (ed. E. Whalley, S.J. Jones, and L.W. Gold), pp. 28–41. Royal Society of Canada, Ottawa.[18, 259, 267]

Kamb, B. and Davis, B.L. (1964). Ice VII, the densest form of ice. *Proceedings of the National Academy of Sciences of the United States of America*, **52**, 1433–9. [262, 273]

Kamb, B. and Prakash, A. (1968). Structure of ice III. *Acta Crystallographica*, **B24**, 1317–27. [272]

Kamb, B., Prakash, A., and Knobler, C. (1967). Structure of ice V. *Acta Crystallographica*, **22**, 706–15. [256, 272]

Kamb, B., Hamilton, W.C., LaPlaca, S.J., and Prakash, A. (1971). Ordered proton configuration in ice II, from single-crystal neutron diffraction. *Journal of Chemical Physics*, **55**, 1934–45. [268]

Kamb, B., Raymond, C.F., Harrison, W.D., Engelhardt, H., Echelmeyer, K.A., Humphrey, N. *et al.* (1985). Glacier surge mechanism: 1982–1983 surge of Variegated Glacier, Alaska. *Science*, **227**, 469–79. [300]

Kawada, S. (1972). Dielectric dispersion and phase transition of KOH doped ice. *Journal of the Physical Society of Japan*, **32**, 1442. [257]

Kawada, S. (1978). Dielectric anisotropy in ice Ih. *Journal of the Physical Society of Japan*, **44**, 1881–6. [92, 95–7]

Kawada, S. (1979). Dielectric properties of heavy ice Ih (D_2O ice). *Journal of the Physical Society of Japan*, **47**, 1850–6. [96]

Kawada, S. (1981). Representations of configurational entropy and dielectric properties of ice Ih. *Journal of the Physical Society of Japan*, **50**, 1233–40. [71]

Kawada, S. (1989*a*). Dielectric properties of KOH-doped D_2O ice. *Journal of the Physical Society of Japan*, **58**, 295–300. [109, 145]

Kawada, S. (1989*b*). Acceleration of dielectric relaxation by KOH-doping and phase transition in ice Ih. *Journal of Physics and Chemistry of Solids*, **50**, 1177–84. [109, 262]

Kawada, S. and Tutiya, R. (1997). Dielectric properties and annealing effects of as-grown KOH-highly doped ice single crystals. *Journal of Physics and Chemistry of Solids*, **58**, 115–21. [109–10, 262]

Kawada, S., Takei, I. and Abe, H. (1989). Development of a new relaxational process having shortened relaxation time and phase transition in KOH-doped ice single crystal. *Journal of the Physical Society of Japan*, **58**, 54–7. [262]

Kawada, S., Iisaka, H., Kitamura, E., Tutiya, R., and Abe, H. (1992). Successive changes and developments of the dielectric relaxation process toward 72 K phase transition in the alkali-hydroxide-doped ice single crystals. In *Physics and chemistry of ice* (ed. N. Maeno and T. Hondoh), pp. 20–6. Hokkaido University Press, Sapporo. [109]

Kawada, S., Jin, R.G., and Abo, M. (1997). Dielectric properties and 110 K anomalies in KOH- and HCl-doped ice single crystals. *Journal of Physical Chemistry*, **B101**, 6223–5. [110]

Keith, W.D. and Saunders, C.P.R. (1990). Further laboratory studies of the charging of graupel during ice crystal interactions. *Atmospheric Research*, **25**, 455–64. [295]

Kell, G.S. (1967). Precise representation of volume properties of water at one atmosphere. *Journal of Chemical and Engineering Data*, **12**, 66–9. [12]

Kern, C.W. and Karplus, M. (1972). The water molecule. In *Water—A comprehensive treatise*, Vol. 1 (ed. F. Franks), pp. 21–91. Plenum Press, New York. [5]

Ketcham, W.M. and Hobbs, P.V. (1969). An experimental determination of the surface energies of ice. *Philosophical Magazine*, **19**, 1161–73. [182]

Kevan, L. (1981). Electron spin echo studies of solvation structure. *Journal of Physical Chemistry*, **85**, 1628–36. [151]

Khusnatdinov, N.N. and Petrenko, V.F. (1996). Fast-growth technique for ice single crystals. *Journal of Crystal Growth*, **163**, 420–5. [13]

Khusnatdinov, N.N., Petrenko, V.F., and Turanov, A.N. (1990). Intrinsic photoconductivity of hexagonal ice. *Physica Status Solidi* (*a*), **118**, 401–8. [152]

Khusnatdinov, N.N., Petrenko, V.F., and Levey, C.G. (1997). Electrical properties of the ice/solid interface. *Journal of Physical Chemistry*, **B101**, 6212–14. [116, 238–9]

Kirby, S., Durham, W., and Stern, L. (1992). The ice I → II transformation: mechanisms and kinetics under hydrostatic and nonhydrostatic conditions. In *Physics and chemistry of ice* (ed. N. Maeno and T. Hondoh), pp. 456–63. Hokkaido University Press, Sapporo. [269]

Kirkwood, J.G. (1939). The dielectric polarization of polar liquids. *Journal of Chemical Physics*, **7**, 911–19. [71]

Kirshenbaum, I. (1951). *Physical properties and analysis of heavy water*. McGraw-Hill, New York. [15]

Klemens, P.G. (1958). Thermal conductivity and lattice vibrational modes. *Solid State Physics*, **7**, 1–98. [44]

Klinger, J. (1975). Low-temperature heat conduction in pure monocrystalline ice. *Journal of Glaciology*, **14**, 517–28. [43, 45]

Klinger, J. (1983). Extraterrestrial ice. A review. *Journal of Physical Chemistry*, **87**, 4209–14. [307]

Klinger, J. and Rochas, G. (1983). Influence of ageing on the heat conduction coefficient of hexagonal ice. *Journal of Physical Chemistry*, **87**, 4155–6. [43–5]

Klug, D.D. and Whalley, E. (1972). Optical spectra of orientationally disordered crystals. IV. Effects of short range correlation of orientations. *Journal of Chemical Physics*, **56**, 553–62. [47]

Klug, D.D. and Whalley, E. (1984). The uncoupled O—H stretch in ice VII. The infrared frequency and integrated intensity up to 189 kbar. *Journal of Chemical Physics*, **81**, 1220–8. [284]

Klug, D.D., Mishima, O., and Whalley, E. (1987). High density amorphous ice. IV. Raman spectrum of uncoupled O—H and O—D oscillators. *Journal of Chemical Physics*, **86**, 5323–8. [284]

Klug, D.D., Tse, J.S., and Whalley, E. (1991*a*). The longitudinal-optic–transverse-optic mode splitting in ice Ih. *Journal of Chemical Physics*, **95**, 7011–12. [58]

Klug, D.D., Whalley, E., Svensson, E.C., Root, J.H., and Sears, V.F. (1991*b*). Densities of vibrational states and heat capacities of crystalline and amorphous H_2O ice determined by neutron scattering. *Physical Review*, **B44**, 841–4. [41]

Knight, C.A. (1962). Studies of Arctic lake ice. *Journal of Glaciology*, **4**, 319–35. [288]

Knight, C.A. (1966). The contact angle of water on ice. *Journal of Colloid and Interface Science*, **25**, 280–4. [243]

Knight, C.A. (1971). Experiments on the contact angle of water on ice. *Philosophical Magazine*, **23**, 153–65. [243]

Knight, C.A. (1996*a*). A simple technique for growing large, optically "perfect" ice crystals. *Journal of Glaciology*, **42**, 585–7. [13]

Knight, C.A. (1996*b*). Surface layers on ice. *Journal of Geophysical Research*, **101**(D8), 12921–8. See also pp. 12929–36. [248]

Knight, C.A. and Knight, N.C. (1968*a*). Spongy hailstone growth criteria I. Orientation fabrics. *Journal of the Atmospheric Sciences*, **25**, 445–52. [294]

Knight, C.A. and Knight, N.C. (1968*b*). Spongy hailstone growth criteria II. Microstructures. *Journal of the Atmospheric Sciences*, **25**, 453–9. [294]

Knight, C.A. and Knight, N.C. (1971). Hailstones. *Scientific American*, **224**(4), 96–103. [294]

Knuts, S., Ojamäe, L., and Hermansson, K. (1993). An *ab-initio* study of the OH stretching frequencies in ice-II, ice-VIII, and ice-IX. *Journal of Chemical Physics*, **99**, 2917–28. [269, 285]

Kobayashi, T. (1961). The growth of snow crystals at low supersaturations. *Philosophical Magazine*, **6**, 1363–70. [12, 294]

Kobayashi, T. and Furukawa, Y. (1975). On twelve branched snow crystals. *Journal of Crystal Growth*, **28**, 21–8. [181]

Kobayashi, T. and Furukawa, Y. (1991). *Snow crystals*. Museum book of the Snow Crystal Museum, Asahikawa, Japan, In Japanese. [293]
König, H. (1943). Eine kubische Eismodifikation. *Zeitschrift für Kristallographie*, **105**, 279–86. [276]
Kouchi, A., Furukawa, Y., and Kuroda, T. (1987). X-ray diffraction pattern of quasi-liquid layer on ice crystal surface. *Journal de Physique*, **48**, Colloque C1, 675–7. [229]
Kozack, R.E. and Jordan, P.C. (1992). Polarizability effects in a four-charge model for water. *Journal of Chemical Physics*, **96**, 3120–30. [8, 30]
Kroes, G.-J. (1992). Surface melting on the (0001) face of TIP4P ice. *Surface Science*, **275**, 365–82. [249]
Kröger, F.A. (1974). *The chemistry of imperfect crystals*, Vol. 2 (2nd edn), Chap. 18. North Holland, Amsterdam. [104, 155]
Kuhn, W. and Thürkauf, M. (1958). Isotopentrennung beim Gefrieren von Wasser und Diffusionskonstanten von D und ^{18}O im Eis. *Helvetica Chimica Acta*, **41**, 938–71. [132]
Kuhs, W.F. and Knupfer, M. (1992). The structure of ice Ih on approaching the melting point. In *Physics and chemistry of ice* (ed. N. Maeno and T. Hondoh), pp. 50–3. Hokkaido University Press, Sapporo. [24]
Kuhs, W.F. and Lehmann, M.S. (1986). The structure of ice-Ih. *Water Science Reviews*, **2**, 1–65. [23–4, 31–2, 59]
Kuhs, W.F. and Lehmann, M.S. (1987). The geometry and orientation of the water molecule in ice Ih. *Journal de Physique*, **48**, Colloque C1, 3–8. [33]
Kuhs, W.F., Finney, J.L., Vettier, C., and Bliss, D.V. (1984). Structure and hydrogen ordering in ices VI, VII, and VIII by neutron powder diffraction. *Journal of Chemical Physics*, **81**, 3612–23. [256, 262, 264, 266–7]
Kuhs, W.F., Bliss, D.V., and Finney, J.L. (1987). High-resolution neutron powder diffraction study of ice Ic. *Journal de Physique*, **48**, Colloque C1, 631–6. [256, 277]
Kuhs, W.F., Ahsbahs, H., Londono, D., and Finney, J.L. (1989). In-situ crystal growth and neutron four-circle diffractometry under high pressure. *Physica*, **B156 & 157**, 684–7. [267]
Kuhs, W.F., Chazallon, B., Radaelli, P.G., and Pauer, F. (1997). Cage occupancy and compressibility of deuterated N_2-clathrate hydrate by neutron diffraction. *Journal of Inclusion Phenomena and Molecular Recognition in Chemistry*, **29**, 65–77. [283]
Kuhs, W.F., Lobban, C., and Finney, J.L. (1998). Partial H-ordering in high pressure ices III and V. *The Review of High Pressure Science and Technology*, **7**, 1141–3. [272]
Kumai, M. (1961). Snow crystals and the identification of the nuclei in the northern United States of America. *Journal of Meteorology*, **18**, 139–50. [295]
Kunst, M. and Warman, J.M. (1983). Nanosecond time-resolved conductivity studies of pulse-ionized ice. 2. The mobility and trapping of protons. *Journal of Physical Chemistry*, **87**, 4093–5. [151]
Kunst, M., Warman, J.M., de Haas, M.P., and Verberne, J.B. (1983). Nanosecond time-resolved conductivity studies of pulsed-ionized ice. 3. The electron as a probe for defects in doped ice. *Journal of Physical Chemistry*, **87**, 4096–8. [151]
Kuroda, T. and Lacmann, R. (1982). Growth kinetics of ice from the vapor phase and its growth forms. *Journal of Crystal Growth*, **56**, 189–205. [247]
Kuroiwa, D. (1964). Internal friction of ice. I. The internal friction of H_2O and D_2O ice, and the influence of chemical impurities on mechanical damping. *Contributions from the Institute of Low Temperature Science*, **A18**, 1–37. [138]
Kvlividze, V.I., Kiselev, V.F., Kurzaev, A.B., and Ushakova, L.A. (1974). The mobile water phase on ice surfaces. *Surface Science*, **44**, 60–8. [240]

LaChapelle, E.R. (1969). *Field guide to snow crystals.* University of Washington Press, Seattle. [293–4]

Lacmann, R. and Stranski, I.N. (1972). The growth of snow crystals. *Journal of Crystal Growth*, **13/14**, 236–40. [247]

Lampert, M.A. and Mark, P. (1970). *Current injection in solids.* Academic Press, New York. [117]

Landy, M. and Freiberger, A. (1967). Studies of ice adhesion. I. Adhesion of ice to plastics. *Journal of Colloid and Interface Science*, **25**, 231–44. [316]

Lang, A.R. (1959). Studies of individual dislocations in crystals by X-ray diffraction microscopy. *Journal of Applied Physics*, **30**, 1748–55. [163]

Langham, E.J. (1981). Physics and properties of snowcover. In *Handbook of snow: principles, processes, management and use* (ed. D.M. Gray and D.H. Male), pp. 275–337. Pergamon Press, Toronto. [298]

Latimer, W.M. and Rodebush, W.H. (1920). Polarity and ionization from the standpoint of the Lewis theory of valence. *Journal of the American Chemical Society*, **42**, 1419–33. [6]

Leadbetter, A.J. (1965). The thermodynamic and vibrational properties of H_2O ice and D_2O ice. *Proceedings of the Royal Society of London*, **A287**, 403–25. [33, 41]

Leadbetter, A.J., Ward, R.C., Clark, J.W., Tucker, P.A., Matsuo, T., and Suga, H. (1985). The equilibrium low-temperature structure of ice. *Journal of Chemical Physics*, **82**, 424–8. [258]

Lee, C. and Vanderbilt, D. (1993). Proton transfer in ice. *Chemical Physics Letters*, **210**, 279–84. [144]

Lee, C., Vanderbilt, D., Laasonen, K., Car, R., and Parrinello, M. (1993). *Ab initio* studies on the structural and dynamical properties of ice. *Physical Review*, **B47**, 4863–72. [30, 57, 275]

Lekner, J. (1998). Energetics of hydrogen ordering in ice. *Physica*, **B252**, 149–59. [261]

Leung, S.-K., Brodovitch, J.-C., Newman, K.E., and Percival, P.W. (1987). Muonium diffusion in ice. *Chemical Physics*, **114**, 399–409. [148]

Lewis, J.S. (1995). *Physics and chemistry of the solar system.* Academic Press, San Diego, California. [307]

Li, J.-C. (1996). Inelastic neutron scattering studies of hydrogen bonding in ices. *Journal of Chemical Physics*, **105**, 6733–55. [46, 48–9, 52, 55, 57, 285–6]

Li, J.-C. (1997). Effects of potentials on the vibrational dynamics of ice. *Journal of Physical Chemistry*, **B101**, 6237–42. [57]

Li, J.-C. and Adams, M. (1996). Inelastic incoherent neutron scattering study of the pressure dependence of ice VII and VIII. *Europhysics Letters*, **34**, 675–80. [285]

Li, J.-C. and Ross, D.K. (1992). Neutron scattering studies of ice dynamics. Part I—Inelastic incoherent neutron scattering studies of ice Ih (D_2O, H_2O and HDO). In *Physics and chemistry of ice* (ed. N. Maeno and T. Hondoh), pp. 27–34. Hokkaido University Press, Sapporo. [46, 48–9, 57]

Li, J.-C. and Ross, D.K. (1993). Evidence for two kinds of hydrogen bond in ice. *Nature*, **365**, 327–9. [57]

Li, J.-C. and Ross, D.K. (1994). Inelastic neutron scattering studies of defect modes of H in D_2O ice Ih. *Journal of Physics: Condensed Matter*, **6**, 10823–37.[49–51, 57, 59]

Li, J.-C., Ross, D.K., Londono, J.D., Finney, J.L., Kolesnikov, A., and Ponyatovskii, E.G. (1992). Neutron scattering studies of ice dynamics. Part III—Inelastic incoherent neutron scattering studies of ice II, V, VI, VIII and IX. In *Physics and chemistry of ice* (ed. N. Maeno and T. Hondoh), pp. 43–9. Hokkaido University Press, Sapporo. [285]

Li, J.-C., Nield, V.M., Ross, D.K., Whitworth, R.W., Wilson, C.C., and Keen, D.A. (1994). Diffuse neutron-scattering study of ice Ih. *Philosophical Magazine*, **B69**, 1173–81. [18, 33–4]

Li, J.-C., Nield, V.M., and Jackson, S.M. (1995). Spectroscopic measurements of ice XI. *Chemical Physics Letters*, **241**, 290–4. [52, 58, 262]

Li, J., Jacka, T.H., and Budd, W.F. (1996). Deformation rates in combined compression and shear for ice which is initially isotropic and after the development of strong anisotropy. *Annals of Glaciology*, **23**, 247–52. [199]

Lieb, E.H. (1967). Residual entropy of square ice. *Physical Review*, **162**, 162–72. [28]

Line, C.M.B. and Whitworth, R.W. (1996). A high resolution neutron powder diffraction study of D_2O ice XI. *Journal of Chemical Physics*, **104**, 10008–13. [20, 256, 258, 260–1]

Liu, F., Baker, I., Yao, G., and Dudley, M. (1992). Dislocations and grain boundaries in polycrystalline ice: a preliminary study by synchrotron X-ray topography. *Journal of Materials Science*, **27**, 2719–25. [13, 164]

Liu, F., Baker, I., and Dudley, M. (1993). Dynamic observation of dislocation generation at grain boundaries in ice. *Philosophical Magazine*, **A67**, 1261–76. [168]

Liu, F., Baker, I., and Dudley, M. (1995). Dislocation–grain boundary interactions in ice crystals. *Philosophical Magazine*, **A71**, 15–42. [168, 197]

Liu, K., Cruzan, J.D., and Saykally, R.J. (1996). Water clusters. *Science*, **271**, 929–33. [8]

Livingston, F.E., Whipple, G.C., and George, S.M. (1998). Surface and bulk diffusion of HDO on ultrathin single-crystal ice multilayers on Ru(001). *Journal of Chemical Physics*, **108**, 2197–207. [133]

Lliboutry, L. (1965). *Traité de glaciologie*, Vol. 2. Masson, Paris. [299]

Lobban, C., Finney, J.L., and Kuhs, W.F. (1998). The structure of a new phase of ice. *Nature*, **391**, 268–70. [255–6, 270–1, 273]

Lock, G.S.H. (1990). *The growth and decay of ice*. Cambridge University Press, Cambridge. [4, 287]

Londono, D., Finney, J.L., and Kuhs, W.F. (1992). Formation, stability, and structure of helium hydrate at high-pressure. *Journal of Chemical Physics*, **97**, 547–52. [269]

Londono, J.D., Kuhs, W.F., and Finney, J.L. (1993). Neutron diffraction studies of ices III and IX on under-pressure and recovered samples. *Journal of Chemical Physics*, **98**, 4878–88. [256, 272]

Long, E.A. and Kemp, J.D. (1936). The entropy of deuterium oxide and the third law of thermodynamics. Heat capacity of deuterium oxide from 15 to 298 K. The melting point and heat of fusion. *Journal of the American Chemical Society*, **58**, 1829–34. [40]

Loria, A., Mazzega, E., del Pennino, U., and Andreotti, G. (1978). Measurements of the electrical properties of ice Ih single crystals by admittance and thermally stimulated depolarization techniques. *Journal of Glaciology*, **21**, 219–30. [92, 123–4]

Luth, K. and Scheiner, S. (1992). Calculation of barriers to proton transfer using a variety of electron correlation methods. *International Journal of Quantum Chemistry: Quantum Chemistry Symposium*, **26**, 817–35. [144]

Lutz, H.D. and Jung, C. (1997). Water molecules and hydroxide ions in condensed materials; correlation of spectroscopic and structural data. *Journal of Molecular Structure*, **404**, 63–6. [52]

McClung, D. and Schaerer, P. (1993). *The avalanche handbook*. The Mountaineers, Seattle. [297]

McConnel, J.C. (1891). On the plasticity of an ice crystal. *Proceedings of the Royal Society of London*, **49**, 323–43. [184, 186]

McConnel, J.C. and Kidd, D.A. (1888). On the plasticity of glacier and other ice. *Proceedings of the Royal Society of London*, **44**, 331–67. [184]

Mackay, J.R. (1972). The world of underground ice. *Annals of the Association of American Geographers*, **62**, 1–22. [305]

Mader, H.M. (1992). The thermal behaviour of the water-vein system in polycrystalline ice. *Journal of Glaciology*, **38**, 359–74. [182]

Mae, S. (1968). Void formation during non-basal glide in ice single crystals under tension. *Philosophical Magazine*, **18**, 101–14. [192]

Maeno, N. (1973). Measurements of surface and volume conductivities of single ice crystals. In *Physics and chemistry of ice* (ed. E. Whalley, S.J. Jones, and L.W. Gold), pp. 140–3. Royal Society of Canada, Ottawa. [234]

Maeno, N. and Ebinuma, T. (1983). Pressure sintering of ice and its implication to the densification of snow at polar glaciers and ice sheets. *Journal of Physical Chemistry*, **87**, 4103–10. [297]

Maeno, N. and Nishimura, H. (1978). The electrical properties of ice surfaces. *Journal of Glaciology*, **21**, 193–205. [234–5]

Magono, C. and Lee, C. (1966). Meteorological classification of natural snow crystals. *Journal of the Faculty of Science of Hokkaido University*, Series 7, **2**, 321–35. [294]

Maï, C. (1976). Étude par topographie X du comportement dynamique des dislocations dans la glace Ih. *Comptes Rendus de l'Academie des Sciences de Paris*, **B282**, 515–18. [164, 171]

Maï, C., Perez, J., Tatibouet, J., and Vassoille, R. (1978). Vitesse des dislocations dans la glace dopée avec HF. *Journal de Physique Lettres*, **39**, 307–10. [175]

Maidique, M.A., von Hippel, A., and Westphal, W.B. (1971). Transfer of protons through "pure" ice Ih single crystals. III. Extrinsic versus intrinsic polarization; surface versus volume conduction. *Journal of Chemical Physics*, **54**, 150–60. [234]

Makkonen, L. (1997). Surface melting of ice. *Journal of Physical Chemistry*, **B101**, 6196–200. [243, 248]

Malmberg, C.G. and Maryott, A.A. (1956). Dielectric constant of water from 0° to 100°. *Journal of Research of the National Bureau of Standards*, **56**, 1–8. [63]

Manley, M.E. and Schulson, E.M. (1997). On the strain-rate sensitivity of columnar ice. *Journal of Glaciology*, **43**, 408–10. [201]

Marchi, M., Tse, J.S., and Klein, M.L. (1986). Lattice vibrations and infrared absorption of ice Ih. *Journal of Chemical Physics*, **85**, 2414–18. [58]

Maréchal, Y. (1991). Infrared spectra of water. I. Effect of temperature and of H/D isotopic dilution. *Journal of Chemical Physics*, **95**, 5565–73. [50–1]

Mason, B.J. (1971). *The physics of clouds* (2nd edn). Clarendon Press, Oxford. [293]

Materer, N., Starke, U., Barbieri, A., Van Hove, M.A., Somorjai, G.A., Kroes, G.-J. *et al.* (1997). Molecular surface structure of ice (0001): dynamical low-energy electron diffraction, total-energy calculations and molecular dynamics simulations. *Surface Science*, **381**, 190–210. [228]

Matsuda, M. and Wakahama, G. (1978). Crystallographic structure of polycrystalline ice. *Journal of Glaciology*, **21**, 607–20. [180]

Matsuo, T., Tajima, Y., and Suga, H. (1986). Calorimetric study of a phase transition in D_2O ice Ih doped with KOD: ice XI. *Journal of Physics and Chemistry of Solids*, **47**, 165–73. [40–1, 254, 257]

Matsuoka, O., Clementi, E., and Yoshimine, M. (1976). CI study of the water dimer potential surface. *Journal of Chemical Physics*, **64**, 1351–61. [9]

Matsuoka, T., Fujita, S., and Mae, S. (1996). Effect of temperature on dielectric properties of ice in the range 5–39 GHz. *Journal of Applied Physics*, **80**, 5884–90. [216–17]

Matsuoka, T., Fujita, S., Morishima, S., and Mae, S. (1997). Precise measurement of dielectric anisotropy in ice Ih at 39 GHz. *Journal of Applied Physics*, **81**, 2344–8. [95, 218]

Mayer, E. and Brüggeller, P. (1982). Vitrification of pure liquid water by high pressure jet freezing. *Nature*, **298**, 715–18. [279]

Mayer, E. and Pletzer, R. (1986). Astrophysical implications of amorphous ice—a microporous solid. *Nature*, **319**, 298–301. [278]

Mazzega, E., del Pennino, U., Loria, A., and Mantovani, S. (1976). Volta effect and liquid-like layer at the ice surface. *Journal of Chemical Physics*, **64**, 1028–31. [240]

Mellor, M. (1975). A review of basic snow mechanics. In *Snow mechanics. Proceedings of the Grindelwald Symposium, April 1974*, pp. 251–91. International Association of Hydrological Sciences Publication No. 114. [297–8]

Mellor, M. and Cole, D.M. (1982). Deformation and failure of ice under constant stress or constant strain-rate. *Cold Regions Science and Technology*, **5**, 201–19. [190, 198, 201]

Michel, B. (1978). *Ice mechanics*. Les Presses de l'Université Laval, Quebec. [184, 287]

Michel, B. and Ramseier, R.O. (1971). Classification of river and lake ice. *Canadian Geotechnical Journal*, **8**, 36–45. [287]

Minagawa, I. (1981). Ferroelectric phase transition and anisotropy of dielectric constant in ice Ih. *Journal of the Physical Society of Japan*, **50**, 3669–76. [71]

Minagawa, I. (1990). Phase transition of ice Ih–XI. *Journal of the Physical Society of Japan*, **59**, 1676–85. [71, 261]

Minagawa, I. (1992). Low temperature phases of ice Ih. In *Physics and chemistry of ice* (ed. N. Maeno and T. Hondoh), pp. 14–19. Hokkaido University Press, Sapporo. [261]

Minceva-Sukarova, B., Sherman, W.F., and Wilkinson, G.R. (1984). The Raman spectra of ice (Ih, II, III, V, VI and IX) as functions of pressure and temperature. *Journal of Physics C: Solid State Physics*, **17**, 5833–50. [284]

Minceva-Sukarova, B., Slark, G.E., and Sherman, W.F. (1986). The Raman spectra of ice V and ice VI and evidence of partial ordering at low temperatures. *Journal of Molecular Structure*, **143**, 87–90. [272]

Mishima, O. (1996). Relationship between melting and amorphization of ice. *Nature*, **384**, 546–9. [280]

Mishima, O., Calvert, L.D., and Whalley, E. (1984). 'Melting ice' I at 77 K and 10 kbar: a new method of making amorphous solids. *Nature*, **310**, 393–5. [279]

Mishima, O., Calvert, L.D., and Whalley, E. (1985). An apparently first-order transition between two amorphous phases of ice induced by pressure. *Nature*, **314**, 76–8. [279]

Mitzdorf, U. and Helmreich, D. (1971). Elastic constants of D_2O ice and variation of intermolecular forces on deuteration. *Journal of the Acoustical Society of America*, **49**, 723–8. [56]

Mizuno, Y. and Hanafusa, N. (1987). Studies of surface properties of ice using nuclear magnetic resonance. *Journal de Physique*, **48**, Colloque C1, 511–17. [240–2]

Mogensen, O.E. and Eldrup, M. (1978). Vacancies in pure ice studied by positron annihilation techniques. *Journal of Glaciology*, **21**, 85–99. [136]

Molina, M.J. (1996). Polar ozone depletion (Nobel lecture). *Angewandte Chemie International Edition in English*, **35**, 1778–85. [296]

Moore, J.C. and Paren, J.G. (1987). A new technique for dielectric logging of Antarctic ice cores. *Journal de Physique*, **48**, Colloque C1, 155–60. [304]

Moore, J., Paren, J., and Oerter, H. (1992). Sea salt dependent electrical conduction in polar ice. *Journal of Geophysical Research*, **97**(B13), 19803–12. [111–12, 304]

Moore, J.C., Wolff, E.W., Clausen, H.B., Hammer, C.U., Legrand, M.R., and Fuhrer, K. (1994). Electrical response of the Summit-Greenland ice core to ammonium, sulphuric acid, and hydrochloric acid. *Geophysical Research Letters*, **21**, 565–8.[112]

Morgan, V.I. (1991). High-temperature ice creep tests. *Cold Regions Science and Technology*, **19**, 295–300. [202]

Morgenstern, M., Müller, J., Michely, T., and Comsa, G. (1997). The ice bilayer on Pt(111): nucleation, structure and melting. *Zeitschrift für Physikalische Chemie*, **198**, 43–72. [228]

Morse, M.D. and Rice, S.A. (1982). Tests of effective pair potentials for water: Predicted ice structures. *Journal of Chemical Physics*, **76**, 650–60. [9, 30]

Mounier, S. and Sixou, P. (1969). A contribution to the study of conductivity and dipolar relaxation in doped ice crystals. In *Physics of ice* (ed. N. Riehl, B. Bullemer, and H. Engelhardt), pp. 562–70. Plenum Press, New York. [93]

Muguruma, J. (1969). Effects of surface condition on the mechanical properties of ice crystals. *British Journal of Applied Physics* (*Journal of Physics D*), **2**, 1517–25. [188]

Muguruma, J. and Higashi, A. (1963). Observation of etch channels on the (0001) plane of ice crystal produced by nonbasal glide. *Journal of the Physical Society of Japan*, **18**, 1261–9. [162]

Muguruma, J., Mae, S., and Higashi, A. (1966). Void formation by non-basal glide in ice single crystals. *Philosophical Magazine*, **13**, 625–9. [192]

Muller, S.W. (1947). *Permafrost or permanently frozen ground and related engineering problems*. J.W. Edwards Inc., Ann Arbor, Michigan. [305]

Mulvaney, R., Wolff, E.W., and Oates, K. (1988). Sulphuric acid at grain boundaries in Antarctic ice. *Nature*, **331**, 247–9. [183]

Nadgornyi, E. (1988). Dislocation dynamics and mechanical properties of crystals. *Progress in Materials Science*, **31**, 1–536. [172]

Nagano, Y., Miyazaki, Y., Matsuo, T., and Suga, H. (1993). Heat capacities and enthalpy of fusion of heavy oxygen water. *Journal of Physical Chemistry*, **97**, 6897–901. [15]

Nagle, J. (1966). Lattice statistics of hydrogen bonded crystals. I The residual entropy of ice. *Journal of Mathematical Physics*, **7**, 1484–91. [27]

Nagle, J.F. (1974). Dielectric constant of ice. *Journal of Chemical Physics*, **61**, 883–8. [69, 71]

Nagle, J.F. (1978). Configurational statistics. *Journal of Glaciology*, **21**, 73–83. [69, 71]

Nagle, J.F. (1979). Theory of the dielectric constant of ice. *Chemical Physics*, **43**, 317–28. [71–2]

Nagle, J.F. (1983). Relevance of ice studies to bioenergetics. *Journal of Physical Chemistry*, **87**, 4086–8. [146]

Nagle, J.F. (1992). Proton transfer in condensed matter. In *Proton transfer in hydrogen-bonded systems*, NATO Advanced Science Institutes Series B Vol. 291 (ed. T. Bountis), pp. 17–28. Plenum Press, New York. [77, 146]

Nakamura, T. and Jones, S.J. (1970). Softening effect of dissolved hydrogen chloride in ice crystals. *Scripta Metallurgica*, **4**, 123–6. [191]

Nakawo, M. and Sinha, N.K. (1981). Growth rate and salinity profile of first-year sea ice in the high Arctic. *Journal of Glaciology*, **27**, 315–30. [292]

Nakaya, U. (1954). *Snow crystals—natural and artificial*. Harvard University Press, Cambridge, Massachusetts. [12, 293–4]

Nakaya, U. (1958). *Mechanical properties of single crystals of ice. Part I. Geometry of deformation*, US Army Snow Ice and Permafrost Research Establishment Research Report no. 28. [13, 186–7]

Narten, A.H., Venkatesh, C.G., and Rice, S.A. (1976). Diffraction pattern and structure of amorphous solid water at 10 and 77 K. *Journal of Chemical Physics*, **64**, 1106–21. [280]

Nasello, O., Di Prinzio, C., and Levi, L. (1992). Grain boundary migration in bicrystals of ice. In *Physics and chemistry of ice* (ed. N. Maeno and T. Hondoh), pp. 206–11. Hokkaido University Press, Sapporo. [181]

Nelmes, R.J., Loveday, J.S., Wilson, R.M., Besson, J.M., Pruzan, Ph., Klotz, S. *et al.* (1993). Neutron diffraction study of the structure of deuterated ice VIII to 10 GPa. *Physical Review Letters*, **71**, 1192–5. [264]

Nelmes, R.J., Loveday, J.S., Marshall, W.G., Besson, J.M., Klotz, S., and Hamel, G. (1998*a*). Structures of ice VII and ice VIII to 20 GPa. *The Review of High Pressure Science and Technology*, **7**, 1138–40. [265]

Nelmes, R.J., Loveday, J.S., Marshall, W.G., Hamel, G., Besson, J.M., and Klotz, S. (1998*b*). Multi-site disordered structure of ice VII to 20 GPa. *Physical Review Letters*, **81**, 2719–22. [264]

Nemat-Nasser, S., and Horii, H.J. (1982). Compression-induced nonplanar crack extension with application to splitting, exfoliation, and rockburst. *Journal of Geophysical Research*, **87**(B8), 6805–21. [208]

Nersesova, Z.A. (1950). Temperature dependence of ice content in ground. *Doklady Akademii Nauk SSSR*, **75**, 845–6. In Russian. [306]

Nickolayev, O. and Petrenko, V.F. (1995). SFM studies of the surface morphology of ice. *Materials Research Society Symposium Proceedings*, **355**, 221–6. [242]

Nickolayev, O.Y. and Schulson, E.M. (1995). Grain-boundary sliding and across-column cracking in columnar ice. *Philosophical Magazine Letters*, **72**, 93–7. [196]

Nield, V.M. and Whitworth, R.W. (1995). The structure of ice Ih from analysis of single-crystal neutron diffuse scattering. *Journal of Physics: Condensed Matter*, **7**, 8259–71. [34]

Nilsson, G., Christensen, H., Pagsberg, P., and Nielson, S.O. (1972). Transient electrons in pulse-irradiated crystalline water and deuterium oxide ice. *Journal of Physical Chemistry*, **76**, 1000–8. [151]

Nixon, W.A. and Schulson, E.M. (1987). A micromechanical view of the fracture toughness of ice. *Journal de Physique*, **48**, Colloque C1, 313–9. [207]

Noll, G. (1978). The influence of the rate of deformation on the electrical properties of ice monocrystals. *Journal of Glaciology*, **21**, 277–89. [92, 94, 114–15]

Nye, J.F. (1952). The mechanics of glacier flow. *Journal of Glaciology*, **2**, 82–93. [300]

Nye, J.F. (1953). The flow law of ice from measurements in glacier tunnels, laboratory experiments and the Jungfraufirn borehole experiment. *Proceedings of the Royal Society of London*, **A219**, 477–89. [199]

Nye, J.F. (1957). *Physical properties of crystals*. Clarendon Press, Oxford. [37]

Nye, J.F. (1965). The flow of a glacier in a channel of rectangular, elliptic or parabolic cross-section. *Journal of Glaciology*, **5**, 661–90. [300]

Nye, J.F. (1992). Water veins and lenses in polycrystalline ice. In *Physics and chemistry of ice* (ed. N. Maeno and T. Hondoh), pp. 200–5. Hokkaido University Press, Sapporo. [182]

Odutola, J.A. and Dyke, T.R. (1980). Partially deuterated water dimers: microwave spectra and structure. *Journal of Chemical Physics*, **72**, 5062–70. [7]

Oguro, M. (1988). Dislocations in artificially grown single crystals. In *Lattice defects in ice crystals* (ed. A. Higashi), pp. 27–47. Hokkaido University Press, Sapporo. [13, 163–4]

Oguro, M. and Higashi, A. (1973). Stacking-fault images in NH_3-doped ice crystals revealed by X-ray diffraction topography. In *Physics and chemistry of ice* (ed.

E. Whalley, S.J. Jones, and L.W. Gold), pp. 338–43. Royal Society of Canada, Ottawa. [179]

Oguro, M. and Higashi, A. (1981). The formation mechanism of concentric dislocation loops in ice single crystals grown from the melt. *Journal of Crystal Growth*, **51**, 71–80. [164]

Oguro, M. and Hondoh, T. (1988). Stacking faults in ice crystals. In *Lattice defects in ice crystals* (ed. A. Higashi), pp. 49–67. Hokkaido University Press, Sapporo. [179]

Oguro, M. and Whitworth, R.W. (1991). Dielectric observations of the transformation of single crystals of KOH-doped ice Ih to ice XI. *Journal of Physics and Chemistry of Solids*, **52**, 401–3. [109–10, 261–2]

Oguro, M., Hatano, K., and Kato, S. (1982). Orientation dependence of internal friction in artificial single crystals of ice. *Cold Regions Science and Technology*, **6**, 29–35. [138]

Oguro, M., Hondoh, T., and Azuma, K. (1988). Interactions between dislocations and point defects in ice crystals. In *Lattice defects in ice crystals* (ed. A. Higashi), pp. 97–128. Hokkaido University Press, Sapporo. [134, 164, 179–80]

Ohtomo, M., Ahmad, S., and Whitworth, R.W. (1987). A technique for the growth of high quality single crystals of ice. *Journal de Physique*, **48**, Colloque C1, 595–8. [13]

Ojamäe, L., Hermansson, K., Dovesi, R., Roetti, C., and Saunders, V.R. (1994). Mechanical and molecular properties of ice VIII from crystal-orbital *ab initio* calculations. *Journal of Chemical Physics*, **100**, 2128–38. [265]

Oksanen, P. and Keinonen, J. (1982). The mechanism of friction of ice. *Wear*, **78**, 315–24. [318, 320]

Onsager, L. (1967). Ferroelectricity of ice? In *Ferroelectricity* (ed. E.F. Weller), pp. 16–19. Elsevier, Amsterdam. [261]

Onsager, L. and Dupuis, M. (1960). The electrical properties of ice. In *Termodinamica dei processi irreversibili*, Rendiconti della Scuola Internazionale di Fisica "Enrico Fermi", Corso X, Varenna, 1959, pp. 294–315. Nicola Zanichelli, Bologna. [79]

Onsager, L. and Dupuis, M. (1962). The electrical properties of ice. In *Electrolytes* (ed. B. Pesce), pp. 27–46. Pergamon Press, Oxford. [68, 76, 79]

Onsager, L. and Runnels, L.K. (1963). Mechanism for self-diffusion in ice. *Proceedings of the National Academy of Sciences of the United States of America*, **50**, 208–10. [138, 142, 147]

Onsager, L. and Runnels, L.K. (1969). Diffusion and relaxation phenomena in ice. *Journal of Chemical Physics*, **50**, 1089–103. [88, 147]

Parsonage, N.G. and Staveley, L.A.K. (1978). *Disorder in crystals*, Chap. 8. Clarendon Press, Oxford. [283]

Pastori Parravicini, G. and Resca, L. (1973). Electronic states and optical properties in cubic ice. *Physical Review*, **B8**, 3009–23. [225]

Paterson, W.S.B. (1994). *The physics of glaciers* (3rd edn). Pergamon/Elsevier Science, Oxford. [299, 301]

Pauling, L. (1935). The structure and entropy of ice and other crystals with some randomness of atomic arrangement. *Journal of the American Chemical Society*, **57**, 2680–4. [15, 25]

Pauling, L. (1960). *The nature of the chemical bond* (3rd edn). Cornell University Press, Ithaca. [6, 19]

Penny, A.H.A. (1948). A theoretical determination of the elastic constants of ice. *Proceedings of the Cambridge Philosophical Society*, **44**, 423–39. [56]

Perez, J., Maï, C., and Vassoille, R. (1978). Cooperative movement of H_2O molecules and dynamic behaviour of dislocations in ice Ih. *Journal of Glaciology*, **21**, 361–74. [175]

Perez, J., Maï, C., Tatibouet, J., and Vassoille, R. (1979). Etude des joints de grains dans la glace Ih par mesure du frottement intérieur. *Physica Status Solidi* (*a*), **52**, 321–30. [182, 205]

Perez, J., Maï, C., Tatibouët, J., and Vassoille, R. (1980). Dynamic behaviour of dislocations in HF-doped ice Ih. *Journal of Glaciology*, **25**, 133–49. [175]

Perovich, D.K. (1996). *Optical properties of sea ice*, US Army Corps of Engineers Cold Regions Research and Engineering Laboratory Monograph 96-1. [292]

Perovich, D.K. and Govoni, J.W. (1991). Absorption coefficients of ice from 250 to 400 nm. *Geophysical Research Letters*, **18**, 1233–5. [220–1]

Perovich, D.K. and Gow, A.J. (1996). A quantitative description of sea ice inclusions. *Journal of Geophysical Research*, **101**(C8), 18327–43. [290]

Peter, T. (1997). Microphysics and heterogeneous chemistry of polar stratospheric clouds. *Annual Review of Physical Chemistry*, **48**, 785–822. [296]

Peter, T. (1998). Polar stratospheric clouds on Earth. In *Solar System ices* (ed. B. Schmitt, C. de Bergh, and M. Festou), pp. 443–75. Kluwer Academic Publishers, Dordrecht. [296]

Peterson, S.W. and Levy, H.A. (1957). A single-crystal neutron diffraction study of heavy ice. *Acta Crystallographica*, **10**, 70–6. [18, 19]

Petit, J.R., Basile, I., Leruyuet, A., Raynaud, D., Lorius, C., Jouzel, J. *et al.* (1997). Four climate cycles in Vostok ice core. *Nature*, **387**, 359–60. [301]

Petrenko, V.F. (1993*a*). On the nature of electrical polarization of materials caused by cracks. Application to ice electromagnetic emission. *Philosophical Magazine*, **B67**, 301–15. [212]

Petrenko, V.F. (1993*b*). *Electrical properties of ice*, US Army Corps of Engineers Cold Regions Research and Engineering Laboratory Special Report 93-20. [94]

Petrenko, V.F. (1994*a*). *The surface of ice*, US Army Corps of Engineers Cold Regions Research and Engineering Laboratory Special Report 94-22. [227]

Petrenko, V.F. (1994*b*). The effect of static electric fields on ice friction. *Journal of Applied Physics*, **76**, 1216–19. [320]

Petrenko, V.F. (1997). Study of the surface of ice, ice/solid and ice/liquid interfaces with scanning force microscopy. *Journal of Physical Chemistry*, **B101**, 6276–81. [242–3, 317]

Petrenko, V.F. (1998). Effect of electric fields on adhesion of ice to mercury. *Journal of Applied Physics*, **84**, 261–7. [317]

Petrenko, V.F. and Chesnakov, V.A. (1990*a*). Recombination injection of H_3O^+ and OH^- charge-carriers during electrolysis of water. *Doklady Akademii Nauk SSSR*, **314**, 359–61. [*Physics—Doklady*, **35**, 822–3.] [121]

Petrenko, V.F. and Chesnakov, V.A. (1990*b*). Nature of carriers in ice. *Fizika Tverdogo Tela*, **32**, 2368–72. [*Soviet Physics—Solid State*, **32**, 1374–77.] [62]

Petrenko, V.F. and Chesnakov, V.A. (1990*c*). Investigations of the physical properties of ice–electronic conductor interfaces. *Fizika Tverdogo Tela*, **32**, 2655–60. [*Soviet Physics—Solid State*, **32**, 1539–42.] [112]

Petrenko, V.F. and Chesnakov, V.A. (1990*d*). Recombination injection of charge carriers in ice. *Fizika Tverdogo Tela*, **32**, 2947–52. [*Soviet Physics—Solid State*, **32**, 1711–14.] [121–3]

Petrenko, V.F. and Colbeck, S.C. (1995). Generation of electric fields by ice and snow friction. *Journal of Applied Physics*, **77**, 4518–21. [238, 320]

Petrenko, V.F. and Khusnatdinov, N.N. (1994). On the nature of photo charge carriers in ice. *Journal of Chemical Physics*, **100**, 9096–105. [152]

Petrenko, V.F. and Maeno, N. (1987). Ice field transistor. *Journal de Physique*, **48**, Colloque C1, 115–19. [115–16, 238]

Petrenko, V.F. and Ryzhkin, I.A. (1984*a*). Dielectric properties of ice in the presence of space charge. *Physica Status Solidi* (*b*), **121**, 421–7. [85–7, 114]

Petrenko, V.F. and Ryzhkin, I.A. (1984*b*). Space-charge-limited currents in ice. *Zhurnal Eksperimental'noi i Teoreticheskoi Fiziki*, **87**, 558–69. [*Journal of Experimental and Theoretical Physics*, **60**, 320–6.] [118]

Petrenko, V.F. and Ryzhkin, I.A. (1984*c*). Theory of anelastic relaxation of ice. *Fizika Tverdogo Tela*, **26**, 2681–8. [*Soviet Physics—Solid State*, **26**, 1624–8.] [138]

Petrenko, V.F. and Ryzhkin, I.A. (1993). Electron energy spectrum of ice. *Physical Review Letters*, **71**, 2626–9. [225]

Petrenko, V.F. and Ryzhkin, I.A. (1997). Surface states of charge carriers and electrical properties of the surface layer of ice. *Journal of Physical Chemistry*, **B101**, 6285–9. [116, 238–9, 250]

Petrenko, V.F. and Schulson, E.M. (1992). The effect of static electric fields on protonic conductivity of ice single crystals. *Philosophical Magazine*, **B66**, 341–53. [120–1]

Petrenko, V.F. and Schulson, E.M. (1993). Action of electric fields on the plastic deformation of pure and doped ice single crystals. *Philosophical Magazine*, **A67**, 173–85. [176, 317]

Petrenko, V.F. and Whitworth, R.W. (1983). Electric currents associated with dislocation motion in ice. *Journal of Physical Chemistry*, **87**, 4022–4. [176]

Petrenko, V.F., Whitworth, R.W., and Glen, J.W. (1983). Effects of proton injection on the electrical properties of ice. *Philosophical Magazine*, **B47**, 259–78. [97, 117–19]

Petrenko, V.F., Ebinuma, T., and Maeno, N. (1986). Protonic photoconductivity of ice. *Physica Status Solidi* (*a*), **93**, 695–702. [153]

Péwé, T.L. (1983). The periglacial environment in North America during Wisconsin time. In *Late-quaternary environments of the United States*, Vol. 1. The Late Pleistocene (ed. S.C. Porter), pp. 157–89. Longman, London. [306]

Picu, R.C. and Gupta, V. (1995*a*). Crack nucleation in columnar ice due to elastic anisotropy and grain boundary sliding. *Acta Metallurgica et Materialia*, **43**, 3783–9. [211]

Picu, R.C. and Gupta, V. (1995*b*). Observations of crack nucleation in columnar ice due to grain boundary sliding. *Acta Metallurgica et Materialia*, **43**, 3791–7. [211]

Pilcher, C.B., Chapman, C.R., Lebofsky, L.A., and Kieffer, H.H. (1970). Saturn's rings: identification of water frost. *Science*, **215**, 1372–3. [309]

Pisani, C., Casassa, S., and Ugliengo, P. (1996). Proton-ordered ice structures at zero pressure. A quantum-mechanical investigation. *Chemical Physics Letters*, **253**, 201–8. [30]

Pistorius, C.W.F.T., Rapoport, E., and Clark, J.B. (1968). Phase diagrams of H_2O and D_2O at high pressures. *Journal of Chemical Physics*, **48**, 5509–14. [254]

Pittenger, B., Cook, D.J., Slaughterbeck, C.R., and Fain, S.C. (1998). Investigation of ice–solid interfaces by force microscopy: Plastic flow and adhesive forces. *Journal of Vacuum Science and Technology*, **A16**, 1832–7. [317]

Pitzer, K.S. and Polissar, J. (1956). The order–disorder problem for ice. *Journal of Physical Chemistry*, **60**, 1140–2. [29]

Plummer, P.L.M. (1992). Structural studies and molecular dynamics simulations of defects in ice. In *Physics and chemistry of ice* (ed. N. Maeno and T. Hondoh), pp. 54–61. Hokkaido University Press, Sapporo. [143]

Plummer, P.L.M. (1997). Quantum mechanical studies of the energetics of ionic defects in icelike systems. *Journal of Physical Chemistry*, **B101**, 6247–50. [144]

Polian, A. and Grimsditch, M. (1984). New high-pressure phase of H_2O: ice X. *Physical Review Letters*, **52**, 1312–14. [274]

Poole, P.H., Essmann, U., Sciortino, F., and Stanley, H.E. (1993). Phase diagram for amorphous solid water. *Physical Review*, **E48**, 4605–10. [280]

Price, P.B. and Bergström, L. (1997). Enhanced Rayleigh scattering as a signature of nanoscale defects in highly transparent solids. *Philosophical Magazine*, **A75**, 1383–90. [220]

Proctor, T.M. (1966). Low-temperature speed of sound in single-crystal ice. *Journal of the Acoustical Society of America*, **39**, 972–7. [39]

Pruppacher, H.R. and Klett, J.D. (1997). *Microphysics of clouds and precipitation* (2nd edn), Kluwer Academic Publishers, Dordrecht. [293–4]

Pruzan, Ph., Chervin, J.C., and Canny, B. (1992). Determination of the D_2O ice VII–VIII transition line by Raman scattering up to 51 GPa. *Journal of Chemical Physics*, **97**, 718–21. [264, 266]

Pruzan, Ph., Chervin, J.C., and Canny, B. (1993). Stability domain of the ice VIII proton-ordered phase at very high pressure and low temperature. *Journal of Chemical Physics*, **99**, 9842–6. [266]

Pruzan, Ph., Wolanin, E., Gauthier, M., Chervin, J.C., Canny, B., Häusermann, D. *et al.* (1997). Raman scattering and X-ray diffraction of ice in the megabar range. Occurrence of a symmetric disordered solid above 62 GPa. *Journal of Physical Chemistry*, **B101**, 6230–3. [274–5]

Rabideau, S.W., Finch, E.D., and Denison, A.B. (1968). Proton and deuteron NMR of ice polymorphs. *Journal of Chemical Physics*, **49**, 4660–5. [32]

Rabier, J. and George, A. (1987). Dislocations and plasticity in semiconductors. II. The relation between dislocation dynamics and plastic deformation. *Revue de Physique Appliquée*, **22**, 1327–51. [191]

Ramseier, R.O. (1967). Self-diffusion of tritium in natural and synthetic ice monocrystals. *Journal of Applied Physics*, **38**, 2553–6. [132]

Raraty, L.E. and Tabor, D. (1958). The adhesion and strength properties of ice. *Proceedings of the Royal Society of London*, **A245**, 184–201. [315]

Ratcliffe, E.H. (1962). Thermal conductivity of ice. New data on the temperature coefficient. *Philosophical Magazine*, **7**, 1197–203. [43]

Ravi-Chandar, K., Adamson, B., Lazo, J., and Dempsey, J.P. (1994). Stress-optic effect in ice. *Applied Physics Letters*, **64**, 1183–5. [222]

Raynaud, D., Jouzel, J., Barnola, J.M., Chappellaz, J., Delmas, R.J., and Lorius, C. (1993). The ice record of greenhouse gases. *Science*, **259**, 926–34. [301]

Read, W.T. (1953). *Dislocations in crystals*. McGraw-Hill, New York. [156, 168, 181]

Readey, D.W. and Kingery, W.D. (1964). Plastic deformation of single crystal ice. *Acta Metallurgica*, **12**, 171–8. [190]

Reinhard, R. (1986). The Giotto encounter with comet Halley. *Nature*, **321**, 313–18. [313]

Renker, B. (1973). Lattice dynamics of hexagonal ice. In *Physics and chemistry of ice* (ed. E. Whalley, S.J. Jones, and L.W. Gold), pp. 82–6. Royal Society of Canada, Ottawa. [55, 57]

Reusch, E. (1864). Beiträge zur Lehre vom Eis. *Annalen der Physik und Chemie* (Poggendorff), **121**, 573–8. [184]

Reynolds, S.E., Brook, M., and Gourley, M.F. (1957). Thunderstorm charge separation. *Journal of Meteorology*, **14**, 426–36. [295]

Rice, S.A., Bergren, M.S., Belch, A.C., and Nielson, G. (1983). A theoretical analysis of the OH stretching spectra of ice Ih, liquid water, and amorphous solid water. *Journal of Physical Chemistry*, **87**, 4295–308. [58]

Richardson, C. (1976). Phase relationships in sea ice as a function of temperature. *Journal of Glaciology*, **17**, 507–19. [291]

Rickman, H. (1998). Composition and physical properties of comets. In *Solar system ices* (ed. B. Schmitt, C. de Bergh, and M. Festou), pp. 395–417. Kluwer Academic Publishers, Dordrecht. [313]

Rigsby, G.P. (1958). Effect of hydrostatic pressure on velocity of shear deformation of single ice crystals. *Journal of Glaciology*, **3**, 273–8. [191]

Rigsby, G.P. (1968). The complexities of the three-dimensional shape of individual crystals in glacier ice. *Journal of Glaciology*, **7**, 233–51. [300]

Rist, M.A. (1997). High-stress ice fracture and friction. *Journal of Physical Chemistry*, **B101**, 6263–6. [191, 321]

Rist, M.A. and Murrell, S.A.F. (1994). Ice triaxial deformation and fracture. *Journal of Glaciology*, **40**, 305–18. [201, 211–12]

Rist, M.A., Sammonds, P.R., Murrell, S.A.F., Meredith, P.G., Oerter, H., and Doake, C.S.M. (1996). Experimental fracture and mechanical properties of Antarctic ice: preliminary results. *Annals of Glaciology*, **23**, 284–92. [301]

Ritzhaupt, G., and Devlin, J.P. (1980). Direct spectroscopic observation of proton exchange and Bjerrum defect migration in cubic ice. *Journal of Chemical Physics*, **72**, 6807–8. [141]

Ritzhaupt, G., Thornton, C., and Devlin, J.P. (1978). Infrared spectrum of D_2O vibrationally decoupled in H_2O ice Ic. *Chemical Physics Letters*, **59**, 420–2. [140]

Rogers, R.R. and Yau, M.K. (1989). *A short course in cloud physics* (3rd edn). Butterworth-Heinemann, Oxford. [293]

Rothery, D.A. (1992). *Satellites of the outer planets*. Clarendon Press, Oxford. [Second edition 2000.] [307–8, 312]

Röttger, K., Endriss, A., Ihringer, J., Doyle, S., and Kuhs, W.F. (1994). Lattice constants and thermal expansion of H_2O and D_2O ice Ih between 10 and 265 K. *Acta Crystallographica*, **B50**, 644–8. [20, 22, 41–2, 256]

Rowland, B. and Devlin, J.P. (1991). Spectra of dangling OH groups at ice cluster surfaces and within pores of amorphous ice. *Journal of Chemical Physics*, **94**, 812–13. [229]

Rowland, B., Kadagathur, N.S., Devlin, J.P., Buch, V., Feldman, T., and Wojcik, M.J. (1995). Infrared spectra of ice surfaces and assignment of surface-localized modes from simulated spectra of cubic ice. *Journal of Chemical Physics*, **102**, 8328–41.[229]

Ruepp, R. (1973). Electrical properties of ice Ih single crystals. In *Physics and chemistry of ice* (ed. E. Whalley, S.J. Jones, and L.W. Gold), pp. 179–86. Royal Society of Canada, Ottawa. [92]

Ruepp, R. and Käss, M. (1969). Dielectric relaxation, bulk and surface conductivity of ice single crystals. In *Physics of ice* (ed. N. Riehl, B. Bullemer, and H. Engelhardt), pp. 555–61. Plenum Press, New York. [234]

Ruocco, G., Sette, F., Bergmann, U., Krisch, M., Masciovecchio, C., Mazzacurati, V. *et al.* (1996). Equivalence of the sound velocity in water and ice at mesoscopic wavelengths. *Nature*, **379**, 521–3. [47]

Ryzhkin, I.A. (1985). Superionic transition in ice. *Solid State Communications*, **56**, 57–60. [128]

Ryzhkin, I.A. and Petrenko, V.F. (1997). Physical mechanisms responsible for ice adhesion. *Journal of Physical Chemistry*, **B101**, 6267–70. [318]

Ryzhkin, I.A. and Whitworth, R.W. (1997). The configurational entropy in the Jaccard theory of the electrical properties of ice. *Journal of Physics: Condensed Matter*, **9**, 395–402. [84, 102]

Sagdeev, R.Z., Blamont, J., Galeev, A.A., Moroz, V.I., Shapiro, V.D., Shevchenko, V.I. *et al.* (1986). Vega spacecraft encounters with comet Halley. *Nature*, **321**, 259–62. [313]

St Lawrence, W.F. and Cole, D.M. (1982). Acoustic emissions from polycrystalline ice. *Cold Regions Science and Technology*, **5**, 183–99. [198]

Sakata, M., Takata, M., Oshizumi, H., Goto, A., and Hondoh, T. (1992). Electron density distribution of ice Ih obtained by the maximum entropy method. In *Physics and chemistry of ice* (ed. N. Maeno and T. Hondoh), pp. 62–8. Hokkaido University Press, Sapporo. [224–5]

Salisbury, J.W., D'Aria, D.M., and Wald, A. (1994). Measurements of thermal infrared spectral reflectance of frost, snow, and ice. *Journal of Geophysical Research*, **99**(B12), 24235–340. [299]

Salm, B. (1982). Mechanical properties of snow. *Reviews of Geophysics and Space Physics*, **20**, 1–19. [297]

Sanderson, T.J.O. (1988). *Ice mechanics—risks to offshore structures*. Graham and Trotman, London. [184, 212, 290]

Sarid, D. (1994). *Scanning force microscopy: with applications to electric, magnetic, and atomic forces* (Revised edn). Oxford University Press, New York. [242]

Saunders, C.P.R. (1993). A review of thunderstorm electrification processes. *Journal of Applied Meteorology*, **32**, 642–55. [295]

Saunders, C.P.R. (1994). Thunderstorm electrification laboratory experiments and charging mechanisms. *Journal of Geophysical Research*, **99**(D5), 10773–9. [295]

Saunders, C.P.R., Keith, W.D., and Mitzeva, R.P. (1991). The effect of liquid water on thunderstrom charging. *Journal of Geophysical Research*, **96**(D6), 11007–17. [295]

Sayles, F.H. (1988). State of the art: mechanical properties of frozen soil. In *Ground freezing 88. Proceedings of the Fifth International Symposium on Ground Freezing, Nottingham, 26–28 July 1988*, Vol. 1 (ed. R.H. Jones and J.T. Holden), pp. 143–65. A.A. Balkema, Rotterdam. [307]

Sceats, M.G. and Rice, S.A. (1980). The water–water pair potential near the hydrogen bonded equilibrium configuration. *Journal of Chemical Physics*, **72**, 3236–47. [58]

Scheiner, S. (1981). Proton transfers in hydrogen-bonded systems. Cationic oligomers of water. *Journal of the American Chemical Society*, **103**, 315–20. [144]

Scheiner, S. (1992). Extraction of the principles of proton transfer by *ab initio* methods. In *Proton transfer in hydrogen-bonded systems*, NATO Advanced Science Institutes Series B Vol. 291 (ed. T. Bountis), pp. 29–47. Plenum Press, New York. [144–5]

Scheiner, S. and Nagle, J.F. (1983). Ab initio molecular orbital estimates of charge partitioning between Bjerrum and ionic defects in ice. *Journal of Physical Chemistry*, **87**, 4267–72. [103]

Scherer, J.R. and Snyder, R.G. (1977). Raman intensities of single crystal ice Ih. *Journal of Chemical Physics*, **67**, 4794–811. [47, 49]

Schiller, P. (1958). Die mechanische Relation in reinen Eiseinkristallen. *Zeitschrift für Physik*, **153**, 1–15. [137–8]

Schmitt, B., de Bergh, C., and Festou, M. (ed.) (1998*a*). *Solar system ices* (Astrophysics and Space Sciences Library, Vol. 227). Kluwer Academic Publishers, Dordrecht. [307]

Schmitt, B., Quirico, E., Trotta, F., and Grundy, W.M. (1998*b*). Optical properties of ices from UV to infrared. In *Solar system ices* (ed. B. Schmitt, C. de Bergh, and M. Festou), pp. 199–240. Kluwer Academic Publishers, Dordrecht. [309]

Schneider, J. and Zeyen, C. (1980). Elastic diffuse neutron scattering due to D–D correlation functions seen in Ih ice. *Journal of Physics C: Solid State Physics*, **13**, 4121–6. [18, 34]

Schulson, E.M. (1987). The fracture of ice Ih. *Journal de Physique*, **48**, Colloque C1, 207–20. [206]

Schulson, E.M. (1990). The brittle compressive fracture of ice. *Acta Metallurgica et Materialia*, **30**, 1963–76. [185, 209–10]
Schulson, E.M. (1997). The brittle failure of ice under compression. *Journal of Physical Chemistry*, **B101**, 6254–8. [207]
Schulson, E.M. and Buck, S.E. (1995). The ductile-to-brittle transition and ductile failure envelopes of orthotropic ice under biaxial compression. *Acta Metallurgica et Materialia*, **43**, 3661–8. [212]
Schulson, E.M. and Hibler, W.D. (1991). Fracture of ice on large scale and small: Arctic leads and wing cracks. *Journal of Glaciology*, **37**, 319–22. [292]
Schulson, E.M., Hoxie, S.G., and Nixon, W.A. (1989). The tensile strength of cracked ice. *Philosophical Magazine*, **A59**, 303–11. [207]
Schulson, E.M., Kuehn, G.A., Jones, D.E., and Fifolt, D.A. (1991). The growth of wing cracks and the brittle compressive failure of ice. *Acta Metallurgica et Materialia*, **39**, 2651–5. [208]
Schwander, J., Neftel, A., Oeschger, H., and Stauffer, B. (1983). Measurement of direct current conductivity on ice samples for climatological applications. *Journal of Physical Chemistry*, **87**, 4157–60. [94]
Schwarz, J. and Weeks, W.F. (1977). Engineering properties of sea ice. *Journal of Glaciology*, **19**, 499–531. [290]
Schweizer, K.S. and Stillinger, F.H. (1984). High pressure phase transitions and hydrogen-bond symmetry in ice polymorphs. *Journal of Chemical Physics*, **80**, 1230–40. [266, 275]
Schwerdtfeger, P. (1963). The thermal properties of sea ice. *Journal of Glaciology*, **4**, 789–807. [292]
Sciortino, F. and Corongiu, G. (1993). Evaluation of experimental quantities via analysis of molecular dynamics data. In *Methods and techniques in computational chemistry* (ed. E. Clementi), pp. 47–79. STEF, Cagliari. [57]
Seki, M., Kobayashi, K., and Nakahara, J. (1981). Optical spectra of hexagonal ice. *Journal of the Physical Society of Japan*, **50**, 2643–8. [223–4]
Sellars, C.M. (1978). Recrystallization of metals during hot deformation. *Philosophical Transactions of the Royal Society of London*, **A288**, 147–58. [204]
Sergent, C., Pougatch, E., Sudul, M., and Bourdelles, B. (1993). Experimental investigation of optical snow properties. *Annals of Glaciology*, **17**, 281–7. [299]
Shapiro, L.H., Johnson, J.B., Sturm, M., and Blaisdell, G.L. (1997). *Snow mechanics. Review of the state of knowledge and applications*, US Army Corps of Engineers Cold Regions Research and Engineering Laboratory Special Report 97-3. [297]
Sharp, R.P. (1988). *Living ice: understanding glaciers and glaciation*. Cambridge University Press, Cambridge. [299]
Shawyer, R.E. and Dean, P. (1972). Atomic vibrations in orientationally disordered systems: II. Hexagonal ice. *Journal of Physics C: Solid State Physics*, **5**, 1028–37. [58]
Shearwood, C. and Whitworth, R.W. (1989). X-ray topographic observations of edge dislocation glide on non-basal planes in ice. *Journal of Glaciology*, **35**, 281–3. [166–7]
Shearwood, C. and Whitworth, R.W. (1991). The velocity of dislocations in ice. *Philosophical Magazine*, **A64**, 289–302. [168, 170–4]
Shearwood, C. and Whitworth, R.W. (1992). The velocity of dislocations in crystals of HCl-doped ice. *Philosophical Magazine*, **A65**, 85–9. [175]
Shearwood, C. and Whitworth, R.W. (1993). Novel processes of dislocation multiplication observed in ice. *Acta Metallurgica et Materialia*, **41**, 205–10. [168]
Shibaguchi, T., Onuki, H., and Onaka, R. (1977). Electronic structures of water and ice. *Journal of the Physical Society of Japan*, **42**, 152–8. [223–4]

Shimizu, H., Nabetani, T., Nishiba, T., and Sasaki, S. (1996). High-pressure elastic properties of the VI and VII phase of ice in dense H_2O and D_2O. *Physical Review*, **B53**, 6107–10. [286]

Shoji, H. and Langway, C.C.J. (1982). Air hydrate inclusions in fresh ice core. *Nature*, **298**, 548–50. [299]

Shubin, V.N., Zhigunov, V.A., Zolotarevsky, V.I., and Dolin, P.I. (1966). Pulse radiolysis of crystalline ice and frozen crystalline aqueous solutions. *Nature*, **212**, 1002–35. [151]

Shumskii, P.A. (1964). *Principles of structural glaciology* (trans. D. Kraus). Dover, New York. [287]

Siegbahn, K. (1974). Electron spectroscopy—an outlook. *Journal of Electron Spectroscopy and Related Phenomena*, **5**, 9–37. [223–4]

Siegle, G. and Weithase, M. (1969). Spin-Gitter-Relaxation der Protonen in hexagonalem Eis. *Zeitschrift für Physik*, **219**, 364–80. [147]

Sinha, N.K. (1977). Dislocations in ice as revealed by etching. *Philosophical Magazine*, **36**, 1385–404. [162]

Sinha, N.K. (1978*a*). Short-term rheology of polycrystalline ice. *Journal of Glaciology*, **21**, 457–73. [38]

Sinha, N.K. (1978*b*). Rheology of columnar-grained ice. *Experimental Mechanics*, **18**, 464–70. [205]

Sinha, N.K. (1984). Intercrystalline cracking, grain-boundary sliding, and delayed elasticity at high temperatures. *Journal of Materials Science*, **19**, 359–76. [205]

Sirota, N.N. and Zhapparov, K.T. (1994). Phase diagram of heavy ice at low temperatures and high pressures. *Doklady Akademii Nauk SSSR*, **334**, 577–80. [*Physics—Doklady*, **39**, 99–102.] [254]

Sivakumar, T.C., Rice, S.A., and Sceats, M.G. (1978). Raman spectroscopic studies of the OH stretching region of low density amorphous solid water and of polycrystalline ice Ih. *Journal of Chemical Physics*, **69**, 3468–76. [48, 51]

Slack, G.A. (1980). Thermal conductivity of ice. *Physical Review*, **B22**, 3065–71. [43–5]

Slater, J.C. (1941). Theory of the transition in KH_2PO_4. *Journal of Chemical Physics*, **9**, 16–33. [68]

Slaughterbeck, C.R., Kukes, E.W., Pittenger, B., Cook, D.J., Williams, P.C., Eden, V.L. *et al.* (1996). Electric field effects on force curves for oxidized silicon tips and ice surfaces in a controlled environment. *Journal of Vacuum Science and Technology*, **A14**, 1213–18. [242–3, 317]

Sloan, E.D. (1990). *Clathrate hydrates of natural gases* (Chemical Industries Series Vol. 39). Marcel Dekker, New York. [281]

Smith, R.C. and Baker, K.S. (1981). Optical properties of the clearest natural waters (200–800 nm). *Applied Optics*, **20**, 177–84. [220]

Smith, T.F. and White, G.K. (1975). The low-temperature thermal expansion and Grüneisen parameter of some tetrahedrally bonded solids. *Journal of Physics C: Solid State Physics*, **8**, 2031–42. [41]

Sokoloff, J.B. (1973). Absence of Hall effect in ice crystals. *Physical Review Letters*, **31**, 90–2. [98]

Solomon, S. (1990). Progress towards a quantitative understanding of Antarctic ozone depletion. *Nature*, **347**, 347–54. [296]

Soper, A.K. (1994). Orientational correlation function for molecular liquids: The case of liquid water. *Journal of Chemical Physics*, **101**, 6888–901. [280]

Souchez, R.A. and Lorrain, R.D. (1991). *Ice composition and glacier dynamics*, (Springer series in physical environment, Vol. 8). Springer-Verlag, Berlin. [301]

Steckel, F. and Szapiro, S. (1963). Physical properties of heavy oxygen water. Part 1—Density and thermal expansion. *Transactions of the Faraday Society*, **59**, 331–43. [15]

Steinemann, A. (1957). Dielektrische Eigenschaften von Eiskristallen. II—Dielektrische Untersuchungen an Eiskristallen mit eingelagerten Fremdatomen. *Helvetica Physica Acta*, **30**, 581–610. [99, 105, 111]

Steinemann, S. (1954). Results of preliminary experiments on the plasticity of ice crystals. *Journal of Glaciology*, **2**, 404–13. [14]

Steinemann, S. (1958). *Experimentelle Untersuchungen zur Plastizität von Eis*, Beiträge zur Geologie der Schweiz 10. Kümmerly and Frey, Bern. [193–204]

Stillinger, F.H. (1982). Low frequency dielectric properties of liquid and solid water. In *The liquid state of matter: Fluids, simple and complex*, (Studies in Statistical Mechanics 8, ed. E.W. Montroll and J.L. Lebowitz), pp. 341–431. North-Holland, Amsterdam. [71]

Stillinger, F.H. and Rahman, A. (1974). Improved simulation of liquid water by molecular dynamics. *Journal of Chemical Physics*, **60**, 1545–57. [8]

Stillinger, F.H. and Schweizer, K.S. (1983). Ice under pressure: Transition to symmetrical hydrogen bonds. *Journal of Physical Chemistry*, **87**, 4281–8. [273, 275]

Stoneham, A.M. and Tasker, P.W. (1985). Metal–non-metal and other interfaces: the role of image interactions. *Journal of Physics C: Solid State Physics*, **18**, L543–548. [318]

Su, X., Lianos, L., Shen, Y.R., and Somorjai, G.A. (1998). Surface-induced ferroelectric ice on Pt(111). *Physical Review Letters*, **80**, 1533–6. [228, 240]

Suga, H. (1997). Ultra-slow relaxation in frozen-in disordered crystals. *Cryo-Letters*, **18**, 55–64. [282]

Suga, H., Matsuo, T., and Yamamuro, O. (1992). Slow dynamics of ordering processes in ice Ih and clathrate hydrates. In *Physics and chemistry of ice* (ed. N. Maeno and T. Hondoh), pp. 1–8. Hokkaido University Press, Sapporo. [258, 282]

Sutton, A.P. (1984). Grain-boundary structure. *International Metals Reviews*, **29**, 377–402. [181]

Suzuki, S. and Kuroiwa, D. (1972). Grain-boundary energy and grain-boundary groove angles in ice. *Journal of Glaciology*, **11**, 265–77. [181]

Svishchev, I.M. and Kusalik, P.G. (1994). Crystallization of liquid water in a molecular dynamics simulation. *Physical Review Letters*, **73**, 975–8. [277]

Svishchev, I.M. and Kusalik, P.G. (1995). Spatial structure in low-temperature amorphous phases of water. *Chemical Physics Letters*, **239**, 349–53. [280]

Svishchev, I.M. and Kusalik, P.G. (1996). Quartzlike polymorph of ice. *Physical Review*, **B53**, R8815–8817. [255]

Szyrmer, W. and Zawadzki, I. (1997). Biogenic and anthropogenic sources of ice-forming nuclei: a review. *Bulletin of the American Meteorological Society*, **78**, 209–28. [295]

Tabor, D. (1987). Friction and wear—developments over the last fifty years. In *Proceedings of the Institution of Mechanical Engineers International Conference 1987–5 on 'Tribology—Friction, lubrication and wear. Fifty years on'*, Vol. 1, pp. 157–72. Mechanical Engineering Publications, London. [320]

Tajima, Y., Matsuo, T., and Suga, H. (1984). Calorimetric study of phase transition in hexagonal ice doped with alkali hydroxides. *Journal of Physics and Chemistry of Solids*, **45**, 1135–44. [109, 254, 257–8]

Takahashi, T. (1970). Electric surface potential of growing ice crystals. *Journal of the Atmospheric Sciences*, **27**, 453–62. [240]

Takahashi, T. (1973). Electrification of growing ice crystals. *Journal of the Atmospheric Sciences*, **30**, 1220–4. [240]

Takei, I. and Maeno, N. (1984). Dielectric properties of single crystals of HCl-doped ice. *Journal of Chemical Physics*, **81**, 6186–90. [106]

Takei, I. and Maeno, N. (1987). Electrical characteristics of point defects in HCl-doped ice. *Journal de Physique*, **48**, Colloque C1, 121–6. [95–6, 106–7]

Takei, I. and Maeno, N. (1997). Dielectric low-frequency dispersion and crossover phenomena of HCl-doped ice. *Journal of Physical Chemistry*, **B101**, 6234–6. [67]

Tammann, G. (1900). Ueber die Grenzen des festen Zustandes IV. *Annalen der Physik*, Series 4, **2**, 1–31. [10, 252, 255]

Tanaka, H. (1998). Thermodynamic stability and negative thermal expansion of hexagonal and cubic ices. *Journal of Chemical Physics*, **108**, 4887–93. [42]

Tanaka, H. and Okabe, I. (1996). Thermodynamic stability of hexagonal and cubic ices. *Chemical Physics Letters*, **259**, 593–8. [30]

Tatibouet, J., Perez, J., and Vassoille, R. (1983). Study of lattice defects in ice by very-low-frequency internal friction measurements. *Journal of Physical Chemistry*, **87**, 4050–4. [138]

Tatibouet, J., Perez, J., and Vassoille, R. (1986). High-temperature internal friction and dislocations in ice Ih. *Journal de Physique*, **47**, 51–60. [171]

Tatibouet, J., Perez, J., and Vassoille, R. (1987). Study of grain boundaries in ice by internal friction measurement. *Journal de Physique*, **48**, Colloque C1, 197–203. [182]

Taub, I.A. and Eiben, K. (1968). Transient solvated electron, hydroxyl, and hydroperoxy radicals in pulse-irradiated crystalline ice. *Journal of Chemical Physics*, **49**, 2499–513. [151]

Taubenberger, R., Hubmann, M., and Gränicher, H. (1973). Effect of hydrostatic pressure on the dielectric properties of ice Ih single crystals. In *Physics and chemistry of ice* (ed. E. Whalley, S.J. Jones, and L.W. Gold), pp. 194–8. Royal Society of Canada, Ottawa. [139]

Taylor, G.I. (1938). Plastic strain in metals. *Journal of the Institute of Metals*, **62**, 307–24. [193]

Taylor, K., Alley, R., Fiacco, J., Grootes, P., Lamorey, G., Mayewski, P. *et al.* (1992). Ice-core dating and chemistry by direct-current electrical conductivity. *Journal of Glaciology*, **38**, 325–32. [111, 303]

Thibert, E. and Dominé, F. (1997). Thermodynamics and kinetics of the solid solution of HCl in ice. *Journal of Physical Chemistry*, **B101**, 3554–65. [150, 296]

Thibert, E. and Dominé, F. (1998). Thermodynamics and kinetics of the solid solution of HNO_3 in ice. *Journal of Physical Chemistry*, **B102**, 4432–9. [150]

Tse, J.S. (1992). Mechanical instability in ice Ih. A mechanism for pressure-induced amorphization. *Journal of Chemical Physics*, **96**, 5482–7. [280]

Tse, J.S., Klein, M.L., and McDonald, I.R. (1984). Lattice vibrations of ices Ih, VIII, and IX. *Journal of Chemical Physics*, **81**, 6124–9. [57]

Tuckerman, M.E., Marx, D., Klein, M.L., and Parrinello, M. (1997). On the quantum nature of the shared proton in hydrogen bonds. *Science*, **275**, 817–20. [144]

Tulk, C.A., Gagnon, R.E., Kiefte, H., and Clouter, M.J. (1994). Elastic constants of ice III by Brillouin spectroscopy. *Journal of Chemical Physics*, **101**, 2350–4. [286]

Tulk, C.A., Gagnon, R.E., Kiefte, H., and Clouter, M.J. (1996). Elastic constants of ice VI by Brillouin spectroscopy. *Journal of Chemical Physics*, **104**, 7854–9. [286]

Tulk, C.A., Gagnon, R.E., Kiefte, H., and Clouter, M.J. (1997). The pressure dependence of the elastic constants of ice III and VI. *Journal of Chemical Physics*, **107**, 10684–90. [286]

Turner, G.J., Stow, C.D., and Keatinge, R. (1987). The manufacture of large samples of monocrystalline ice under microcomputer control. *Journal of Crystal Growth*, **80**, 463–4. [13]

Tyndall, J. (1858). On some physical properties of ice. *Philosophical Transactions of the Royal Society of London*, **148**, 211–29. [227]

Tyndall, J. (1860). *The glaciers of the Alps*. John Murray, London. [184]

Unwin, P.N.T. and Muguruma, J. (1972). Electron microscope observations on the defect structure of ice. *Physica Status Solidi (a)*, **14**, 207–16. [162]

van der Veen, J.F., Pluis, B., and Denier van der Gon, A.W. (1988). Surface melting. In *Chemistry and physics of solid surfaces*, Vol. 7 (ed. R. Vanselow and R.F. Howe), pp. 455–67. Springer-Verlag, Berlin. [227]

van Oss, C.J., Giese, R.F., Wentzek, R., Norris, J., and Chuvilin, E.M. (1992). Surface tension parameters of ice obtained from contact angle data and from positive and negative particle adhesion to advancing freezing fronts. *Journal of Adhesion Science and Technology*, **6**, 503–16. [244, 250]

Varrot, M., Rochas, G., and Klinger, J. (1978). Thermal conductivity of ice in the temperature range 0.5 to 5.0 K. *Journal of Glaciology*, **21**, 241–5. [43, 45]

Vassoille, R., Maï, C., and Perez, J. (1978). Inelastic behaviour of ice Ih single crystals in the low-frequency range due to dislocations. *Journal of Glaciology*, **21**, 375–84. [171]

Vineyard, G.H. (1957). Frequency factors and isotope effects in solid state rate processes. *Journal of Physics and Chemistry of Solids*, **3**, 121–7. [129]

Vömel, H., Hofmann, D.J., Oltmans, S.J., and Harris, J.M. (1995). Evidence for midwinter chemical ozone destruction over Antarctica. *Geophysical Research Letters*, **22**, 2381–4. [296]

Voitkovsky, K.F., Bozhinsky, A.N., Golubev, V.N., Laptev, M.N., Zhigulsky, A.A., and Slesarenko, Y.Y. (1975). Creep-induced changes in structure and density of snow. In *Snow mechanics. Proceedings of the Grindelwald Symposium, April 1974*, pp. 171–9. International Association of Hydrological Sciences Publication No. 114. [298]

von Hippel, A.R. (1954). *Dielectrics and waves*, Chap. 31. Wiley, New York. [292]

von Hippel, A., Knoll, D.B., and Westphal, W.B. (1971). Transfer of protons through "pure" ice Ih single crystals. I. Polarization spectra of ice. *Journal of Chemical Physics*, **54**, 134–44. [94, 96]

Vonnegut, B. (1947). The nucleation of ice formation by silver iodide. *Journal of Applied Physics*, **18**, 593–5. [295]

von Stackelberg, M. and Müller, H.R. (1954). Feste Gashydrate II. *Zeitschrift für Elektrochemie*, **58**, 25–39. [281, 283]

Vos, W.L., Finger, L.W., Hemley, R.J., and Mao, H. (1993). Novel H_2–H_2O clathrates at high pressures. *Physical Review Letters*, **71**, 3150–3. [283]

Vos, W.L., Finger, L.W., Hemley, R.J., and Mao, H. (1996). Pressure dependence of hydrogen bonding in a novel H_2O–H_2 clathrate. *Chemical Physics Letters*, **257**, 524–30. [283]

Wagner, W., Saul, A., and Pruss, A. (1994). International equations for the pressure along the melting and along the sublimation curve of ordinary water substance. *Journal of Physical and Chemical Reference Data*, **23**, 515–27. [11, 254]

Wakahama, G. (1964). On the plastic deformation of ice. V. Plastic deformation of polycrystalline ice. *Low Temperature Science*, **A22**, 1–24. [195]

Wallqvist, A., Martyna, G., and Berne, B.J. (1988). Behavior of the hydrated electron at different temperatures. *Journal of Physical Chemistry*, **92**, 1721–30. [151]

Walrafen, G.E., Abebe, M., Mauer, F.A., Block, S., Piermarini, G.J., and Munro, R. (1982). Raman and x-ray investigations of ice VII to 36 GPa. *Journal of Chemical Physics*, **77**, 2166–74. [262, 284]

Warman, J.M., de Haas, M.P., and Verberne, J.B. (1980). Decay kinetics of excess electrons in crystalline ice. *Journal of Physical Chemistry*, **84**, 1240–8. [151]

Warren, S.G. (1982). Optical properties of snow. *Reviews of Geophysics and Space Physics*, **20**, 67–89. [298]

Warren, S.G. (1984). Optical constants of ice from the ultraviolet to the microwave. *Applied Optics*, **23**, 1206–25. [216–17, 223]

Watanabe, H. (1991). Thermal dilation of water between 0 °C and 44 °C. *Metrologia*, **28**, 33–43. [15]

Webb, W.W. and Hayes, C.E. (1967). Dislocations and plastic deformation of ice. *Philosophical Magazine*, **16**, 909–25. [161]

Weeks, W.F. and Ackley, S.F. (1982). *The growth, structure and properties of sea ice*, US Army Corps of Engineers Cold Regions Research and Engineering Laboratory Monograph 82-1. [290]

Weeks, W.F. and Ackley, S.F. (1986). The growth, structure, and properties of sea ice. In *The geophysics of sea ice*, NATO Advanced Science Institutes Series B, Vol. 146 (ed. N. Untersteiner), pp. 9–164. Plenum Press, New York. [290]

Weeks, W.F. and Assur, A. (1967). *Mechanical properties of sea ice*, US Army Corps of Engineers Cold Regions Research and Engineering Laboratory Monograph M II-C3. [290]

Weertman, J. (1957). On the sliding of glaciers. *Journal of Glaciology*, **3**, 33–8. [300]

Weertman, J. (1963). The Eshelby–Shoeck viscous dislocation damping mechanism applied to the steady-state creep of ice. In *Ice and snow: properties, processes and applications* (ed. W.D. Kingery), pp. 28–33. MIT Press, Cambridge, Massachusetts. [175]

Weertman, J. (1973). Creep of ice. In *Physics and chemistry of ice* (ed. E. Whalley, S.J. Jones, and L.W. Gold), pp. 320–37. Royal Society of Canada, Ottawa. [184]

Wei, Y. and Dempsey, J.P. (1994). The motion of non-basal dislocations in ice crystals. *Philosophical Magazine*, **A69**, 1–10. [162]

Wei, Y., Adamson, R.M., and Dempsey, J.P. (1996). Ice/metal interfaces: fracture energy and fractography. *Journal of Materials Science*, **31**, 943–7. [316]

Weiss, J. and Grasso, J.-R. (1997). Acoustic emission in single crystals of ice. *Journal of Physical Chemistry*, **B101**, 6113–7. [191]

Weiss, J. and Schulson, E.M. (1995). The failure of fresh-water granular ice under multiaxial compressive loading. *Acta Metallurgica et Materialia*, **43**, 2303–15. [212]

Weithase, M., Noack, F., and von Schütz, J. (1971). Proton spin relaxation in hexagonal ice. II: The $T_{1\rho}$ minimum. *Zeitschrift für Physik*, **246**, 91–6. [147]

Wenk, H.-R., Canova, G., Bréchet, Y., and Flandin, L. (1997). A deformation-based model for recrystallization of anisotropic materials. *Acta Materialia*, **45**, 3283–96. [205]

Whalley, E. (1957). The difference in the intermolecular forces of H_2O and D_2O. *Transactions of the Faraday Society*, **53**, 1578–85. [29]

Whalley, E. (1969). Structure problems of ice. In *Physics of ice* (ed. N. Riehl, B. Bullemer, and H. Engelhardt), pp. 19–43. Plenum Press, New York. [253]

Whalley, E. (1974). The OH distance in ice. *Molecular Physics*, **28**, 1105–8. [31–2]

Whalley, E. (1976). The hydrogen bond in ice. In *The hydrogen bond*, Vol. 3 (ed. P. Schuster, G. Zundel, and C. Sandorfy), pp. 1425–70. North-Holland, Amsterdam. [29, 32–3]

Whalley, E. (1977). A detailed assignment of the O—H stretching bands of ice I. *Canadian Journal of Chemistry*, **55**, 3429–41. [51]

Whalley, E. and Bertie, J.E. (1967). Optical spectra of orientationally disordered crystals. I. Theory of translational lattice vibrations. *Journal of Chemical Physics*, **46**, 1264–70. [47]

Whalley, E., Davidson, D.W., and Heath, J.B.R. (1966). Dielectric properties of ice VII. Ice VIII: a new phase of ice. *Journal of Chemical Physics*, **45**, 3976–82. [264]

Whalley, E., Heath, J.B.R., and Davidson, D.W. (1968). Ice IX: An antiferroelectric phase related to ice III. *Journal of Chemical Physics*, **48**, 2362–70. [272]

Whalley, E., Klug, D.D., Floriano, M.A., Svensson, E.C., and Sears, V.F. (1987). Recent work on high-density amorphous ice. *Journal de Physique*, **48**, Colloque C1, 429–34. [279]

Whatley, L.S. and Van Valkenburg, A. (1966). High pressure optics. In *Advances in high pressure research*, Vol. 1 (ed. R.S. Bradley), pp. 327–71. Academic Press, London. [252]

Whipple, F.L. (1950). A comet model. I. The acceleration of comet Encke. *Astrophysical Journal*, **111**, 375–94. [313]

Whipple, F.L. (1985). *The mystery of comets*. Cambridge University Press, Cambridge. [313]

Whitworth, R.W. (1975). Charged dislocations in ionic crystals. *Advances in Physics*, **24**, 203–304, §15.5. [175]

Whitworth, R.W. (1980). Influence of the choice of glide plane on the theory of the velocity of dislocations in ice. *Philosophical Magazine*, **A41**, 521–8. [161, 174]

Whitworth, R.W. (1983). Velocity of dislocations in ice on {0001} and $\{10\bar{1}0\}$ planes. *Journal of Physical Chemistry*, **87**, 4074–8. [174]

Whitworth, R.W., Paren, J.G., and Glen, J.W. (1976). The velocity of dislocations in ice—a theory based on proton disorder. *Philosophical Magazine*, **33**, 409–26. [174]

Wilen, L.A., Wettlaufer, J.S., Elbaum, M., and Schick, M. (1995). Dispersion-force effects in interfacial premelting of ice. *Physical Review*, **B52**, 12426–33. [250, 318]

Wilhelms, F., Kipfstuhl, J., Miller, H., Heinloth, K., and Firestone, J. (1998). Precise dielectric profiling of ice cores: a new device with improved guarding and its theory. *Journal of Glaciology*, **44**, 171–4. [304]

Williams, P.J. and Smith, M.W. (1989). *The frozen earth: Fundamentals of geocryology*. Cambridge University Press, Cambridge. [305]

Wilson, C.J.L. and Zhang, Y. (1994). Comparison between experiment and computer modelling of plane-strain simple-shear ice deformation. *Journal of Glaciology*, **40**, 46–55. [195]

Wilson, C.J.L. and Zhang, Y. (1996). Development of microstructure in the high-temperature deformation of ice. *Annals of Glaciology*, **23**, 293–302. [195]

Wilson, G.J., Chan, R.K., Davidson, D.W., and Whalley, E. (1965). Dielectric properties of ices II, III, V, and VI. *Journal of Chemical Physics*, **43**, 2384–91. [268, 272]

Wolff, E.W. and Paren, J.G. (1984). A two-phase model of electrical conduction in polar ice sheets. *Journal of Geophysical Research*, **89**(B11), 9433–8. [183]

Wolff, E.W., Moore, J.C., Clausen, H.B., Hammer, C.U., Kipfstuhl, J., and Fuhrer, K. (1995). Long-term changes in the acid and salt concentrations of the Greenland Ice Core Project ice core from electrical stratigraphy. *Journal of Geophysical Research*, **100**(D8), 16249–63. [302, 304]

Wolff, E.W., Miners, W.D., Moore, J.C., and Paren, J.G. (1997). Factors controlling the electrical conductivity of ice from the polar regions—A summary. *Journal of Physical Chemistry*, **B101**, 6090–4. [304]

Wollan, E.O., Davidson, W.L., and Shull, C.G. (1949). Neutron diffraction study of the structure of ice. *Physical Review*, **75**, 1348–52. [18]

Wong, P.T.T. and Whalley, E. (1975). Optical spectra of orientationally disordered crystals. V. Raman spectrum of ice Ih in the range 4000–350 cm^{-1}. *Journal of Chemical Physics*, **62**, 2418–25. [46–7]

Wong, P.T.T. and Whalley, E. (1976). Optical spectra of orientationally disordered crystals. VI. The Raman spectrum of the translational lattice vibrations of ice Ih. *Journal of Chemical Physics*, **65**, 829–36. [46–7]

Wooldridge, P.J. and Devlin, J.P. (1988). Proton trapping and defect energetics in ice from FT-IR monitoring of photoinduced isotopic exchange of isolated D_2O. *Journal of Chemical Physics*, **88**, 3086–91. [141]

Wörz, O. and Cole, R.H. (1969). Dielectric properties of ice I. *Journal of Chemical Physics*, **51**, 1546–51. [96]

Xantheas, S.S. (1996*a*). Significance of higher-order many-body interaction energy terms in water clusters and bulk water. *Philosophical Magazine*, **B73**, 107–15. [8]

Xantheas, S.S. (1996*b*). On the importance of the fragment relaxation energy terms in the estimation of the basis set superposition error correction to the intermolecular interaction energy. *Journal of Chemical Physics*, **104**, 8821–4. [8]

Xantheas, S.S. and Dunning, T.H.J. (1993). *Ab initio* studies of cyclic water clusters $(H_2O)_n$, $n = 1$–6. I. Optimal structures and vibrational spectra. *Journal of Chemical Physics*, **99**, 8774–92. [5, 8]

Xie, Y.M., Remington, R.B., and Schaefer, H.F. (1994). The protonated water dimer: extensive theoretical studies of $H_5O_2^+$. *Journal of Chemical Physics*, **101**, 4878–84. [144]

Yamamoto, K. (1980). Supercooling of the coexisting state of ice VII and water within ice VI region observed in diamond–anvil pressure cells. *Japanese Journal of Applied Physics*, **19**, 1841–5. [252]

Yamamuro, O. and Suga, H. (1989). Thermodynamic studies of clathrate hydrates. *Journal of Thermal Analysis*, **35**, 2025–64. [281]

Yamamuro, O., Oguni, M., Matsuo, T., and Suga, H. (1987). High-pressure calorimetric study on the ice XI–Ih transition. *Journal of Chemical Physics*, **86**, 5137–40. [260]

Yamamuro, O., Oguni, M., Matsuo, T., and Suga, H. (1988). Calorimetric study of pure and KOH-doped tetrahydrofuran clathrate hydrate. *Journal of Physics and Chemistry of Solids*, **49**, 425–34. [282]

Yamamuro, O., Matsuo, T., Suga, H., David, W.I.F., Ibberson, R.M., and Leadbetter, A.J. (1995). A neutron diffraction study of tetrahydrofuran and acetone clathrate hydrates. *Physica*, **B213 & 214**, 405–7. [282–3]

Yen, Y.C. (1981). *Review of thermal properties of snow, ice, and sea ice*, US Army Corps of Engineers Cold Regions Research and Engineering Laboratory Report 81-10. [292]

Yen, Y.C., Cheng, K.C., and Fukusako, S. (1991). Review of intrinsic thermophysical properties of snow, ice, sea ice, and frost. In *Third international symposium on cold regions heat transfer, Fairbanks, Alaska, June 1991* (ed. J.P. Zarling and S.L. Faussett), pp. 187–218. University of Alaska, Fairbanks. [292]

Zarcomb, S. and Brill, R. (1956). Solid solutions of ice and NH_4F and their dielectric properties. *Journal of Chemical Physics*, **24**, 895–902. [111]

Zaretskii, A.V., Petrenko, V.F., Ryzhkin, I.A., and Trukhanov, A.V. (1987*a*). Theoretical and experimental study of ice in the presence of a space charge. *Journal de Physique*, **48**, Colloque C1, 93–8. [114]

Zaretskii, A.V., Petrenko, V.F., Trukhanov, A.V., Aziev, E.A., and Tonkonogov, M.P. (1987*b*). Theoretical and experimental study of pure and doped ice Ih by the method of thermally stimulated depolarization. *Journal de Physique*, **48**, Colloque C1, 87–91. [123–5]

Zaretskii, A.V., Petrenko, V.F., and Chesnakov, V.A. (1988). The protonic conductivity of heavily KOH-doped ice. *Physica Status Solidi* (*a*), **109**, 373–81. [110]

Zaretskii, A.V., Howe, R., and Whitworth, R.W. (1991). Dielectric studies of the transition of ice Ih to ice XI. *Philosophical Magazine*, **B63**, 757–68. [92, 262]

Zhang, J.L. and Hibler, W.D. (1997). On an efficient numerical method for modelling sea ice dynamics. *Journal of Geophysical Research*, **102**(C4), 8691–702. [292]

Zhang, Y., Hobbs, B.E., and Jessell, M.W. (1994*a*). The effect of grain-boundary sliding on fabric development in polycrystalline aggregates. *Journal of Structural Geology*, **16**, 1315–25. [195]

Zhang, Y., Hobbs, B.E., and Ord, A. (1994*b*). A numerical simulation of fabric development in polycrystalline aggregates with one slip system. *Journal of Structural Geology*, **16**, 1297–313. [195–6]

Zolotaryuk, A.V., Savin, A.V., and Economou, E.N. (1998). Dichotomous collective proton dynamics in ice. *Physical Review*, **B57**, 234–45. [146]

Index